T0135695

Anna-Lena Leistner

Biocompatible Photochromic Diketopiperazine-based Supramolecular Hydrogelator Systems

Logos Verlag Berlin

λογος

Bibliografische Information der Deutschen Nationalbibliothek

Die Deutsche Nationalbibliothek verzeichnet diese Publikation in der Deutschen Nationalbibliografie; detaillierte bibliografische Daten sind im Internet über http://dnb.d-nb.de abrufbar.

ISBN 978-3-8325-5707-2

Logos Verlag Berlin GmbH
Georg-Knorr-Str. 4, Geb. 10,
D-12681 Berlin
Germany

Tel.: +49 (0)30 / 42 85 10 90
Fax: +49 (0)30 / 42 85 10 92
http://www.logos-verlag.de

German Title of this Thesis

Biokompatible Photochrome Supramolekulare Hydrogelsysteme basierend auf dem Diketopiperazin Strukturmotiv

Preface

Some of the presented results were published or submitted during the preparation of this thesis (*vide infra*).

If applicable, each chapter includes a list of authors with the individual contributions described briefly. Additionally, if some of the presented results have already been partly discussed in other theses, it is stated at the beginning of the respective chapters.

A.-L. Leistner, S. Kirchner, J. Karcher, T. Bantle, M. L. Schulte, P. Gödtel, C. Fengler, Z. L. Pianowski, *Chem. Eur. J.* **2021**, *27*, 8094.

DOI: 10.1002/chem.202005486

Fluorinated Azobenzenes Switchable with Red Light

A.-L. Leistner, D. G. Kistner, C. Fengler and Z. L. Pianowski, *RSC Adv.*, **2022**, 12, 4771

DOI: 10.1039/D1RA09218A

Reversible photodissipation of composite photochromic azobenzene-alginate supramolecular hydrogels

A.-L. Leistner, Z. L. Pianowski, Cover Feature: Smart Photochromic Materials Triggered with Visible Light (Eur. J. Org. Chem. 19/2022). *Eur. J. Org. Chem.* ***2022**, e202101271.*

DOI: 10.1002/ejoc.202101271

Smart Photochromic Materials Triggered with Visible Light

S. Kirchner, **A.-L. Leistner**, Z. L. Pianowski, Photoswitchable Peptides and Proteins. In *Molecular Photoswitches*, **2022**; pp 987-1013.

DOI: 10.1002/9783527827626.ch40

Photoswitchable Peptides and Proteins

F. Hoffmann, **A. L. Leistner**, S. Kirchner, B. Luy, C. Muhle-Goll*, Z. L. Pianowski*, *Eur. J. Org. Chem.* 2023, e202300227.

Cargo encapsulation in photochromic supramolecular hydrogels depends on specific guest-gelator supramolecular interactions.

Table of contents

Kurzzusammenfassung / Abstract in German

Gele werden durch einen physikalischen Zustand charakterisiert, der sowohl Eigenschaften von Feststoffen als auch von Flüssigkeiten aufweist. Sie werden aus einem Geliermittel und einer flüssigen Komponente gebildet. Eine wichtige Klasse stellen die Hydrogele dar, die sich aufgrund ihres hohen Wassergehalts und ihrer Ähnlichkeit in mechanischer Zusammensetzung zu natürlichem Gewebe hervorragend für biomedizinische Anwendungen eignen. Fortschritte wurden bei der Entwicklung supramolekularer Hydrogele erzielt, die entweder aus Gelatoren mit niedrigem Molekulargewicht oder Makromolekülen bestehen. Diese arrangieren sich spontan und oft synergistisch durch Selbstorganisation, basierend auf dynamischen, intermolekularen nicht-kovalenten Bindungen. Diese Bindungen setzen sich aus Wasserstoffbrückenbindungen, Metall-Ligand-Koordination, π-π-, Wirt-Gast-, elektrostatischen und/oder Van-der-Waals-Wechselwirkungen zusammen. Die Selbstorganisation und Auflösung von Gelen kann durch verschiedene Parameter wie elektrische Spannung, Ultraschall, pH-Wert oder Licht ausgelöst werden. Letzteres gewinnt immer mehr an Aufmerksamkeit, da es präzise, räumlich und zeitlich kontrolliert, nicht kontaminierend und unabhängig von anderen Einflüssen eingesetzt werden kann. Diese auf Licht reagierenden Hydrogele, auch photochrome Hydrogele genannt, sind mit "photochemischen Schaltern" ausgestattet, die durch Licht reversible molekulare Reaktionen, typischerweise *E*/*Z*-Isomerisierungen oder perizyklische Reaktionen, ermöglichen. In dieser Arbeit wurde eine Reihe von photochromen supramolekularen Hydrogelen untersucht, die aus biokompatiblen zyklischen Dipeptiden mit niedrigem Molekulargewicht bestehen und lichtmodulierte Polarität aufweisen. Es wurde eine Reihe von Azobenzolderivaten verwendet, die hauptsächlich mit sichtbarem Licht schaltbar sind. Insbesondere wurden die bekannten Strukturmotive der tetra-*ortho*-chloro- und dichloro-di-fluor-substituierte Azobenzole, die mit rotem Licht isomerisiert werden, erfolgreich in das Grunddesign des Gelators (DKP-Lys) integriert. Darüber hinaus wurde die unerwartete Entdeckung gemacht, dass tetra-*ortho*-fluorierte Azobenzole (TFABs) – die ursprünglich mit grünem Licht aktiviert wurden – nach der Substitution mit konjugierten ungesättigten Substituenten auch für rotes Licht empfindlich werden, was das konjugierte TFAB-Chromophor zu einem wertvollen Baustein für die Einbindung in Biomaterialien oder in photopharmakologische Wirkstoffe macht. Eine neue Methode zur Stabilisierung von Hydrogelatoren auf DKP-Lys-Basis wurde durch kooperative Gelierung des basischen DKP-Lys mit dem saurem Polymer Alginat entwickelt. Durch Austausch des photochromen Gelators mit Ca^{2+}-Ionen kommt es zur Bildung ganz natürlicher Hydrogele. Darüber hinaus wurde ein Gelierungsprotokoll entwickelt, das ohne Hitze auskommt, was für die Beladung der Hydrogele mit hitzeempfindlicher Fracht (Proteine, lebende Zellen) wichtig ist. Zuletzt wurde die Wasserstoffbrückenbindungen fördernde peptidische Struktur, die im DKP-Motiv zu finden ist, mit dem Fluorophor Naphthalimid verknüpft, um „AIE-gens“ zu erhalten. Fluoreszenzmessungen zeigten das Potenzial zur Bildung von Nanoaggregaten, vermittelt durch äußere Einflüsse wie z.B. Licht.

Abstract

Gels represent a physical state that exhibits properties of both solids and liquids. They are formed from a gelling agent and fluid component. Their important class are hydrogels, which are suitable for biomedical applications due to their high-water content and their potential compositional and mechanical similarities to native soft tissues. Advances were made in the development of supramolecular hydrogels, formed either by low-molecular-weight gelators or macromolecules in a spontaneous and often synergistic manner by self-assembly through dynamic networks of intermolecular non-covalent bonds. These interactions include π–π stacking, hydrogen bonding, metal-ligand coordination, host-guest, electrostatic and/or van der Waals interactions.

Self-assembly and dissipation of gels can be triggered by various parameters such as electrical voltage, ultrasound, pH, or light. The latter is gaining increasing attention, as it can be applied precisely with spatial and temporal control in a non-contaminating manner and orthogonally to other components of the system. These light-triggered hydrogels, also known as photochromic hydrogels, are equipped with 'photochemical switches'. Such switches give reversible molecular responses, typically in the form of *E*/*Z*-isomerizations or pericyclic reactions, triggered by light.

In this thesis, a collection of photochromic supramolecular hydrogels comprised of biocompatible low-MW cyclic dipeptide gelators with light-modulated polarity was explored. A number of photochromic azobenzene derivatives, mainly switchable with visible light was used. In particular, previously reported tetra-*ortho*-chloro- and di-chloro-di-fluoro-substituted azobenzenes triggered with red light were successfully incorporated into the basic gelator design (DKP-Lys). Moreover, an unexpected discovery was made, that tetra-*ortho* fluorinated azobenzenes (TFABs) – originally activated with green light – upon substitution with conjugated unsaturated substituents become sensitive to red light as well, which makes the conjugated TFAB chromophore a valuable building block for incorporation into biomaterials or into photopharmacology agents.

A new method for stabilization of DKP-Lys-based hydrogelators was provided by composite gelation of the basic cyclic dipeptide with polymeric acidic alginate. There, replacement of the photochromic gelator by equilibration with Ca^{2+} ions leads to all-natural hydrogels. In addition, a gelation protocol without the necessity of heat was developed, which is important for loading the hydrogels with heat-sensitive cargo (proteins, living cells).

Finally, the hydrogen bond-promoting peptidic structure, which can be found in the DKP motif, was linked with the fluorophore naphthalimide to obtain aggregation-induced emission luminogens (AIE-gens). Fluorescence measurement revealed the potential to form nanoaggregates, mediated by external factors such as light.

1 Introduction

1.1 Molecular photoswitches

In the course of the last twenty years, the research interest in 'photochemical switches' highly increased. They react on light, which can be controlled with excellent spatiotemporal precision, yet it is orthogonal towards most elements of chemical and biochemical systems. Light, or specifically photons, do not contaminate the irradiated material.[1] Toxic effects mostly occur only upon irradiation with short-wave UV-light, along with increased scattering of light in the ultraviolet frequencies on cell components.[2] Therefore, light is an attractive trigger to control or study biological function[3] or functional materials[4]. Moreover, the increased interest in photoswitches can be attributed to the emerging field of photopharmacology, in which problems of classical pharmacotherapy such as adverse effects, environmental toxicity, and the appearance of drug resistance can be reduced due to spatiotemporally targeted activation of the drug.[5] An example for the concept of photopharmacology is demonstrated by Wanner *et al.* in 2014 in form of azobenzene derivatives of the mGAT1 inhibitor nipecotic acid. By isomerization of the azobenzene the activity of active membrane transporters was regulated.[6] Another, recently introduced by Pianowski group is an *E/Z*-isomerization of a hemipiperazine photochrome inside plinabulin derivatives, which can increase its cytotoxicity by three orders of magnitude.[7]

The functional core of light-triggered systems discussed in this work are molecular photoswitches. Compounds of this class give reversible molecular responses, typically in the form of *E/Z*-isomerizations[8] or pericyclic reactions[9], triggered by light. The former primarily result in distinct structural changes, whereas the latter are characterized by large changes in electronic properties and/or polarity. The transformation between two forms with distinct absorption spectra upon irradiation is also referred to as photochromism and was first described by J. Fritzsche when he observed the reversible photodimerization of anthracene in 1866.[10] Another classification of photoswitches is based on their back reaction to the thermodynamically stable form after initial excitation: In T-type switches this process is driven thermally, whereas in P-type switches it is exclusively photochemically.[9] Furthermore, in the context of T-type switches we distinguish between the positive photochromism - when the color of the switch fades in the absence of light - or the negative photochromism - when the color reversibly fades upon light irradiation and is restored in the dark.[11] These characteristics classify photoswitches as valuable tools for optical manipulation of various materials with spatiotemporal precision, thereby allowing for a wide variety of applications.

1.1.1 Most commonly used photoswitches

A plethora of photoswitchable molecules were developed over the last decades. Some of them have become firmly established and are frequently applied due to the extensive knowledge about them. The following chapter describes a selection of the most commonly used photoswitches.

Indigoids

The structure of indigoids is derived from the dye indigo (ἰνδικόν, ancient Greek for "blue dye from India") which is one of the oldest known pigments (Scheme 1 a)). Initial production was by extraction of the glycoside indican from the leaves of *Indigofera tinctoria*, followed by fermentation to indoxyl and final oxidation to indigo on air[12]. *E*/*Z*-photoisomerization of the double bond in unsubstituted Indigo (**1**) is not possible because of excited-state proton transfer[13] (ESPT), though after the discovery of hemiindigo (**3**) photoisomerization in 1999[14], a large spectrum of indigo derivatives was synthesized and investigated. Still, the photochemistry of parent hemiindigo (**3,** Scheme 1 b)) is constraint by unwanted side reactions such as [2+2] cycloadditions or triplet generation.[15] ESPT in the parent indigo (**1**) inhibits photoisomerization[13], consequently disubstitution of the indigo NH (**2**) paves the way to new indigo derived visible light switches. The properties of these compounds are highly tunable depending on the substituents.

a)

1 *Indigo*
(R = H)

2 *N,N-disubstituted indigos*
(R = Alkyl, Ar)

b)

stilbene part

indigo part

3 *Hemiindigo (X = NR; Y = CH)*

4 *Hemithioindigo (X = S; Y = CH)*

5 *Iminothioindoxyl (X = S; Y = N)*

R' = various substituents

Scheme 1: Indigoid photoswitches. a) the structures of non-photochromic indigo **1** and it's *N,N*-disubstituted photochromic derivatives. b) Related photochromic compounds originating from the indoxyl and thioindoxyl core, the *E/Z*-photoisomerization is demonstrated.

Tremendous improvement was achieved in 2017 by the group of Dube through the generation of push-pull systems *via* donor substitution, with the indigo carbonyl assuming the acceptor part.[16] These new hemiindigo (**3**) derivatives are characterized by high bistability of up to 83 years for some derivatives at room temperature, solvent-independent visible light photoswitching and high photostationary states (PSS).

Hemithioindigos (**4**) - the combination of thioindigos and stilbenes (Scheme 1) can be isomerized with visible light if strong donors are installed in *para* position on the stilbene unit.[17] The problem of low thermal stability of these derivatives is solved by additional donor-*para*-substitution related to the sulfur in the thioindigo part.

Iminothioindoxyls (**5**) are the hybrid of thioindigos and azobenzenes and are characterized by rapid isomerization, a short thermal half-life in the range of milliseconds, and photoisomer band separation of over 100 nm. They stand out for switching in solid state and high solvent tolerance, rendering them suitable for various applications.[18]

Unlike most photoswitches, indigoids are characterized by high photostability and have the advantage that their core structure absorbs light in the visible region of the spectrum. Furthermore, their fast and efficient photoreaction between the isomers, accompanied by large changes in geometry, classify indigoids as interesting chromophores in supramolecular and biological chemistry.[19]

Stilbenes

Stilbenes (**6**) have been well investigated for 75 years as model systems for "condensed phases" comprising liquid and solid state and furthermore they are representative systems for *E*/*Z*-isomerizations.[20] *E*-stilbene absorbs ultraviolet (UV) light and isomerizes to *Z*-stilbene (Scheme 2 a)). The *Z*-form, however, is not stable and can either isomerize back to its *E*-isomer or undergo a 6π-electrocyclization forming dihydrophenanthrene (**7**).[21] This cyclic intermediate reverts thermally to stilbene (**6**) in darkness and in absence of oxygen. In the presence of oxygen, an irreversible conversion from dihydrophenanthrene (**7**) to phenanthrene (**8**) takes place, limiting the applicability of this simple photoswitch.[21] The introduction of cyclic stilbenes (stiff-stilbenes **9**, Scheme 2)) enables clean, high quantum yield *E*/*Z*-photoisomerization with excellent thermal stability of the *Z*-isomer.[22]

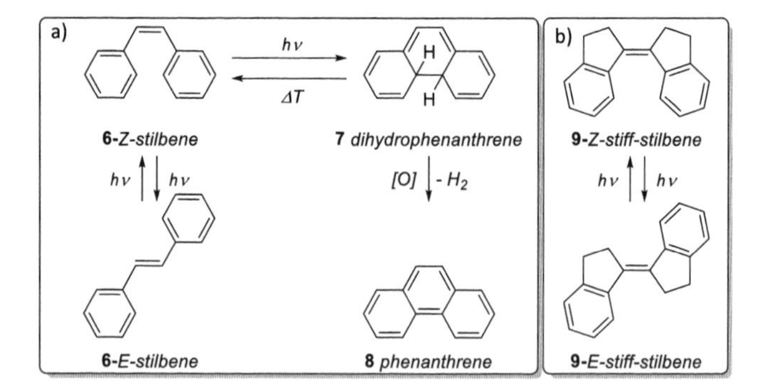

Scheme 2: a) *E*/*Z*-photoisomerization and photocyclization of stilbene (**6**). In the presence of oxygen, phenanthrene (**8**) is irreversibly formed from dihydrophenanthrene. b) Reversible *E*/*Z*-photoisomerization of a stiff-stilbene.

Diarylethenes (DAE).

The underlying photoisomerization mechanism of diarylethenes (**10**) has already been described in the 6π-electrocyclization observed in plain stilbenes (**6**). Methyl substitution at the 2- and 6-position of the respective stilbene phenyl rings hinders the hydrogen elimination, enabling reversible photoisomerization even in the presence of oxygen.[23] Replacement of the phenyl rings by thiophene strongly increases the lifetime of the dihydrointermediate from a few minutes[24] to several hours[25]. Further substitution of these thiophene-containing diarylethenes with maleic acid at the bridging double bond finally provided one of the first examples for thermally irreversible photochromic diarylethenes with additional inhibited *E*/*Z*-isomerization.[24] Therefore, diarylethenes (**10**) are characterized as P-type photochromic compounds with outstanding high fatigue resistance, fast coloration/decoloration (in the picosecond range) and high quantum yields.[26]

$h\nu_1$ / $h\nu_2$

open form — *6π-electrocyclization* — *closed form*

10 *Diarylethenes*

Scheme 3: P-type photochromic diarylethenes (**10**) undergo 6π-electrocyclization under irradiation.

The classical DAE (**10**, Scheme 3) requires UV light activation for photocyclization. For visible light switching, several strategies are exploited.[27] Examples are π-system extension[28], backbone modification[29], triplet sensitization[30], upconverting nanoparticles[31] or multiphoton absorption[32].

Spiropyrans

The multistimuli-responsive spiropyrans (**11**) exhibit photochromism, thermochromism[33], acidochromism[34], mechanochromism[35] and electrochromism[36]. The reversible photoinduced isomerization of spiropyrans is based on the interconversion between the colorless closed spiropyran form and the open colored zwitterionic (thermally unstable) merocyanine form. Synthetic access is achieved by condensation of salicylaldehydes with Fischer bases (methylene indolines).[37] Bidirectional photoisomerization occurs in the picosecond range[38] with high quantum yields. However, spiropyrans are seldom addressable with light >400 nm. Here, other strategies have to come into effect, such as oxidative dimerization, which offers an alternative switching mechanism.[39]

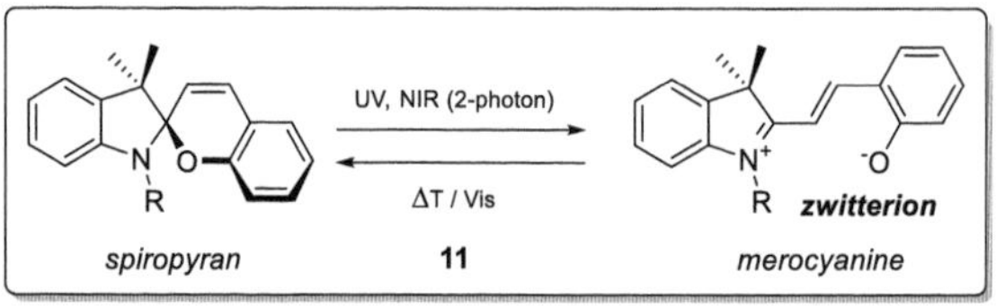

Scheme 4: Photoinduced isomerization between the colorless closed spiropyran form and the open colored zwitterionic (thermally unstable) merocyanine form.

Azobenzenes

Azobenzenes (ABs,**12**) are the most frequently applied molecular photoswitches (Scheme 5). Upon irradiation with appropriate wavelengths, they undergo *E*/*Z*-isomerization. Unsubstituted ABs photoisomerize from *E* to *Z* under UV light and revert from *Z* to *E* either under blue light irradiation or thermally.

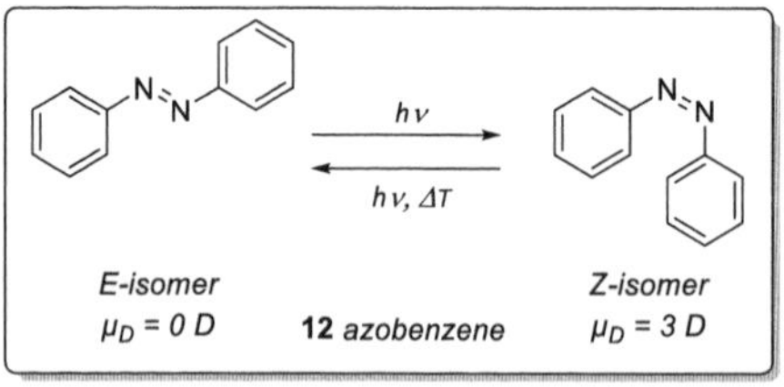

Scheme 5: The non-polar and planar *E*-azobenzene isomerizes under UV irradiation to the polar and bent *Z*-azobenzene.

E-AB has a planar conformation and is non-polar (*E*: μ_D = 0 D), whereas the *Z*-AB is strongly bent, which leads to a higher dipole moment (*Z*: μ_D = 3 D).[40] ABs are applied in diverse fields of research like biomolecules[41], molecular machines[42] or material sciences[43] and, to fit the respective necessary requirements, their photochromic properties, such as half-life, absorption wavelength, switching rates, quantum yields or photostationary states of ABs are tunable. These tuning strategies include introduction of heteroatoms or various substitution patterns and will be discussed in a later section.

1.1.2 Emerging photoswitches

Besides the previous described photoswitches, novel types of photochromic molecules are reported every year. A brief selection of these recently developed switches is discussed in the following chapter.

Hydrazones

Photoisomerization of hydrazones (**13**) is not a new phenomenon and was already reported in a series of publications by Courtot *et al.*, describing his investigations on the photoisomerization of a series of 1,2,3-tricarbonyl-2-arylhydrazones derived from β-diketones and β-ketoesters.[44] The C=N bond in hydrazones can *E*→*Z* isomerize under UV light, although the obtained *Z*-isomers are very short-lived species. An intermolecular H-bond can stabilize the *Z*-isomer and is essential for reversible photoisomerization, but also offers a further adjustment point for chemical isomerization, for example by pH tuning (Scheme 6).[45]

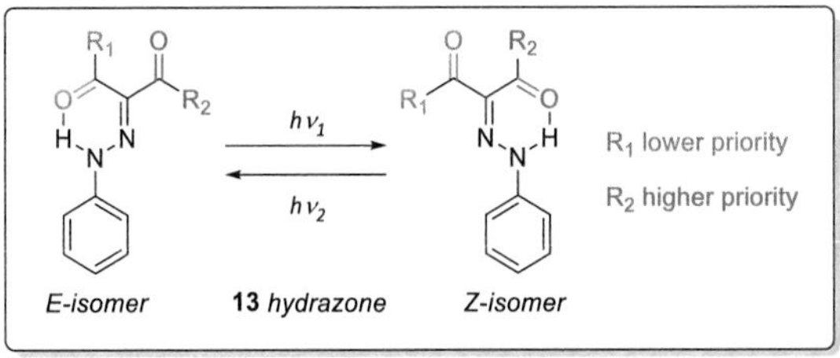

Scheme 6: Photoisomerization and H-bond stabilization of hydrazones.

Due to their promising photoswitching properties with regard to quantum yields, PSS and photostability, APRAHAMIAN group decided to drive their investigation forward and is pioneering the development of new hydrazone switches.[46] For example, the design of push-pull systems enables photoisomerization solely with visible light.[47]

Donor-Acceptor Stenhouse adducts

Donor-Acceptor Stenhouse adducts (DASAs, **14**) are a recently (2014)[48] reported T-type class of visible light photoswitches with negative photochromism (Scheme 7). They are characterized by a rapid and modular synthesis based on furfural – a building block accessible from renewable resources[49] – which is condensed with 1,3-dicarbonyls. In presence of primary or secondary amines, the furan ring is opened and the Stenhouse salt is formed. Upon visible light irradiation, this colored hydrophobic linear triene forms the ring-closed and colorless zwitterionic cyclopentanone form.[48] The switching mechanism is based on *Z*→*E* photoisomerization followed by conrotatory 4π-electrocyclization.[50]

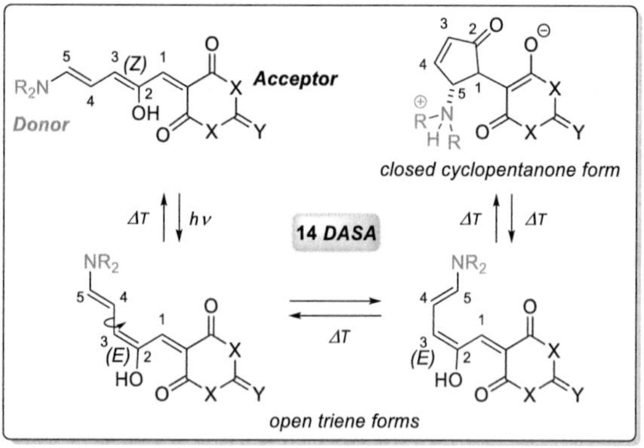

Scheme 7: Proposed mechanism for DASA photoisomerization. An initial *Z*→*E*-photoisomerization around C_2–C_3 is followed by a rotation around C_3–C_4, to facilitate a subsequent thermal cyclization with additional proton transfer.[51]

Dihydropyrenes

The photochromic behavior of dihydropyrenes (DHPs , **15**) has been known for quite some time[52], however this class of photoswitch was disregarded until recently when a simplified synthetic access towards *t*-butyl-DHP was reported.[53] DHPs undergo a 6π-cycloreversion to colorless cyclophanedienes (CPDs) upon irradiation with visible light. Back-conversion to the fully aromatic form occurs thermally (T-type photoswitch, Scheme 8).

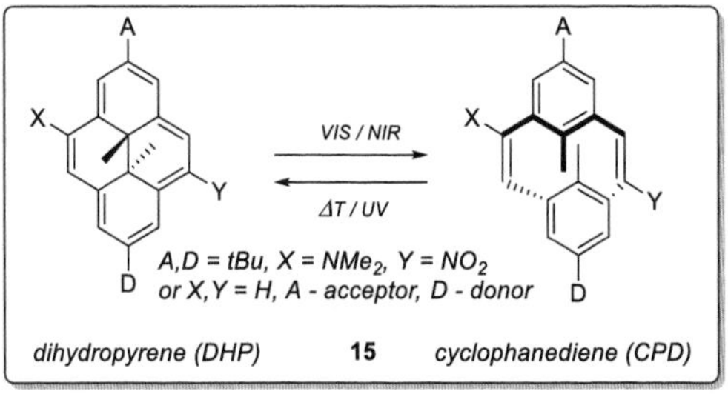

Scheme 8: DHPs are usually negative photochromic photoswitches triggered with visible or NIR light, depending on donor-acceptor functionalization. The photoisomerization is based on a 6π-cycloreversion to colorless CPDs with long thermal lifetimes (often longer than 24 h).

Hemipiperazines

Hemipiperazines (HPIs, **16**) are a new class of photoswitches based on *E*/*Z*-isomerizations driven by visible light and characterized by high bistability. HPIs consist of an arylidene (hemi-stilbene) moiety comparable to hemiindigo attached to the cyclic dipeptide (2,5-diketopiperazine, see Scheme 9). Discovery of this novel photoswitchable motif occurred during investigation of the non-small cell lung cancer drug plinabulin by PIANOWSKI group Characteristic behavior of HPIs is a weak solvatochromism and a strong acidochromism, which also drastically influences the lifetime. The photochromic properties of the HPI motif are tunable by modification and substitution of the stilbene unit, exemplary a *para* or *ortho* substitution with methoxy groups enables conjugation and leads to a slight bathochromic shift.[7]

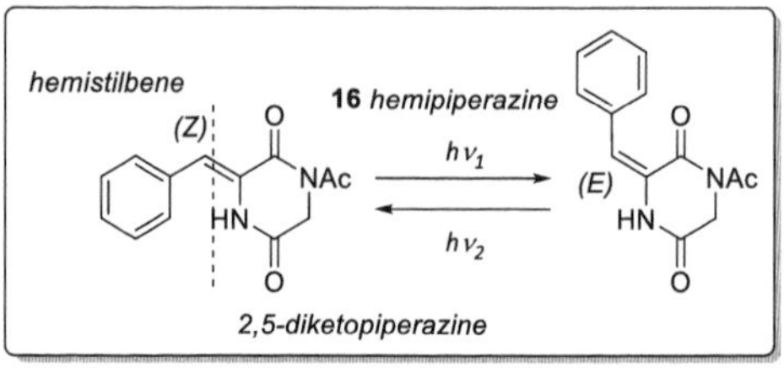

Scheme 9: *E*/*Z*-isomerization of HPIs. The effect was first observed in the non-small cell lung cancer drug plinabulin.

1.1.3 A deeper insight into azobenzenes

The photoswitch of choice in this work is the well-known azobenzene (**12**). There is a tremendous amount of knowledge about this type of photoswitch and it is already being employed in a variety of applications. These often take advantage of the large change in polarity and geometry upon isomerization and are either settled in diverse biological contexts, where the azobenzene moiety is functionalized to be suitable for implementation into peptides[54] or deoxyribonucleic acids[55] (DNA), for example, but they are also attractive scaffolds in materials science.[56] The properties of azobenzenes are easily tunable by different approaches, the influence of substitution patterns on the photophysical properties is well-understood and broadly described, and their synthesis is thoroughly investigated as well.

1.1.3.1 Synthetic approaches

Over time, a large variety of synthetic strategies for azobenzenes were discovered and developed. The discussion of all possible methods would go beyond the scope of this thesis. Therefore, a selection of the most important synthetic methods and the ones most relevant to this thesis is provided below.

Azo coupling

The most commonly used method for synthesizing unsymmetrical azobenzenes is the classic azo coupling (Peter Griess, 1858, Scheme 10)[57]. An aromatic primary amine (**17**) is diazotized at low temperatures to form a weak electrophile (**18**). The electrophilicity of the diazonium salt (**18**) is enhanced by electron withdrawing groups (like, e.g. nitro-groups), so they react with electron-rich aromatic nucleophiles (**19**, usually substituted with electron donating groups (EDG); e.g. ethers or amines) in an electrophilic aromatic substitution reaction, although other mechanisms for the coupling are subject of ongoing discussions[58].

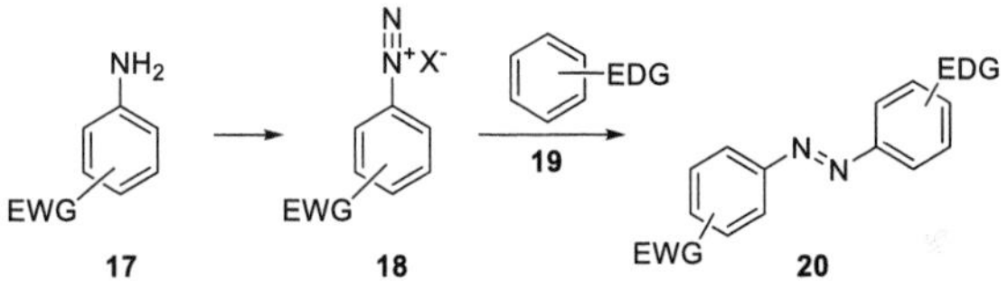

Scheme 10: Classic azo coupling using diazonium salts. After diazotization of a primary aromatic amine **17**, it reacts with a second electron-rich aromatic nucleophile **19** in an electrophilic aromatic substitution reaction.

The EDGs also determine the direction of substitution, which is favored in *para* position to the EDG due to low steric hindrance combined with pronounced stabilization of the positive charge in the σ-complex. If the *para* position is not available, substitution in *ortho* position occurs. Another important parameter of the reaction is the pH. Firstly, the diazonium cations' stability is pH dependent, as they are stable only under acidic and moderately basic conditions. Moreover, the reaction requires either acidic conditions in case of amines as reactants or mildly basic conditions to enhance the nucleophilicity of phenols. Electron poor diazonium salts can also dimerize to azo compounds when treated with copper metal and an acid or with copper(I) salts over an electron transfer mechanism involving radicals.[59]

Oxidative coupling of anilines to (symmetrical) azobenzenes

In 1972, electrolytic oxidation of aromatic amines was reported as a new method to form symmetric azobenzenes. However, described yields were low with 48% as the highest reported value for the coupling of 2,4,5-trichloroaniline to 2,2',4,4',6,6'-hexachloroazobenzene, while unsubstituted azobenzene was obtained with a yield of only 21%.[60] Later, different oxidation agents were described, such as Mn-based reagents (Scheme 11)[61], Ag_2CO_3[62], or hypervalent iodides for example *tert*-butyl hypoiodite, to name a few. Furthermore, metal catalyzed oxidative couplings using O_2 or air as oxidant were reported. Here, gold nanoparticles at higher pressures and temperatures[63], but also inexpensive CuBr[64] under mild conditions is applied. The dependency on the substitution pattern and, in many cases, low yields restrict the versatile application of these oxidants. Consequently, the need for development of improved systems and methods was promoted.[65]

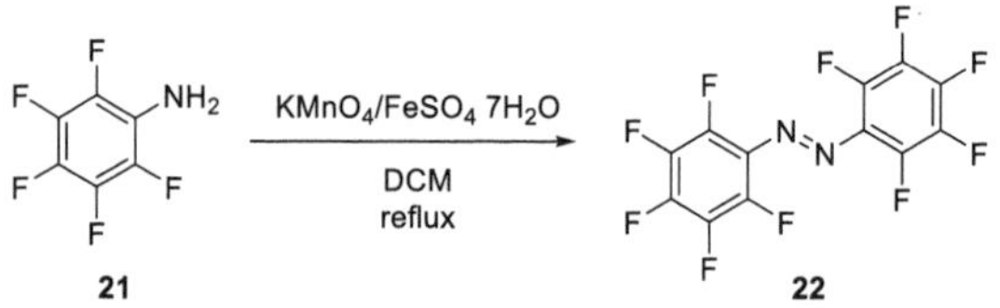

$KMnO_4/FeSO_4\ 7H_2O$, DCM, reflux: **21** → **22**

Scheme 11: Synthesis of decafluoroazobenzene (**22**) applying the oxidative coupling strategy. The yield of the reaction with electron poor anilines is low (20-35%) and variation of solvents ($CHCl_3$, CCl_4 or toluene) or oxidizing agents (^{t}BuOI, copper(I)/pyridine with air) provides no improvement.[61a]

Mills-reaction between nitrosoarenes and anilines

In the Mills reaction[66], anilines react with aromatic nitroso compounds, usually employing glacial acetic acid as solvent.[67] Mechanistically, the Mills reaction is described by the nucleophilic attack of the amine function on the nitroso group, followed by dehydration (Scheme 12). This method provides access to unsymmetrical azobenzenes.

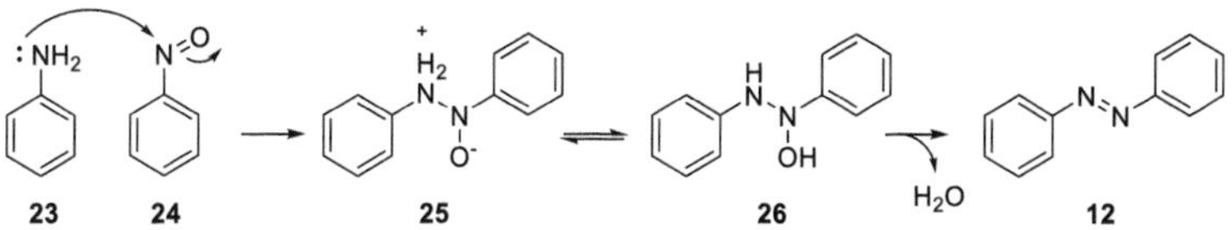

Scheme 12: Azo-bond formation between aniline and an aromatic nitroso compound. The product is formed after dehydration.

The nitroso starting material (**24**) is accessible by various synthetic procedures, usually oxidation reactions. Some examples are oxidation of aromatic methylhydroxylamine with *tert*-butyl hypochlorite[68], *m*-chloroperbenzoic acid[69], 2,3-dichloro-5,6-dicyanobenzoquinone[70] or $KMnO_4$[71]. These homogeneous reactions are often slow and low yielding due to side reactions or low product stability. Therefore, a heterogenous system like the reaction of Oxone® with aniline in Dichloromethane (DCM)/H_2O is more convenient.[68] Coincidentally, the biphasic system also ensures separation of the product from possible side products and substrates.

Reductive coupling of nitroarenes

Another method for symmetrical azobenzenes is the reductive coupling of nitroarenes with various reducing agents (for example $LiAlH_4$[72], $NaBH_4$[73], $SnCl_2$/NaOH[74], $TiCl_4$/$LiAlH_4$[75], etc.). The reduction also works electrochemically: Magnesium electrodes have been used for the synthesis of 4,4'-dimethylazobenzene in very good yield (85%).[76] The presence of acid or base is indispensable for the proposed mechanism[77] (Scheme 13), in which the nitro compound (**27**) is first reduced to the respective nitroso (**24**) and hydroxylamine (**28**). Both intermediates are then converted into radical anions (**29**), which dimerize to a *N,N'*-dihydroxy intermediate (**30**). The following azoxy-intermediate (**31**) formation by dehydration is either acid or base catalyzed and is the rate determining step of the reaction.[77] Finally, azobenzene (**12**) is formed after reduction of the azoxy-intermediate (**31**).

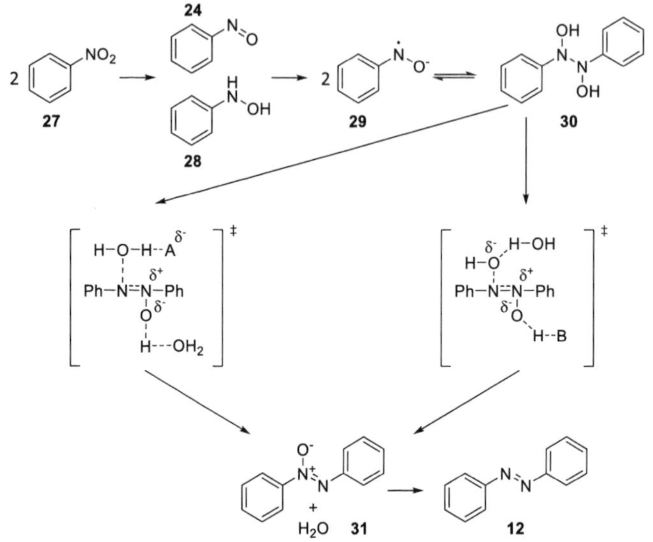

Scheme 13: Reductive coupling of nitroarenes to symmetric azobenzenes adapted after Merino[68]. The transition state varies, depending on the catalyst which is either an acid (A) or base (B).

Azo bond formation by addition of lithium organyls to diazonium salts

The synthesis of sterically demanding ABs, for example tetra-*ortho*-chloroazobenzene, is challenging with conventional methods like azo coupling or Mills reaction. Therefore, different synthetic pathways were employed: FERINGA group proposed *ortho*-lithiation of aromatic substrates followed by coupling reaction with aryldiazonium salts, also derived from chloroarenes.[78] Symmetric as well as asymmetric ABs are accessible with this novel method. Suitable substrates for the reaction are 1,3-disubstituted arenes, in particular those bearing methoxy-, fluoro- or chloro-substituents, because of their efficient *ortho*-lithiation and since they produce visible-light addressable ABs. Feringa's work is based on literature examples where Grignard or organozinc reagents[79] are coupled with aryldiazonium salts. Furthermore, HERGES group reported procedures in which lithiated compounds reacted with diazonium salts to form heterocyclic azobenzenes.[80]

Homocoupling of anilines with *N*-chlorosuccinimide

Many of the methods described so far do not have a high substrate tolerance or have long reaction times. To avoid these problems, novel reactions have been developed, one of which was a novel modification of the oxidative homocoupling of anilines, reported by Lin *et al.* There, anilines are treated with *N*-chlorosuccinimide (NCS) as well as the organic base 1,8-diazabicyclo[5.4.0]undec-7-ene (DBU) and react through oxidative coupling at low temperatures (–78 °C) within minutes to the desired symmetrical azobenzenes.[81] This one-pot synthesis provides good yields (56-93%) within short reaction times and has therefore also been extensively applied during the preparation of this thesis.

1.1.3.2 Visible light activated azobenzenes

Photoswitches are applied in various fields, particularly frequently in biology[3] and recently in photopharmacology[7, 82]. Here, adverse effects occur upon irradiation with highly energetic UV-light as the light is scattered on cellular components or absorbed by biomolecules. Therefore, bathochromic shifts of the absorption maxima are desired to allow bidirectional switching with visible light frequencies, and the scientific community has been and still is putting a tremendous effort into the development of new photoswitch designs to fulfill this requirement amongst others. In the following chapter, modifications of the azobenzene scaffold are presented.

Push-pull systems

The concept of push-pull systems relies on substitution with electron donating (usually dialkylamines) and withdrawing groups (such as nitro groups) attached in *para* position to the azo bond. These ABs have a bathochromically shifted π–π^* absorption band[83], though their polarized structure leads to pronounced solvato- and acidochromism as well as rapid thermal $Z \rightarrow E$ isomerization. The short half-lives of the *Z*-isomers are advantageous for photomodulation of fast biological processes such as the visual process[84]. However, change of the switching mechanism from linear transition (i.e. inversion) to rotation[85] and the short half-lives hamper the photophysical analysis of push-pull ABs. Moreover, the absorbance intensity is strongly dependent on the solvent and another limiting factor for some applications (e.g., intramolecular crosslinkers in peptide chains) is the missing symmetry.

Tetra-*ortho* substituted azobenzenes

Substitution with electronegative substituents (such as the heteroatoms O, N, S) in *ortho* position lowers the energetic barrier for isomerization due to their contribution of a fraction of their nonbonding electron density to the aromatic systems.[86] Bulky substituents in *ortho* position additionally distort the planarity of the *E*-isomer, affecting the energy of both π and π^* molecular orbitals and resulting in the splitting of $n \rightarrow \pi^*$ bands of both isomers. In the unsubstituted AB, the $n \rightarrow \pi^*$ bands that appear in the visible light range are overlapping and cannot be addressed separately, which is, however, possible in the tetra-*ortho* substituted species with visible light frequencies (Figure 1).

Earliest reported *ortho*-substituted ABs have 2,2'-aminoalkyl substituents (**32**) which enable isomerization at wavelengths up to 530 nm, combined with good photochemical yields and stability in reducing aqueous environments. However, with increasing electron density, these compounds suffer from photobleaching.[87] One solution to this problem was reported by Woolley *et. al* in the form of fully visible light switchable tetra-*ortho* methoxy ABs (**33**) which destabilize the n-orbital of *Z*-azobenzene by repulsive interactions between O and N lone pair electrons. Usually, *E*→*Z* isomerization occurs under green light irradiation (PSS: 80%) and revert from *Z*→*E* under blue light (PSS: 85%). Though, these compounds were found to be sensitive to glutathione (GSH, $t_{1/2}$ ≈1.5 h in 10 mM) and therefore limited in their applicability in a biological context.[88]

Another approach to overcome the problem of photobleaching is the replacement of the *ortho* oxygen atoms by sulfur (*S*-ethyl groups, **34**). These compounds *E*→*Z* isomerize with both blue and green light (selected derivatives also with red light), but due to overlap of the spectra of *E*- and *Z*-isomers, the photostationary state of the *Z*-isomers is limited to 75% and their thermal relaxation is faster ($t_{1/2}$ (dimethyl sulfoxide, DMSO, 25 °C) in the range of minutes) compared to the methoxy compounds ($t_{1/2}$ (DMSO, 25 °C) ≈ 14 d).[86b]

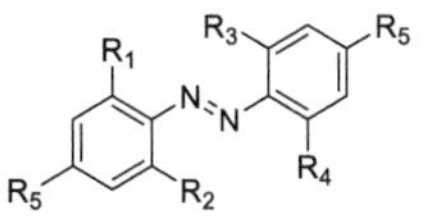

	R_1	R_2	R_3	R_4	R_5	$\lambda_{E\rightarrow Z}$
32	1-methyl-piperazine		1-methyl-piperazine			530 nm
33	OMe	OMe	OMe	OMe	H	530 nm
34	SEt	SEt	SEt	SEt	NH_2	518 nm
35	F	F	F	F	H	530 nm
36	F	F	F	F	CHO	>630 nm
37	Cl	Cl	Cl	Cl	H	635 nm
38	F	F	Cl	Cl	H	530-660 nm

Figure 1: Examples of *ortho*-substituted azobenzenes and their irradiation wavelength for *E*→*Z* isomerization.

Further improvement is achieved by substitution with halogens, which reduce the electron density in the nearby N=N bond and thereby lower the n-orbital energy. On this basis, the HECHT group synthesized tetra-*ortho* fluoro ABs (**35**) which isomerize under visible light with high photoconversions (green light 90% *E*→*Z* and blue light 97% *Z*→*E*) and are remarkably bistable ($t_{1/2}$ (DMSO, 25 °C) >700 d).[89] Shortly afterwards, WOOLLEY group published the tetra-*ortho* chloro ABs (**37**) with even further red shifted n→π* transitions. These compounds are neither sensitive to photobleaching nor glutathione reduction.[90] Using the tetra-*ortho* fluoro ABs, PIANOWSKI group demonstrated that their supplementary functionalization with sp^2 substituents (–CHO, –C=C–) in the *para* position results in an additional bathochromic shift. These derivatives can also be switched with high photoconversions (660 nm 82 % *E*→*Z*, 407 nm 85 % *Z*→*E* for the *para*-bisaldehyde, **36**), although at the cost of decreased thermal stability and glutathione resistance ($t_{1/2}$ ≈10 h).[91] Recent publications combine the enhanced n→π* band separation (*E*→*Z*: 530 nm 86%, 660 nm 59%) with the high bistability ($t_{1/2}$ (DMSO, 70 °C) ≈ 16 h) by combining two fluorine and two chlorine atoms in *ortho* position ("dfdc-AB", **38**).[92] Many of the substitution strategies discussed in this paragraph have already been applied in biological context[93] as well as in other functional materials.[56c, 94] In particular, the halogen substitution strategy is pursued in this thesis, due to its excellent performance in the biological environment.

Azonium switches

Amino-substituted ABs (such as methyl orange) have the capability to form azonium ions, but their low pKa values in the range of 1.5–3.5 prevent the azonium species from being formed at neutral pH.[95] Furthermore, the thermal relaxation rate of most azonium ions is fast with half-lives in the µs range, thus production of significant concentrations of the *Z*-isomer is difficult. An explanation for this characteristic is the decreased double-bond character of the azo-bond. Therefore, an improved substitution pattern capable of forming azonium ions is presented by additional tetra-*ortho* substitution with methoxy groups, which allow H-bonding with the azonium proton, with additional resonance stabilization.[96] This structural design enables formation of red-light addressable azonium ions at neutral pH.[97] Changing the substituent in *para* position by variation of the amine influences the degree of delocalization of the *para*-amino nitrogen lone pairs into the ring system. Moreover, methoxy substitution in *meta* position was predicted to further induce a bathochromic shift of the absorbance according to the distribution rule of auxochromes.[98] Several compounds following the described substitution pattern (as depicted in Figure 2) undergo single photon photoswitching with far-red- and near-IR light.[96, 99]

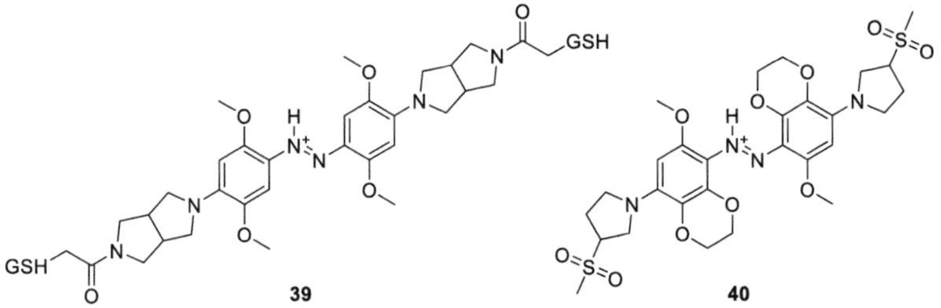

Figure 2: Compound **39** absorbs in the near-IR region and switches at physiological pH values with a life-time (*Z*-isomer) of 10 µs. DOM-azo **40**), is stable for months in neutral aqueous solutions, undergoes *E*-to-*Z* photoswitching with 720 nm light, and thermally reverts to the stable *E*-isomer with a half-life near 1 s.[96, 99]

Azoheteroarenes

Many visible-light-shifted ABs are characterized by short (µs-ms) lifetimes, e.g. push-pull systems or protonated ABs. Azoheteroarenes have one or both phenyl rings replaced by aromatic heterocycles and are far less studied than other AB derivatives.Their previous applications were focusing on the development of systems with strong nonlinear optic and solvatochromic properties.[100] So far, reported examples are featuring pyrimidine, pyridine, pyrrole, imidazole, pyrazole, indazole or indole scaffolds.[101] Depending on the heterocyclic building block, diverse half-lives of the *Z*-isomers are reported and range from few nanoseconds (azopyrimidine[102]) up to 1000 days (*N*-methylated pyrazole derivative[100]). However, only few of the azoheteroarenes feature visible-light-driven isomerization. One notable example are azobenzazoles by Kennedy *et al.* with increased lifetime and bathochromically shifted absorption upon protonation. In the literature azobenzazoles photoisomerize (*E*⟶*Z)* with blue light (488 nm) and after protonation with yellow (591 nm) light), while also being highly fatigue resistant.[103] The effect of prolonged half-live is surprising, considering the opposite effect of protonation on the half-lives of non-heterocyclic ABs, but can be explained by the formation of an intramolecular hydrogen bond.

Diazocines – bridged azobenzenes

The extraordinary photoswitching properties of diazocines – also known as bridged azobenzenes – were first reported by HERGES group in 2009. An AB was bridged in *ortho* position by an ethylene linker, and the resulting target molecule **41** exhibits excellent switching efficiencies at visible-light-shifted wavelengths.[104] Due to the strained geometry of the diazocine structure, new conformational preferences emerged: In contrast to the parent AB, the *Z*-isomer is the thermally stable form in diazocines (Scheme 14). From here, a series of further improved diazocines was developed, and one focus of further investigation was the influence of heteroatoms in the bridge. HERGES group described *E/Z*-photoisomerization with red light (660 nm) for sulfur- (**44**, $t_{1/2}$ = 3.5 d) and oxygen-containing bridges (**43**, $t_{1/2}$ = 89 s).[105] Moreover, depending on their further substitution, nitrogen atoms enable photoisomerization with near-IR light (**45**, NMe-CH_2-bridged diazocine, 740 nm)[106] or photoisomerization under aqueous conditions (**42**, NAc-CH_2-bridged diazocine, 520–600 nm).[106-107] For all derivatives the *Z*→*E* photoisomerization occurs usually under violet or UV light.

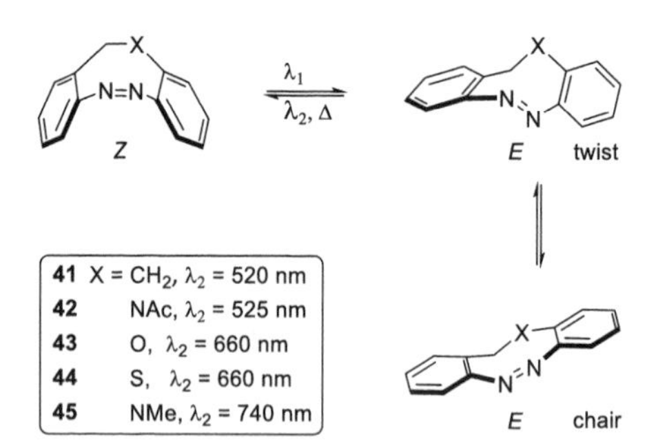

Scheme 14: The energetically higher *E*-diazocine can adopt a twisted or chair form and isomerizes under visible or NIR light irradiation (depending on the functionalization of the bridge) to the thermally stable *Z*-isomer.

1.1.3.3 Azobenzenes in biological applications

Photocontrol of biomolecules is a challenging research area with various demands on the biological material, but even more on the photoswitching unit. They must be stable under physiological conditions, as well as being able to isomerize efficiently and at biocompatible wavelengths. A central player in biomolecules are nucleic acids, which in form of desoxyribonucleic acid (DNA), store genetic information or as ribonucleic acid (RNA) play a central role in the transfer of genetic information into proteins. Five different approaches are known to synthetically modify DNA or RNA with photoswitches: (1) base replacement, (2) base modification, (3) sugar modification, (d) phosphodiester modification and (5) 5′-caps.[108] The latter is the synthetically most facile method, as the photoswitch is attached to the presynthesized oligonucleotide.[109] Impressive results are obtained when azobenzene moieties are introduced into the DNA strand *via* a threoninol scaffold to reversibly control DNA hybridization (Scheme 15): Depending on the threoninol (L- or D-form) and substitution of the azobenzene, two modes of action are possible. In D-threoninol, the DNA duplex is stabilized by intercalation of planar *E*-azobenzene and destabilized by isomerization to the non-planar *Z*-azobenzene. In contrast, the duplex-stabilizing ability of *E*-azobenzene linked through L-threoninol is less pronounced, in particular for azobenzenes with bulky alkyl groups in *para* position. Consequently, the photoisomerization here has the opposite effect on hybridization and strand separation.[110]

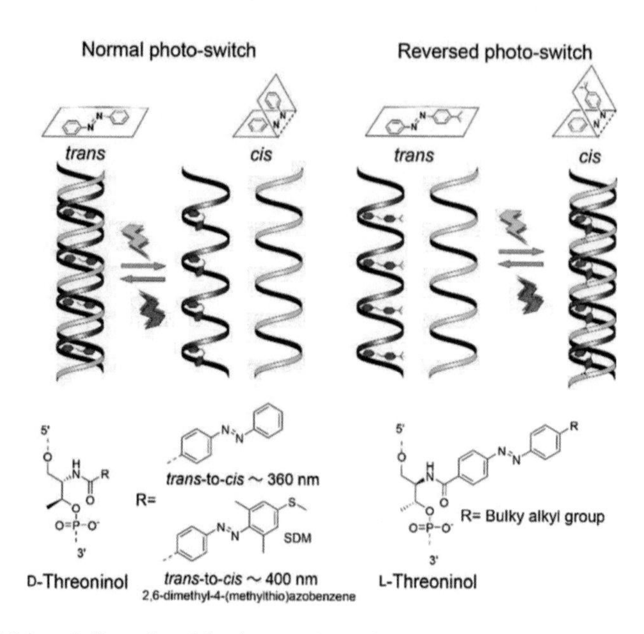

Scheme 15: Schematic illustration of the photo-regulation of DNA hybridization. Left; *Normal* photo-switch with an azobenzene on D-threoninol scaffold. Right; *Reversed* photo-switch with an azobenzene modified with bulky alkyl group on L-threoninol. Reprinted with permission from H. Asanuma, T. Ishikawa, Y. Yamano, K. Murayama, X. Liang, *ChemPhotoChem* 2019, *3*, 418. Copyright (2023), with permission from WILEY.

This concept was applied to photocontrol the isothermal amplification of DNA by template-mediated ligation[111], which is a rapid process that allows for the presence of thermally unstable molecules and provides an alternative approach to replicate nucleic acids (polymerase chain reaction, PCR).

In most cases of photomodulation of biomolecules, a pronounced conformational or geometric change is required to induce a functional change, which is particularly significant in the case of proteins and peptides, since their function and interaction is often based on their 3D structure. Additionally, the photoswitch must be linked to the peptide in a way that isomerization affects the biomolecule. Hence, the first challenge in the development of photochromic peptides is the incorporation of the photochromic building block. In general, there are three different covalent approaches (Scheme 16).

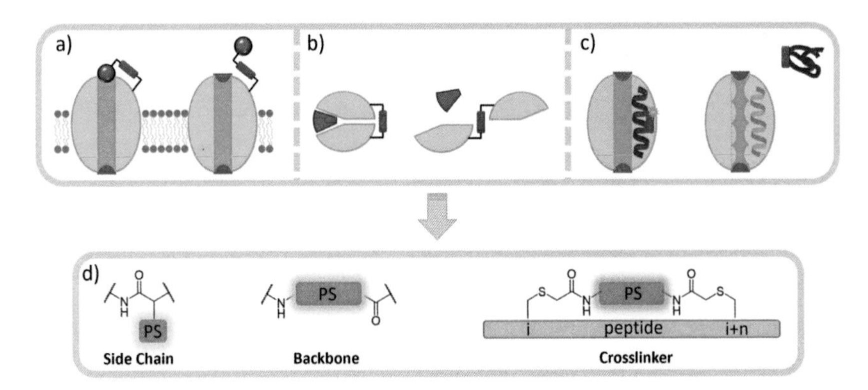

Scheme 16: Schematic illustration of peptide or protein photomodulation by photoswitchable tethered ligands. a) an ion channel is blocked or opened by photoisomerization. b) A functional enzyme is formed after photoinduced geometry change. c) Peptide secondary structure is influenced by photoisomerization. d) Suitable covalent incorporation methods of photoswitches into peptides. PS = photoswitch. In the side chain method the photochromic unit is located at the side chain of an amino acid. In the second method, the photoswitch is incorporated as part of the peptide backbone. Photoswitchable peptides are accessible by linking two separate amino acids by their side chain.

First, there is the side chain method, in which the photochromic unit is located on the side chain of an amino acid. One advantage of these unnatural amino acids is their applicability in standard solid phase synthesis. In the second method, the photoswitch is incorporated as part of the peptide backbone. Here, adaptions are made to enable facile synthesis. For example, A. Aemissegger and D. Hilvert described a practicable protocol towards an azobenzene backbone, compatible with solid phase peptide synthesis using standard 9-fluorenylmethoxycarbonyl (Fmoc,) methods.[112] They inserted azobenzene units into the loop region of a peptide known to fold into a β-hairpin in aqueous solution. As a result, it was possible to reversibly control the conversion of a short open-chain polypeptide from undefined to a structured β-hairpin.[112-113] In this way, azobenzenes have been incorporated into further short linear[114] and also into cyclic peptides[115] for conformational studies and to influence biological signaling pathways further downstream. Another method for building photoswitchable peptides constitutes the linkage of two separate amino acids *via* their side chain. Cross-linkage by photoswitches is challenging and parameters such as the length and thereby flexibility of the attached side chain or distance between the crosslinked amino acids must be taken into consideration. Optimal photocontrol of crosslinked peptides can be achieved when one isomer matches the natural helical state of a peptide while the other does not.[116]

Besides nucleic acids and proteins, another major class of biomolecules are carbohydrates, which are classified according to their degree of polymerization into monosaccharides, oligosaccharides, and polysaccharides (linear or branched). In eucaryotic cells, carbohydrates play a significant role in the glycocalyx, which is a complex component of the extracellular membrane. Since carbohydrates are difficult to analyze, photosensitive azobenzene glycoconjugates are promising tools for studying the glycocalyx. Photoisomerization of the azobenzene unit enables orientational control of carbohydrate presentation in space and on the surface, within glycopolymers, macrocycles or in supramolecular arrangements.[117]

Manipulation of biological material is not constrained to direct synthetic modification of large biopolymers. In addition, strategies employing diverse small molecules ranging from natural products to non-biogenic peptides or respective mimetics have been invented.[118] These small molecules can selectively address individual properties of multifunctional biological structures, allow for kinetic analysis, are readily and rapidly applicable and moreover their effects are reversible in most cases. One large fraction of these small molecules is emerging as photochromic ligands in the ascending field of photopharmacology; the primary objective of which being the restriction of the range of action of a small molecule or drug by local activation to limit side or environmental effects. A striking example of photopharmacology are the azobenzene-derived combretastatin A-4 analogue photostatins (PST) – microtubule targeting drugs developed by TRAUNER group in 2015. These PSTs were applied in cell assays with a human breast cancer cell line and displayed up to 100-fold increased cytotoxic effects under irradiation compared to the dark state.[119] However, UV light irradiation was required for activating photoisomerization of the PSTs. PIANOWSKI group later demonstrated that activation of microtubule targeting drugs by visible light is also possible. A derivative of the non-small cell lung cancer drug plinabulin is activated by cyan light (490 nm) and demonstrates a >1800-fold activity enhancement in cell viability experiments against human colon cancer cells.[7]

1.1.3.4 Azobenzenes in smart materials

As described before, the implementation of photoswitches into materials is frequently performed in a biological context. However, the versatility of photoswitches is not limited to this field of application. Photomodulation of smart materials (causing functional changes in diverse organic and non-organic scaffolds) offers new opportunities and expands their applicability. Examples of such materials are photoresponsive polymers[94, 120] and gels[121], molecular platforms[122], porous organic materials[56c, 123], surfaces[124], and solar thermal energy storage devices[4, 56a, 125].

To give one specific example: HEINKE and HECHT group frequently collaborate to investigate azobenzene-containing metal-organic frameworks (MOFs). Recently, they reported a nanoporous film made of such a framework material. Here, the wettability is controlled by photoswitching of the *ortho*-fluorinated azobenzene moieties and by reversible incorporation of guest molecules with different polarities into the pores.[126] Dynamic control of this interfacial property facilitates the application in sensors (such as coatings for oil/water separation)[127] and is typically investigated by (static) contact angle measurements. This angle is formed at the intersection of the liquid-solid and liquid-vapor interface and serves as a classification criterium (hydrophilic ($\theta_e < 90°$), hydrophobic ($90° < \theta_e < 150°$) or superhydrophobic ($\theta_e > 150°$)).[128] By loading the nanoporous film of HEINKE and HECHT groups with various guest molecules (butanediol, octanol, and hexadecane), the polarity of the guest@MOF-film changes, shifting the water contact angle from ~25° to ~100°. The photomodulation of the surface-mounted metal–organic frameworks (SURMOF,) without guest molecules and the associated change in the contact angle is shown in Figure 3.

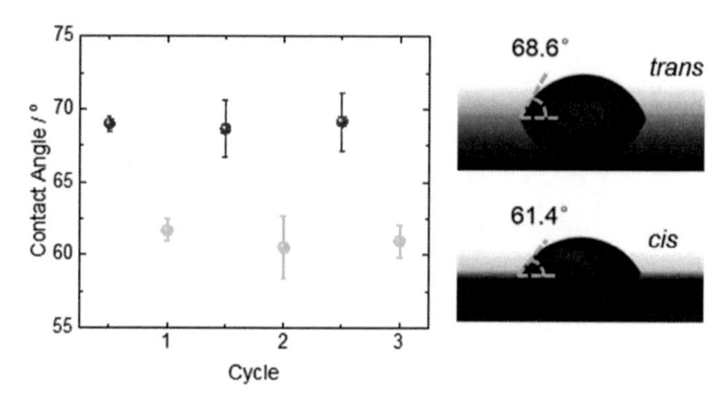

Figure 3: The contact angle of water on the photoswitchable (unloaded) SURMOF. The violet spheres were obtained upon violet light irradiation (*E*) and the green spheres upon green light (*Z*). The irradiation was performed for three cycles. For each point, the contact angle measurements were repeated three times at different positions on the sample and the average values with standard deviations are shown. Photographs of the water droplet on the sample in *trans* and *cis* are shown on the right-hand side. Taken and cited in verbatim from Z. Zhang, D. Chen, D. Mutruc, S. Hecht and L. Heinke, *Chem. Commun.*, 2022, 58, 13963 DOI: 10.1039/D2CC03862E.

Furthermore, there is an increasing interest in stimuli-responsive polymers. Their sensitivity to small changes in the environmental conditions caused by physical, chemical, or biochemical stimuli enables their application in the fields of biotechnology, medicine, and engineering. The largest class of smart polymers is represented by temperature-responsive polymers, featuring lower critical solution temperatures, nevertheless photo-responsive polymers are gaining more and more interest due to the spatiotemporal precision of the stimulus light.[129] Some polymers fall into another class of materials, namely gels, which will be discussed in more depth in the following chapter.

1.2 Gels

The term gel is not easily defined. Gels represent a physical state that exhibits properties of both solids and liquids. Already in 1861, Thomas Graham attempted to describe his observations about gels, since at that time many analytical tools were still missing: "As gelatine appears to be its type, it is proposed to designate substances of the class as *colloids*, and to speak of their peculiar form of aggregation as the *colloidal condition of matter*."[130] Later in 1949, Hermans suggested the following criteria: "(1) gels are coherent colloid disperse systems of at least two components" (gelling agent and fluid component), "(2) gels exhibit mechanical properties characteristic of the solid state and (3) each component extend themselves continuously throughout the whole system".[131] These criteria are still used today, even though deeper knowledge and understanding of the gel forming processes is now available.

As the fluid component is a crucial part in a gel, it also serves as a differentiating characteristic. The most prevalent types are hydrogels and organogels, which use a liquid solvent (water or organic solvent, respectively) as fluid component. However, there exist special cases in which, for example, an ionic liquid serves as the fluid gel part, resulting in ion gels.[132] Here, problems such as evaporation or leakage are avoided.[133] If the fluid component is removed by evaporation or freeze-drying, the gels often shrink and what remains is a dry solid, referred to as xerogel. Upon rewetting with the fluid component, the xerogel swells and reforms the gel.[134] Further classification of gels is dependent on parameters like the charge or preparation method, but also on their mechanical and structural characteristics (Scheme 17).

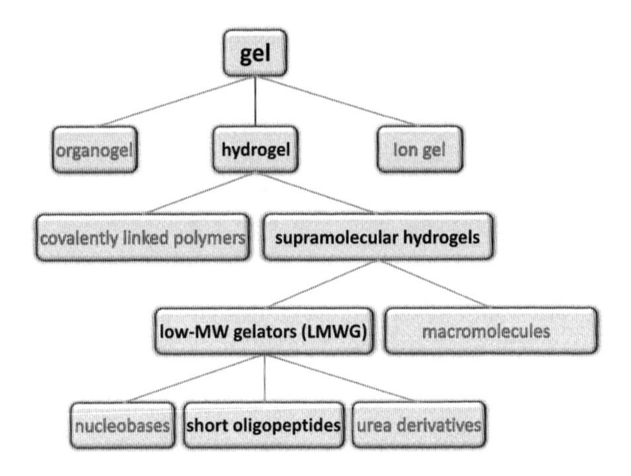

Scheme 17: Hierarchical display of gel classification. The highlighted categories are particularly relevant in this thesis.

1.2.1 Hydrogels

Hydrogels are highly suitable for biomedical applications (controlled therapeutic delivery[135], wound healing[136], tissue engineering[137], tissue adhesion[138] and more) due to their high water content and their potential compositional and mechanical similarities to native soft tissues.[139] Therefore, hydrogels are often referred to as soft materials. As they are the core of this work, they will be discussed in more detail. Hydrogels are either covalently crosslinked polymer networks (gelatin, alginate, agarose, synthesized by conventional polymerization methods[140]) or formed by physical forces as supramolecular hydrogels from macromolecules or low molecular weight gelators (LMWGs).

When developing gels for biomedical applications, the properties of the gel need to be customized to fit the properties of the biological target tissue, such as mechanical strength. This strength and elasticity are required when skin, cartilage, or even vascular conduit[141] are simulated, for which purpose polymeric hydrogels are more suitable. Additionally, polymeric hydrogels of natural origin - like collagen, gelatin and glycosaminoglycans - have a high affinity for proteins, which can be advantageous in biological applications. For instance, in tissue engineering cell attachment is promoted when certain proteins (extracellular matrix proteins) are presented on the gel surface.[142]

Further essential parameters for biomedical applications include self-healing (the ability to rebuild bonds and repair damage to restore the initial gel), stimuli responsiveness (heat, pH, light, redox potential)[143], thixotropy (the reversible, isothermal gel-to-sol transition when a viscous material is under agitation or subjected to increased shear rate[144]) and on-demand reversibility.[145] These requirements are difficult to fulfill for covalently crosslinked polymers without causing polymer degradation and the preparation of synthetic polymers by free radical or condensation chemistry is also not compatible with *in vivo* conditions.

Consequently, other approaches are exploited in which linear polymer chains are crosslinked by physical interactions between the chains. These interactions can be disrupted by shear stress, enabling, for example, injection of these gels.[140] Hydrogels formed by these non-covalent interactions are classified as supramolecular hydrogels. Work in the field of supramolecular chemistry, also termed "chemistry beyond the molecule", was honored in 1987 with the Nobel Prize, which was awarded to Donald J. Cram, Jean-Marie Lehn, and Charles J. Pedersen "for their development and use of molecules with structure-specific interactions of high selectivity".[146] From then on, the scientific interest in studying molecular recognition and high-order assemblies formed by noncovalent interactions increased[147], which is also reflected in the number of publications featuring supramolecular hydrogels (from 2000 to 2022, the number of annual publications increased by a factor of >100)[148]. In more detail, supramolecular hydrogels are formed either by LMWGs (short linear oligopeptides,[149] nucleobases,[150] or urea derivatives[151]) or macromolecules in a spontaneous and often synergistic manner by self-assembly through dynamic intermolecular non-covalent bonds. These are π–π stacking, hydrogen bonding, metal-ligand coordination, host-guest, electrostatic and/or van der Waals interactions.[152]

One excellent example for the crosslinking of macromolecules is the Ca^{2+} dependent gelation of alginate and pectin. Alginate is a naturally occurring polysaccharide composed of β-D-mannuronic and α-L-guluronic acid. It can be found in brown seaweeds and some bacteria, while pectin mainly consists of derivatives of galacturonic acid and is produced in the cell walls of terrestrial plants such as citrus fruits. Due to their natural origin, both polysaccharides are interesting gelling agents for the food industry. The gelation mechanism is based on the classic egg-box model (Figure 4 a)), in which two antiparallel polyuronate chains are stabilized by Ca^{2+} to form egg-box shaped dimers and aggregate in lateral direction to form multimers.[153] Due to the reliable gelation, alginate gels were applied in the concept of double-network gelation (two interpenetrating networks[154]) to increase gel strength and toughness by combining alginate and polyethylene glycol.[155] Even more interesting is the application of a double-network hydrogel composed of sodium alginate and gelatin crosslinked by calcium chloride and microbial transglutaminase, which promotes covalent crosslinks between glutamine and lysine groups, respectively.

Venous and arterial engineered vessels are formed by microfluidic bioprinting of these components and corresponding human cells can be seeded into the inner and outer part of the tube to exhibit key properties of blood vessels (Figure 4 b)).[141]

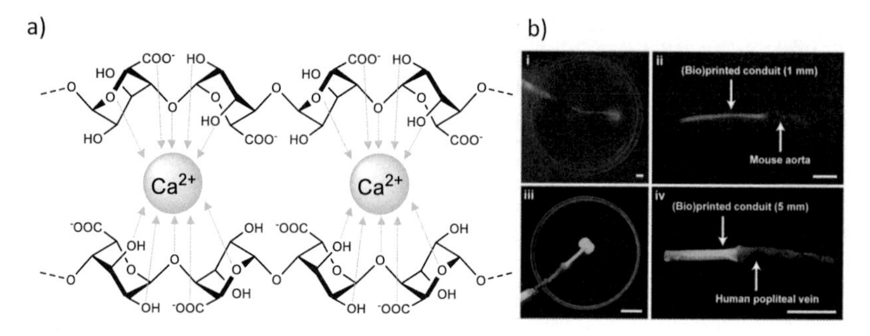

Figure 4: a) schematic representation of the egg-box model in alginate hydrogels. b) *Ex vivo* connection and perfusion of bioprinted vascular conduits linked to native vessels by bioglue. (i and ii) A small-sized printed vascular conduit (1-mm outer diameter) anastomosed with a mouse aorta; (iii and iv) a printed larger-sized (5-mm outer diameter) vascular conduit anastomosed with a human popliteal vein. Scale bars, 20 mm. (adapted from Wang, Di, *et al.* "Microfluidic bioprinting of tough hydrogel-based vascular conduits for functional blood vessels." Science Advances 8.43 (2022): eabq6900. in accordance with the terms of the Creative Commons (CC) license).

1.2.2 Peptides as LMWGs

LMWGs are typically small molecules with a molecular weight below 3000 Da.[156] Their molecular structure is often amphiphilic or bolaamphiphilic, which facilitates their supramolecular arrangement into anisotropic constructs, usually nanofibers. With increasing concentration, these fibers are starting to intertwine and crosslink to three-dimensional networks with the capacity to immobilize the fluid component through surface tension and capillary forces.[145]

The interplay of hydrophilicity and hydrophobicity is crucial for the property of a LMWG to build strong fibers. If balance is not maintained, the LMWG will either dissolve in the solvent or precipitate. Moreover, biocompatibility is essential and therefore artificial derivatives of natural biomolecules, such as amino acids or small peptides, are usually chosen. The functional groups in amino acids or the peptide bond can form intermolecular hydrogen bonds, though they are not sufficient to form strong fibers. Hydrophobic amino acid side chains with π–π stacking ability or capping with aromatic groups provide the required additional stabilization to obtain potent LMWGs.[157] However, these aromatic capping groups may limit biocompatibility. Therefore, a new strategy was followed by NACHTSHEIM group: Two amino acids are combined to the smallest cyclic peptide, diketopiperazine (DKP), rendering the capping group dispensable (Figure 5).[158] Additionally, the rigidity of the DKP ring in the core of the molecule promotes cooperative intermolecular hydrogen bonding, which is predicted to be more favorable than in linear peptide chains.

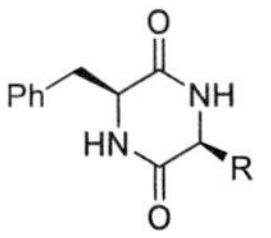

DKP

46 R = H, cyclo(L-Phe-Gly)
47 CH_2-OH, cyclo(L-Phe-L-Ser)
48 CH_2-SH, cyclo(L-Phe-L-Cys)
49 CH_2-CH_2-CO_2H, cyclo(L-Phe-L-Glu)
50 CH_2-(4-imidazolyl), cyclo(L-Phe-L-His)
51 $(CH_2)_4$-NH_2, cyclo(L-Phe-L-Lys)

Figure 5: Cyclic dipeptides by NACHTSHEIM group. The DKP core structure is highlighted in blue.

The NACHTSHEIM group synthesized a small library of DKPs consisting of phenylalanine condensed with one of the hydrophilic amino acids glycine (**46**), serine (**47**), cysteine (**48**), glutamate (**49**), histidine (**50**), and lysine (**51**). All candidates were proven to form hydrogels. The lowest critical gelation concentration was detected for the cysteine derivative, but lysine or glutamate are also interesting due to their pH-dependency and cooperative gelation property when combined.[158]

1.2.3 Photoswitchable hydrogels

Self-assembly and dissipation of gels can be triggered by various parameters. However, light as a trigger is gaining more and more attention, as it can be applied precisely with spatial and temporal control in a non-contaminating manner and orthogonal to other influences, such as pH, redox potential, or mechanical force. Additionally, biocompatibility is ensured when the correct range of wavelengths is chosen. Consequently, efforts were made to develop photoreactive hydrogels and different concepts were applied. Their design ranges from the implementation of photocleavable 2-nitrobenzyl groups into short peptides[159], over gel-to-sol transitions with *E*/*Z*-photoswitchable fumaric-amide motif[160] for example in nano-gel droplets [161] to the installment of fully functional sophisticated photoswitches like spiropyrane[162] or azobenzene[163] into short peptides. Furthermore, photoswitches can be designed as standalone hydrogelators such as arylazopyrozoles, which are the lowest known LMWGs that respond reversibly to light.[164] In the case of photoswitches, the ability of the switches to stack in one of their isomeric forms (*E*-azobenzene/merocyanine) by intermolecular π–π stacking provides the required stabilization to form hydrogels from LMWGs. Both switches experience a large change in polarity upon isomerization into their corresponding isomers and the loss of π–π stacking stabilization leads to gel-to-sol transitions or gel destabilization.

This concept was adapted from Pianowski *et al.* who saw the potential in the DKP based LMWGs of NACHTSHEIM group to form intermolecular π–π stacking with the phenylalanine side chain and elongated the phenyl ring to an azobenzene (Scheme 18).[121b]

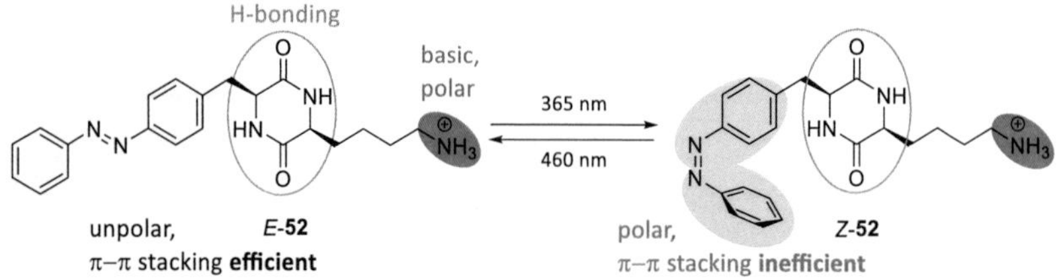

Scheme 18: LMWG **52** developed by Pianowski *et al.* The molecule is built from the natural amino acid L-lysine and an unnatural azobenzene amino acid. The subunits which are critical for gelation are highlighted.[121b]

The LMWG **52** (Scheme 18) can form supramolecular self-healing hydrogels with various additives or dopants (acid, NaCl or DNA) to tune the strength of the gel. The addition of double stranded DNA oligomers (ca. 1300 bp long) increased the melting temperature by over 10 °C compared to the undoped gel due to the cooperative effect of multiple salt bridges. The DNA dopant could be released from the gel by irradiation with UV light (365 nm). Independently, the gel could be transformed to liquid by UV light irradiation and the gel structure was restored by irradiation with blue light (460 nm).[121b] The LMWG **52** itself is stable under UV light (>24 h, 365 nm, 20 W), however, the potential dopant (especially DNA) or the desired application may be sensitive to this highly energetic light. Therefore, the azobenzene properties needed some adjustment and based on the minor structural adaptations required, *ortho* fluorinated azobenzenes were synthesized into corresponding biocompatible LMWGs by PIANOWSKI group[165] These visible-light-addressable LMWGs were successfully applied for drug delivery and antibiotics or anti-inflammatory drugs[165], but also the potent anti-cancer agent plinabulin[135b] was released from the gel by visible light.

Though such LMWGs already exhibit the desired properties such as visible light switching and good stability, there is still potential for optimization. Regarding the synthetic effort, it would be favorable to reduce the critical gelation concentration by modification of the LMWG or investigation of the scope of dopants. Moreover, activation by red or NIR light would further increase biocompatibility and enable deeper penetration of tissue.

1.3 Aggregation induced emission

A common effect in conventional organic dyes or fluorophores is aggregation-caused quenching (ACQ) due to intermolecular π-π stacking, which can limit their applicability in, for example, organic nanodots or organic light emitting diodes (OLED).[166] The opposite effect is aggregation-induced emission (AIE) with weak or negligible emission in dilute solution but strong emission in the aggregate state. This concept is comparatively new and was first formulated in 2001 by Luo *et al.* who described the propeller shaped compound 1-methyl-1,2,3,4,5-pentaphenylsilole[167], which is barely visible as a wet spot on a thin layer chromatography (TLC) plate, but exhibits strong fluorescence as a dry spot. Simply put, the mechanism of AIE relies on the restriction of intramolecular motions/rotations (RIM).[168] In dilute solution, rotation and vibration are rapid, so the nonradiative decay rate k_{nr} is high. In aggregate or solid-state, intermolecular interactions such as π-π stacking restrict intramolecular motions and with k_{nr} suppressed, the radiative decay k_r can compete. However, aggregate formation is not mandatory, taking the mechanism as basis, RIM can also be achieved at low temperatures[169], in highly viscous solutions[169], in rigid matrices[170] or in metal organic frameworks[171]. Further investigation of the AIE effect led to the discovery of another phenomenon, the aggregation induced emission enhancement[172] (AIEE). AIEE-active molecules emit weakly in all physical states and exhibit an emission enhancement under certain conditions. This effect occurs in association with intramolecular planarization, specific aggregation (H- or J-aggregation[173]), twisted intramolecular charge transfer (TICT), ESPT or *E*/*Z*-isomerization.[174] However, none of these mechanisms apply universally and it was agreed to keep RIM as the most reliable mechanistic explanation.[175]

AIE-luminogens (AIE-gen) are promising candidates for application in bioimaging, chemosensing, optoelectronics and stimuli-responsive systems[168], thus the development of new AIE-gens has been vigorously pursued. Here, the potential of 1,8-naphthalimide (**53**) derivatives was soon recognized due to their high fluorescence quantum yields, outstanding photostability[176] and facile synthetic incorporation by *N*-functionalization of 1,8-naphthalic anhydride precursors by Gabriel synthesis.

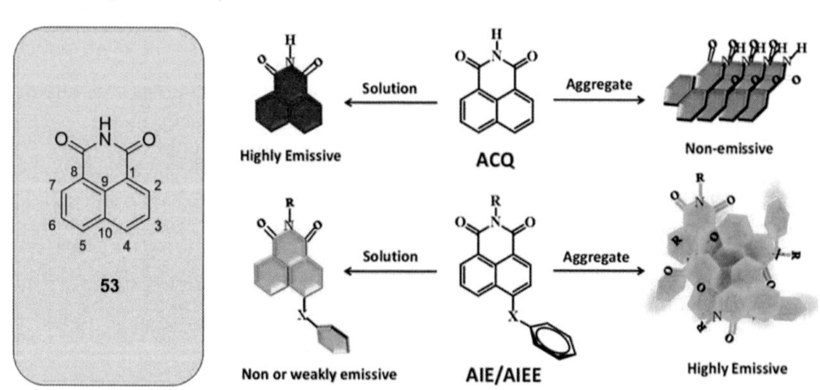

Scheme 19: Structure of 1,8-naphthalimide with reactive positions 3 and 4 highlighted. Representation of the ACQ and AIE/AIEE effect. Reprinted (adapted) with permission from *ACS Appl. Mater. Interfaces* 2018, 10, 15, 12081-12111. Copyright (2023) American Chemical Society.

The naphthalic anhydride precursors and unsubstituted naphthalimide normally suffer from aggregation-caused quenching. Their N-functionalization by alkyl chains, small to medium peptides or amphiphilic π-systems transforms the molecules into AIE-gens. Furthermore, the aromatic core of 1,8-naphthalimides is an electron acceptor and functionalization at the 3 or 4 position (Scheme 19) with donor structures (amine or hydroxyl groups) also results in AIE-gen generation or in solvatochromic effects by intramolecular charge transfer (ICT).[177]

Examples are reported in the literature where the effect of peptide sequences on supramolecular interactions of naphthalimide/peptide conjugates was investigated (Scheme 20). Referring back to Section 1.2.2, small peptides can be capped with aromatic scaffolds to produce functional hydrogelators. Based on the previously described properties, naphthalimide qualifies as such a scaffold and supramolecular hydrogelators with AIE-gen properties containing it were developed.[178] Moreover, the cyclic dipeptide DKP repeatedly demonstrates its qualities in the assembly of supramolecular structures. Under aqueous conditions, strong hydrogen bonding promotes the formation of aggregates, while the addition of DMSO inhibits hydrogen bonding and quenches the fluorescence. Such a system was applied as a chemosensor for the detection of electron-deficient phenolic compounds and drugs.[179]

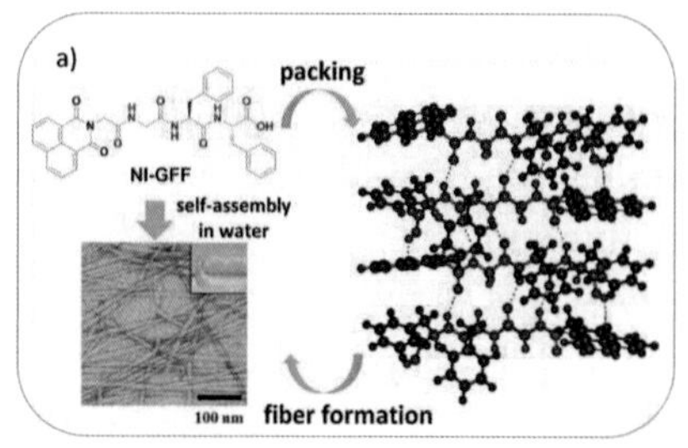

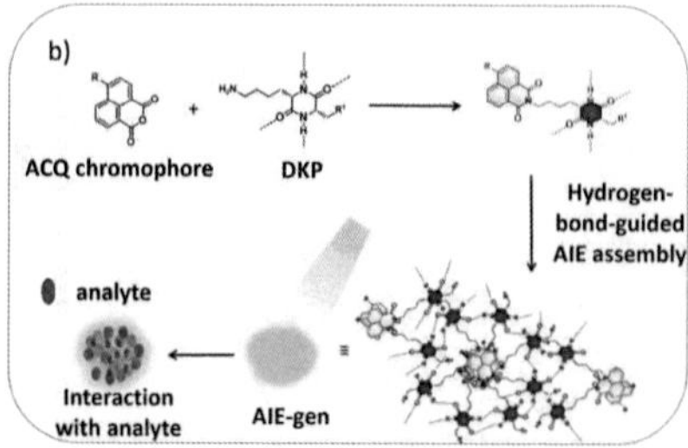

Scheme 20:a) Naphthalimide-tripeptide conjugates self-assemble to fibers in water. Reprinted (adapted) with permission from *Langmuir 2016, 32, 30, 7630-7638*. Copyright (2023) American Chemical Society." b) AIE-gen which assembles due to hydrogen bonding of the DKP units and is applied as chemosensor. Reprinted (adapted) with permission from *J. Org. Chem. 2020, 85, 3, 1525-1536* Copyright (2023) American Chemical Society."

2 Objective

Hydrogels are soft material, which are often applied in biomedical context. Self-healing, thixotropy, on-demand reversibility and stimuli responsiveness are important parameters for extending the applicability of biomedical hydrogels. A promising stimulus is light on which molecular photoswitches react in high spatiotemporal precision. The research described in this thesis focuses on the synthesis of azobenzene photoswitches which react on red light, the development and investigation of novel biocompatible photochromic supramolecular hydrogels and the improvement of existing supramolecular hydrogel systems (Figure 6).

In the first part of this thesis, the isomerization wavelength of *ortho* fluorinated azobenzenes was bathochromically shifted by conjugation with unsaturated substituents. Besides detailed photophysical investigation, the new chromophore was employed in a photochromic smart material (**54**).

In the next part, a non-conjugated version (**55**) of the new *ortho* fluorinated smart material was synthesized. The aim was to assess the influence of the flexibility at the azobenzene-DKP linkage on the gelation behavior.

In the following part, a new photochromic LMWG **56** was synthesized with the aim to serve as model system to investigate the effects of π-π stacking and ionic interactions on gel formation and stability.

Next, the new gelation conditions, which were determined in this work, were applied on existing LMWGs **52** and **57** (developed by the PIANOWSKI group) to reduce the critical gelation concentration, improve the stability and enhance biocompatibility.

Synthetic access to an *ortho* chloro and difluoro-dichloro azobenzene amino acid (**73** and **115**) was developed and the chromophores were applied in novel LMWGs **58** and **59** to operate with light within the therapeutical window.

In the last part, the hydrogen bond promoting peptidic structure, which can be found in the DKP motif, was linked with the fluorophore naphthalimide to obtain AIE-gens **93** and **94**. Fluorescence measurement revealed the potential to form nanoaggregates.

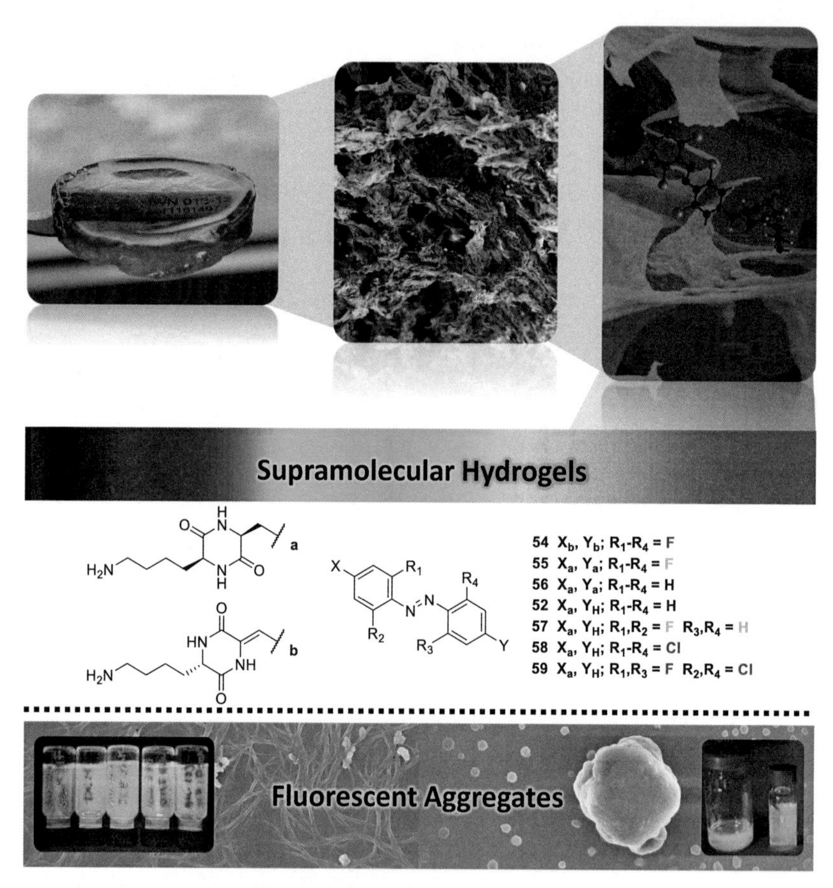

Figure 6: The supramolecular hydrogel project intends to combine various azobenzenes reacting on UV, green or red light with the DKP-Lys gelator motif. The hydrogen bond promoting peptide structure of the hydrogelator was combined with a fluorophore and fluorescent aggregates were formed.

3 Results and discussion

3.1 Fluorinated azobenzenes switchable with red light

Preface

Parts of the following chapter were published in Chemistry—A European Journal (WILEY-VCH)

A.-L. Leistner, S. Kirchner, J. Karcher, T. Bantle, M. L. Schulte, P. Gödtel, C. Fengler, Z. L. Pianowski, *Chem. Eur. J.* **2021**, *27*, 8094.

DOI: 10.1002/chem.202005486

Fluorinated Azobenzenes Switchable with Red Light

Author contributions

The synthesis and characterization was done by the first author, with contributions by J.Karcher, T. Bantle. and M. L. Schulte. All calculations were performed by S. Kirchner with the GAUSSIAN09 program package,[180] employing the B3LYP/6311G* level of theory.[181]

Motivation

Johannes Karcher developed the first photochromic hydrogelators of the PIANOWSKI group. Here the most promising candidate was a tetra-*ortho*-fluorine substituted azobenzene hydrogelator, due to its robustness and addressability with visible light. This compound is biocompatible, can form stable hydrogels under biological conditions (in PBS) and incorporate the potent anti-cancer agent Plinabulin, which is released upon green light irradiation.[135b] However, the preparation of this hydrogelator is tedious, particularly the formation of the chiral fluorinated azobenzene amino acid. To facilitate the synthesis, it was decided to remove the stereogenic center connecting the azobenzene unit to the DKP scaffold and replace it with an achiral double bond. The desired molecule is easily accessible by aldol condensation of aromatic aldehydes with *N*-protected DKPs. Consequently, a tetra-*ortho*-fluorinated azobenzene aldehyde

(**36**) was the target of previous studies.[182] During investigation of the photochromic characteristics of tetra-*ortho*-fluorinated azobenzene *para*-aldehyde **36**, unusual behavior was observed. *E*/*Z*-photoisomerization of the target compound was addressable with red-light. Until then, tetra-*ortho*-fluorinated azobenzenes were known to isomerize exclusively under green light.[89] To further analyze these observations, a small library of aldehyde/alcohol functionalized *ortho*-fluorinated azobenzenes was synthesized (Figure 7). Part of the synthesis and their complete photophysical investigation entered the scope of this PhD thesis. Moreover, the targeted gelator **54** with removed chirality center at the azobenzene-DKP linkage (although in its dimeric form) was synthesized and characterized.

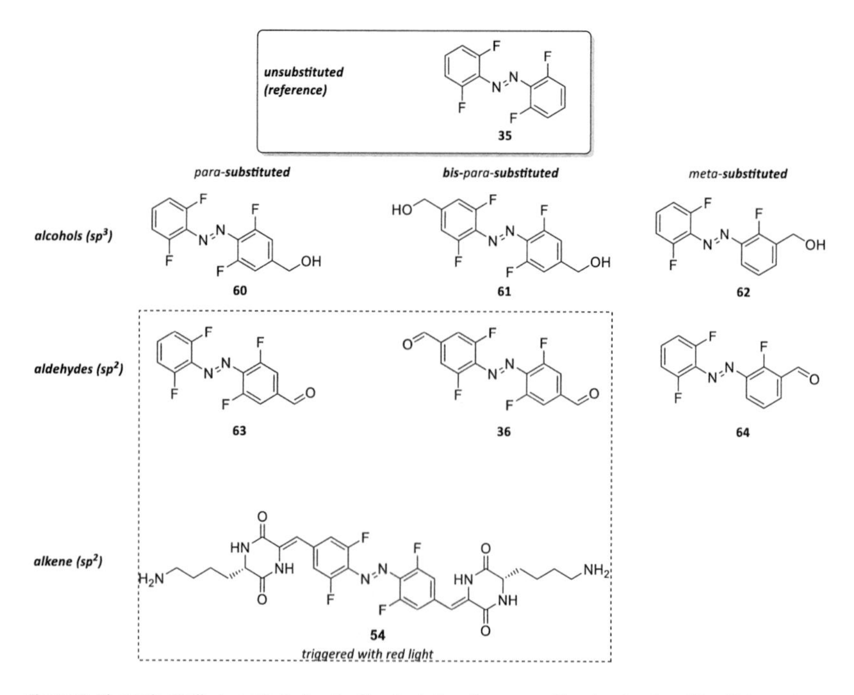

Figure 7: Photophysically investigated *ortho*-fluorinated azobenzenes. The structural motif, which reacts on red light is highlighted in red color.

Synthesis

Part of the discussed structures (**35, 36, 60-64**) were synthesized during the master thesis of Anna-Lena Leistner. The synthesis of bis-*para*-substituted aldehyde **36** over the bis-*para*-substituted alcohol **61** is time-consuming and low-yielding (7-step synthesis, 3% overall yield). An alternative synthetic route (Scheme 21) was developed starting from the cheap building block 2,6-difluoroaniline (**65**) which was first formylated in a copper-catalyzed tandem oxidation/formylation reaction where DMSO serves as solvent and carbonyl carbon source.[183] This was followed by a homocoupling reaction (0) using NCS to receive the target bis-*para*-substituted tetra-*ortho*-fluoroazobenzene **36**. Thereby, the purification was optimized: first flash column chromatography was performed to remove major impurities, and then overoxidized side products (*para* carboxylic acid) were removed by trituration with methanol. In summary, the synthetic effort was reduced to two steps and the overall yield was increased to 16%. With the bis-*para*-substituted tetra-*ortho*-fluoroazobenzene **36**, the core building block for subsequent aldol condensation with the DKP was formed.

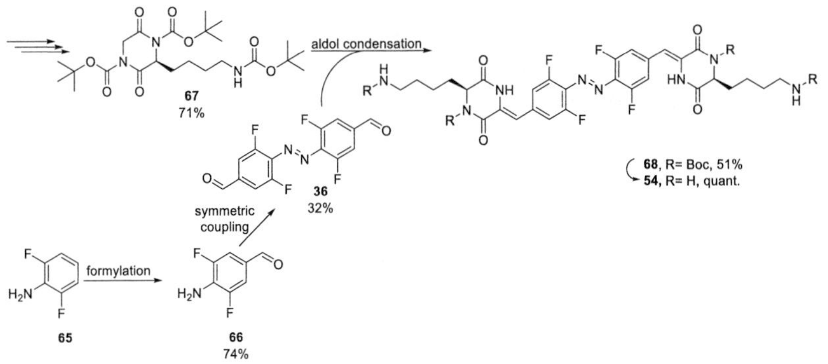

67 71%; aldol condensation; **36** 32%; symmetric coupling; **65**; formylation; **66** 74%; **68**, R= Boc, 51%; **54**, R= H, quant.

Scheme 21: Key step of the modular synthesis towards **54** is the aldol condensation between bis-*para*-substituted tetra-*ortho*-fluoroazobenzene **36** and the functionalized DKP **67**.

The second part of the synthesis was the development of the DKP building block **67** formed from the dipeptide of glycine and lysine, bearing Boc protection groups to avoid side reactions and support the aldol condensation. The reaction enhancing effect of Boc at the DKP amide is comparable to acetylation. This was first observed by Gallina *et al.* (1973), who described that 1,4-diacetylpiperazine-2,5-dione is more reactive than the non-acetylated form.[184] Commercially available Fmoc-Lys(Boc)-OH was condensed with the methyl ester of glycine and subsequently cyclized to the DKP cyclo(Lys(Boc)-Gly) **96**. The amides in the DKP-ring were protected by the Boc-protection group to obtain the DKP building block **67**. In a double aldol condensation reaction, the DKP building block was attached to the azobenzene in both *para*-positions in moderate yield (**68**, 51%). A final quantitative deprotection produced the new hydrogelator bis-(cyclo(Lys)-(2,6-difluoro-4-vinyl-azobenzene) **54**.

Photophysical properties

Photophysical investigation of the azobenzenes **35, 36, 60-64** started by determination of the PSS using ^{1}H NMR spectroscopy measurements (summarized in Table 1). In accordance with the observations during Anna-Lena Leistners master thesis, conjugated aldehydes **36**, **63** and compound **54** produced high $PSS_{E \rightarrow Z}$ under red light irradiation. Furthermore, pronounced photochromism was observed (Scheme 22).

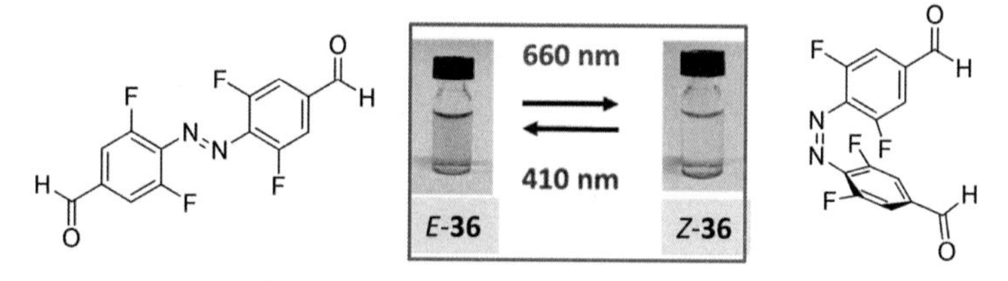

Scheme 22: Photoisomerization of the *bis-para*-aldehyde **36**. The photochromism can be observed with the naked eye as the color of the dissolved compound **36** changes from orange to yellow after *E*→*Z* isomerization.

The negative control compounds which are the non-conjugated aldehyde **64**, alcohols **60**, **61**, **62** or the unsubstituted tetra-*ortho*-fluoroazobenzene **35** did not switch significantly under red light irradiation. Reverse photoisomerization from *Z*→*E* occurred under irradiation with violet light.

Table 1: PSS determined by ^{1}H NMR measurements. (The 660 nm LED was additionally equipped with a 630 nm cut-off filter SCHOTT RG-630 to eliminate the <630 nm component of the emitted light.)

Compound	% of *Z*-x in PSS λ_{max} 407 nm	% of *Z*-x in PSS λ_{max} 523 nm	% of *Z*-x in PSS λ_{max} 660 nm
35	15%	94%	20%
60	15%	92%	19%
61	15%	91%	24%
63	10%	83%	**75%**
36	15%	67%	**82%**
62	18%	85%	29%
64	16%	87%	26%
54	29%[a]	55%	**61%**

π-System elongation is a common tool to vary the absorption bands of photoswitches; consequently, the hypothesis was that the bathochromic shift in the azobenzenes with *sp*^{2}-hybridized conjugated substituents can be attributed to the extended conjugated π-electron system of the chromophore. The shift was observed in the absorption spectra by a peak tail above 600 nm for **36**, **54** and **63**. For example, at 630 nm the molar extinction coefficient $\varepsilon_{630\ nm}$ of *E*-**36** was 11.5 $M^{-1}\ cm^{-1}$ and even higher for *E*-**54** with $\varepsilon_{630\ nm}$ = 18.3 $M^{-1}\ cm^{-1}$. The peak tail originates from the S_0 - S_1 absorption band, or more common n-π* transition, of azobenzenes and requires sufficient separation for the respective isomers to enable selective addressing by visible light irradiation. The band separation was exceptionally large for the *bis-para*-aldehyde **36** (Δ(n-π*) = 51 nm) but already mono *para*-aldehyde substitution (**63**) resulted in larger band separation (Δ(n-π*) = 45 nm) compared to the corresponding alcohols (Δ(n-π*) = 40 nm or 34 nm respectively (Figure 8).

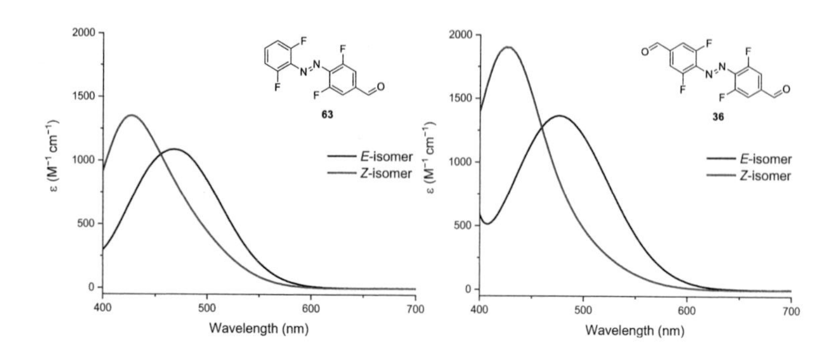

Figure 8: Molar extinction coefficients of compound **63** and **36** between 400 nm and 700 nm. The band separation of the n-π* transitions is clearly visible.

An important parameter of T-type photoswitches is their thermal stability, which was therefore investigated for the bathochromically shifted compounds **36**, **54** and **36** (Figure 63 and Figure 64). Corresponding samples were prepared and measured in MeCN at 60 °C (**36**, **63**) or DMSO (**54**) at 20 °C. The thermal half-life of *Z*-**63** was 10.8 h, which is comparable to other TFAB derivatives, while the half-life of *Z*-**36** was significantly lower with 3.2 h and more similar to unsubstituted azobenzenes.[61a] The half-life of compound **54** (14.8 h at 20 °C in DMSO) was considerably decreased compared to the aldehydes **36** and **63** but still in the range of other red light-switchable azobenzenes (see FK11 by WOOLLEY group: $t_{1/2}$ = 6 h at 37 °C).[90]

To gain deeper insight into the observations, the experimental data were corroborated with theoretical calculations performed by Susanne Kirchner with the GAUSSIAN09 program package,[180] employing the PBE0-D3/def2-TZVP level of theory.[181] The hypothesis that *para*-substitution with aldehydes results in an extended π-conjugation was supported by the results obtained in a ground state optimization. Both HOMO (n) and LUMO (π*) were shifted to lower energies for compounds **36** and **63**. This effect was more pronounced for the LUMO (π*) which resulted in a smaller HOMO-LUMO gap for the *para*-aldehyde substituted compounds (**63**, $\Delta_{HL}(E)$ = 3.838 eV; **36**, $\Delta_{HL}(E)$ = 3.670 eV) compared to the control substances non-conjugated aldehyde **64**

($\Delta_{HL}(E)$ = 4.077 eV), unsubstituted **35** ($\Delta_{HL}(E)$ = 4.077 eV), or sp^3-substituted derivatives **60, 61** or **62** ($\Delta_{HL}(E)$ = 4.047 eV, 4.015 eV or 4.115 eV respectively) (see also 5.3.2 computations). Based on the optimized structures, time-dependent calculations were performed, applying the polarizable continuum model (PCM) for the solvent MeCN (see 5.3.2 computations, Table 11). The calculated spectra agree with the measured spectra (Figure 96 to Figure 98) and the bathochromic shift of compounds **36** and **63** is clearly visible.

Next, the photostability of compound **54** was investigated in view of the desired application as hydrogelator, for which multiple cycles of irradiation are of interest. As proof of photostability, a 1 mM solution in DMSO was cycled (10×) between the $PSS_{660\,nm}$ and $PSS_{470\,nm}$ (Figure 60). Additionally, the HPLC traces of the first and the last cycle can be found in Figure 61 and Figure 62. The compound **54** remained stable during all cycles. Another parameter to be considered in biological context is the stability under reducing conditions. Reduction of azobenzene derivatives to arylhydrazines by thiol groups in biological systems is considered the most serious limitation for their *in vivo* application as photoswitches. To mimic the intracellular reduction potential, compound **54** was added to a glutathione solution (10 mM) and analyzed over the course of 12 h at 25 °C. After 10 h, 50% of the compound was degraded, which is restrictive for some long-time *in vivo* applications.

Compound 54 as hydrogelator

The next step was to verify whether a compound without the chiral center connecting the azobenzene unit to the DKP scaffold is a promising LMWG. The structure of **54** is bolaamphiphilic, as the hydrophilic DKP units with lysine side chain are separated by the hydrophobic planar azobenzene core. Based on the chemical structure fiber formation was assumed. However, LMWGs are sensitive to environmental conditions and the first challenge was the development of a proper gel formation conditions. First, compound **54** was suspended in distilled water and boiled for a short amount of time. At a concentration of 2 wt% an opaque viscous material was formed, which was stable at the inverted vial test and had a melting temperature of 55 °C. No viscous material was observed at lower concentrations. The preliminary work of Johannes Karcher demonstrated increased stability of azobenzene-DKP-Lysine LMWGs with salt as additive. Therefore, the influence of salts on viscous material of compound **54** was investigated (Table 2).

Table 2: Compound **54** formed stable viscous materials with different aqueous fluid components. The melting temperature was determined. Further information is given in 5.3.4. Vial a) – c) demonstrate the inverted vial test. a) non-stable hydrogel flows after light irradiation; b) stable hydrogel; c) stable hydrogel before inversion.

Fluid component	Approx. conc.	T_m °C
Water	2 wt%	55
Ringer's solution	1 wt%	84± 14
200 mM NaCl	1 wt%	93 ± 5

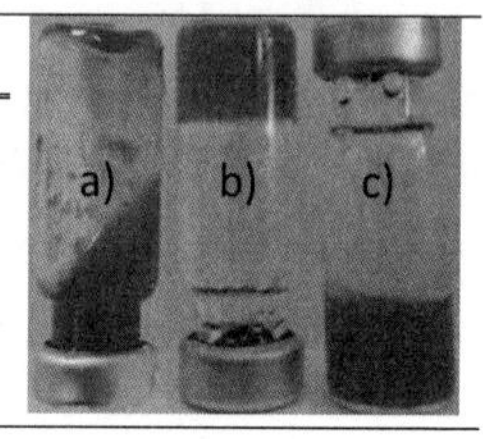

Under physiological conditions with Ringer's solution as fluid component the stability increased. The critical gelation concentration was lowered to 1 wt%, accompanied by an increase of the melting temperature to 84 °C. Further stabilization was observed in 200 mM aqueous NaCl solution. However, the formed viscous material was in all cases of low mechanical stability and collapsed under mechanical stress such as shaking of the vial.

Quantification of the viscoelastic behavior tan δ was done by rheological measurements. The composition of 2 wt% compound **54** in water was analyzed on a strain-controlled rotational rheometer. First the linear viscoelastic regime (LVE) was determined by a strain sweep at 1 Hz. A subsequent frequency sweep at a strain of 0.1% measured the frequency dependence of the storage G' and loss modulus G'' in the range of 0.03 – 100 Hz. The resulting values for the storage modulus G' were higher than for the loss modulus G'' with a $\tan\delta = G''/G'$ of 0.17 at 1 Hz, indicating a gel-like behavior. Compared to the tetra-*ortho*-fluoro LMWG by Johannes Karcher, however, G' of compound **54** (50 Pa) is at least one order of magnitude lower.[135b]

Photoisomerization of compound **54** had already been proven by NMR and UV-Vis experiments. Next, the influence of light irradiation (660 nm with cut-off filter >630 nm) on the viscous material was analyzed. After 5 min of irradiation (56 mW/cm^2), the viscous material dissipated to a non-viscous, opaque fluid (inverted vial test proved liquid flow, Table 2 a)). The viscous material could be restored by short boiling to thermally revert the photoswitch to the *E*-isomer.

Deeper insight into the structure was gained from electron microscopic investigation of the material (Figure 103 to Figure 108), with the results being summarized in Figure 9. In the dark state of the material (a) multiple µm-long thick rigid-fiber-like structures were observed. They are intersecting and additionally supported by smaller fibers.

When the material was irradiated at 660 nm (Figure 9 b) the quantity of thick fibers decreased, and the thin fibers become more prominent. After dilution of this sample and further irradiation at 660 nm a total decay of thick fibers was observed, and a thinner less rigid network remained (d). The initial state of the structure could be restored after shortly boiling the liquified material (c). The xerogel (e) was also examined by scanning electron microscope (SEM) and a sponge-like network was revealed.

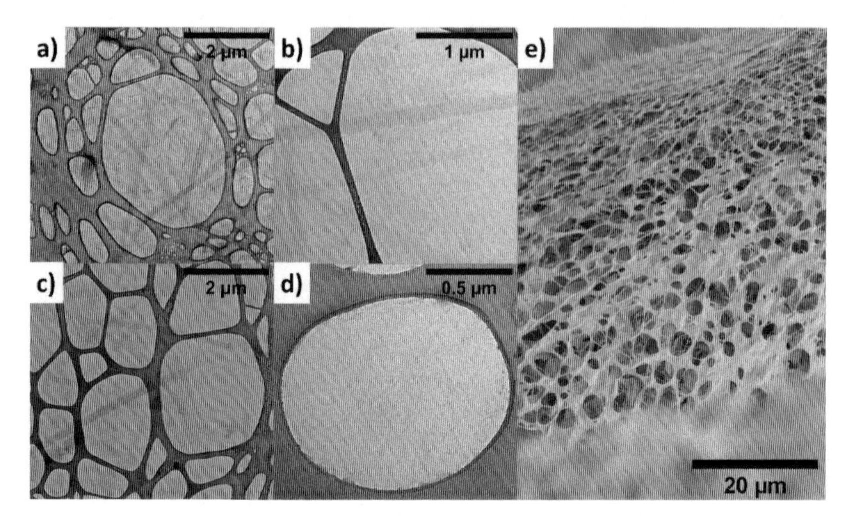

Figure 9: Electron microscopy of the supramolecular assemblies composed from **54** (1 wt%) under aqueous conditions (Ringer's solution+1 % lead citrate). a) The structure is visible as multiple micrometer long and nanometer thick fibers, supported by fibers in nanometer size; b) after irradiation at 660 nm (50 min), the quantity of the described thick fibers decreases and more smaller structures in the nm-scale are formed; c) regeneration of thick fibers occurs after boiling the irradiated gel; d) upon 1 : 10 dilution of b) and irradiation at 660 nm, only small fibers are visible. e) freeze-dried samples of hydrogel from 2 wt% **54** in water. A sponge-like network is revealed.

The conditions in electron microscopy strongly deviate from the natural environment of the viscous material. Therefore, ^{1}H NMR transverse relaxation (T_2) measurements were performed to analyze the microstructure of the material wetted with fluid component by detection of the molecular mobility (1-10 nm). The decay of the transverse magnetization using time-domain NMR techniques is quantitatively linked to the molecular mobility. A magic sandwich echo (MSE) pulse sequence was used to refocus the initial transverse magnetization (100 ms) of rigid components in combination with a Carr-Purcell-Meiboom-Gill (CPMG) pulse sequence to refocus the magnetization of more mobile components up to 1 s. The results of the measurements are shown in Figure 102. During the first 100 ms the FID decayed by 80%. This observation correlates with a highly polycrystalline microstructure and supports the rigid structures proposed by the microscope images.

Conclusion

Removal of the chiral center connecting the azobenzene unit to the DKP scaffold provided a modular synthetic approach towards potential LMWG materials. The synthetic effort was decreased compared to the tetra-*ortho*-fluorine substituted azobenzene hydrogelator of Johannes Karcher additionally, the Boc-protected building block cyclo(Lys(Boc)-Gly) (**67**) can be functionalized with various aromatic aldehydes to produce even more novel materials. However, the investigation of material formed by **54** indicated that a flexible linker between azobenzene and DKP scaffold is essential for efficient self-assembly of fibrous networks.

In conclusion, a new approach to design bathochromically shifted azobenzenes was demonstrated. The new conjugated TFAB chromophore **36** can be incorporated into larger structures. For demonstration, it was applied in a functional material of viscous character in aqueous media under physiological conditions, which can be liquified to a non-viscous fluid by red-light irradiation. The biocompatibility of the material is high enough to elicit effects, for example liquefaction upon irradiation occurs faster than biodegradation, thus rendering the conjugated TFAB chromophore **36** a valuable building block for incorporation into biomaterials or in photopharmacology agents.

3.2 Symmetrical chiral TFAB gelator

Motivation

In chapter 3.1 a new symmetrical LMWG **54** with a conjugated TFAB as core structure was described. The missing flexibility was detrimental for hydrogelation with high mechanical stability. So far, flexible but asymmetric systems consisting of azobenzene and DKP-Lys were demonstrated by Johannes Karcher. The logical consequence is the development of a flexible and symmetric version of the TFAB hydrogelator **54** (Figure 10) to gain deeper insight into the influence of rigidity on the system and to compare the new symmetric system to the asymmetric version.

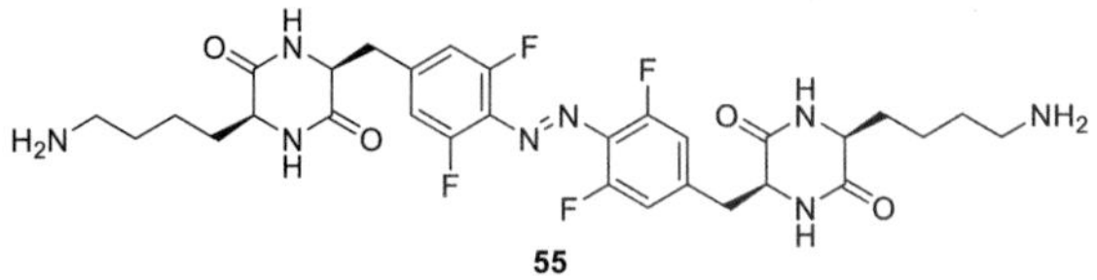

55

Figure 10: Structure of the targeted symmetric TFAB hydrogelator **55**. The structure of the already known asymmetric TFAB hydrogelator system of Johannes Karcher is depicted in blue.

Synthesis

The synthesis towards the new target structure (Figure 10) was based on the strategy developed by Johannes Karcher. The synthetic steps to amino acid building block **97** were performed according to the procedures already published with similar yields.[135b] Next, the new amino acid was symmetrically coupled to form the azobenzene core of the structure (**98**). The following steps apply for both functional ends of the molecule. First, the methyl ester was saponified with excess of lithium hydroxide. To purify the dicarboxylic acid **99**, addition of formic acid (FA) was necessary during column chromatographic purification. Then, the free carboxylic acids reacted in a peptide coupling with commercially available H-Lys(Boc)-OMe·HCl to **100**. The subsequent removal of the Boc protecting groups under acidic conditions proceeded quantitatively. In the last step, the 2,5-DKP was formed by intramolecular cyclization. Simple washing of the crude product with cold MeCN to remove the reagents gave the title compound **55** in excellent yield (94%) and purity (Figure 11).

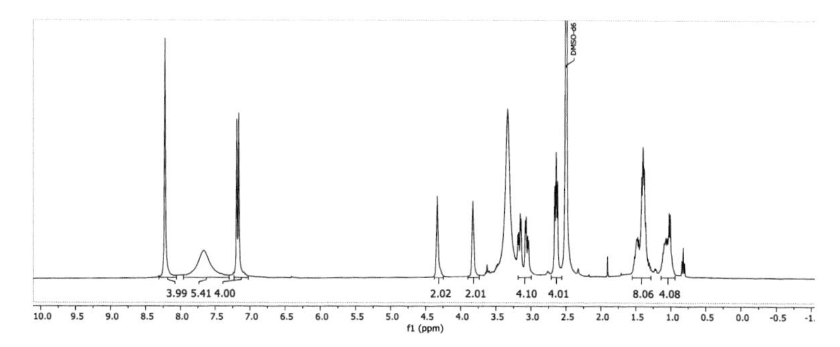

Figure 11: ^{1}H NMR in DMSO-d$_6$ of compound **55**. Minor impurities are visible at 1.91 (AcOH)[185] and 0.82 (H grease)[185].

Compound 55 as hydrogelator

The target compound **55** dissolved well in DMSO and DMF, showed a decreased solubility in MeOH, while being insoluble in toluene, $CHCl_3$, acetone and MeCN. Based on the preliminary work on TFAB-DKP-Lys LMWG **54**, the same procedure to form hydrogel material was applied to the new gelator **55** (Table 15). Since it was assumed that formation of a flexible molecular network is promoted by the flexibility of the chemical structure, gel formation was attempted at comparably low concentration of 2 wt% in distilled water, Ringer's solution, and PBS buffer. Under these conditions, the sample remained liquid and a precipitate was observed in PBS due to the low solubility of **55**. Based on the experiments by NACHTSHEIM group*[158]* PBS buffer at basic pH was selected for the following gelation experiments. In PBS (pH = 8) at a concentration of 2 wt% the sample remained liquid. However, in PBS (pH = 10) an increase of viscosity was observed. Here, at a concentration of 2 wt%, the sample was not homogenous due to precipitate, so the concentration was lowered to 0.5 wt% to fully dissolve the sample, while still observing the increased viscosity.

Conclusion

Compound **55** was synthesized to determine the influence of the flexibility at the azobenzene-DKP linkage on the gelation behavior. No successful gel formation at 2 wt% in aqueous media under physiological conditions (comparable to the conditions where compound **54** formed viscous material) and pH sensitivity suggests that the low mechanical strength of the viscous material based on **54** derives from its bolaamphiphilic structure. The unsaturation and extended conjugation in compound **54** presumably enhanced the π-π stacking ability of the hydrophobic core. This interaction is reduced in the flexible compound **55** which leads to increased influence of the hydrophilic lysine side chains. It was concluded that an environment-sensitive system as provided by **55** is too demanding in biological context and hence no further characterization of the compound was conducted.

3.3 Symmetrical chiral gelator – model system for the addition of dopants

Preface

Parts of the following chapter were published in RSC Advances (Royal Society of Chemistry)

A. Leistner, D. G. Kistner, C. Fengler and Z. L. Pianowski, *RSC Adv.*, 2022, **12**, 4771

DOI: 10.1039/D1RA09218A

Reversible photodissipation of composite photochromic azobenzene-alginate supramolecular hydrogels

Author contributions

The following syntheses towards compound **56**, PSS determination, UV-Vis spectroscopy, Gelation experiments (Table 16 and Table 18) and light induced gel-to-sol transition were performed by David Georg Kistner in his bachelor thesis under the supervision of Anna-Lena Leistner. The complete characterization and interpretation were done by Anna-Lena Leistner. Rheological and NMR relaxation experiments were performed by Christian Fengler.

Motivation

π-π stacking and ionic interactions are important parameters to regulate formation of supramolecular structures in hydrogels. Therefore, a model system azobenzene-DKP gelator was developed to assess both parameters. There are different models to describe π-π stacking of two aromatic systems (Figure 12).

Two electron-rich aromatics can either stack by off-center parallel stacking or by edge-to-face interactions. Face-centered stacking is disfavored due to repulsion between the negatively charged π-electron clouds. In electron-deficient aromatics the electronic situation is reversed, here the π-electron density is drawn away from the aromatic core. Consequently, in mixtures of electron-rich and electron-deficient aromatics the face-centered stacking is favored.[186]

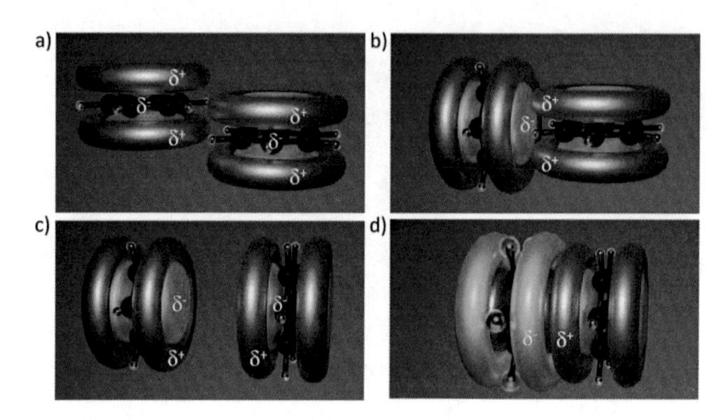

Figure 12: Schematic representation of the π-π stacking models: a) off-center parallel stacking of two electron rich aromatics; b) edge-to-face interaction of two electron rich aromatics; c) face-centered stacking of two electron rich aromatics is disfavored due to repulsion; d) face-centered stacking of electron-rich and electron-deficient aromatics.

Unsubstituted azobenzene serves as electron-rich equivalent and highly fluorinated aromatics (such as decafluoro-azobenzene **22,** Figure 13) may form the electron-deficient part. Incorporation of a highly fluorinated azobenzene into the DKP-Lys scaffold is challenging, so unsubstituted azobenzene was chosen as core of the LMWG, while fluorinated derivatives will be added separately. Application of fluorinated azobenzenes also enables orthogonal manipulation of the system by irradiation with different wavelengths. To address the ionic interactions as second parameter, the bolaamphiphilic structure of symmetric azobenzene-(DKP-Lys)$_2$ **56** is suitable. The resulting target structure is demonstrated in Figure 13.

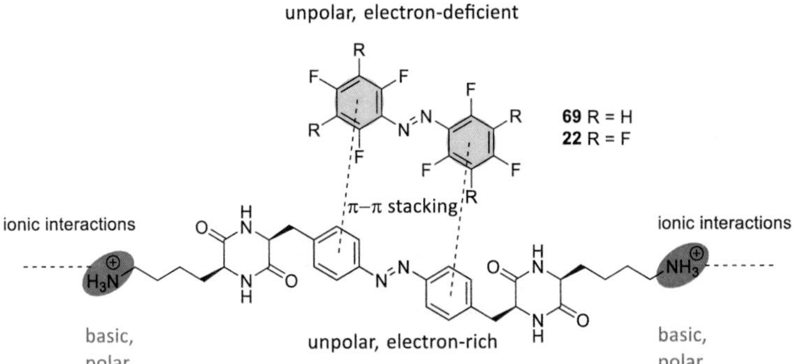

Figure 13: Target structures of this chapter. A model system LMWG **56** with an electron-rich core and basic polar side chains was developed.

As an additive to obtain a charge complementary system, sodium alginate was chosen since it forms aggregates in form of the egg-box model with divalent cations (see 1.2.1). Due to the basic lysine side chains of the target structure **56**, a stabilizing interaction with the carboxylic acids of the alginate is assumed.

Synthesis

Synthesis towards the LMWG target structure **56** started from the natural amino acid phenylalanine (**70**, Phe) and was performed by David Georg Kistner during his Bachelor Thesis under supervision of Anna-Lena Leistner. First, the nitrogen, which is required for the azo bond formation in later steps, was introduced by nitration in *para*-position. Then protecting groups were installed at the amino functionality and carboxylic group to avoid side reactions and facilitate the purification. Subsequently, the nitro-group in *para*-position was reduced to a primary amine using hydrogen gas in combination with a heterogenous catalyst (Pd/C). Silica gel chromatography was not required in the first four steps and the reactions were done on multi-gram scale. Next, the unnatural and protected amino acid **105** was coupled symmetrically to form the corresponding azobenzene **106.** The following steps were similar to the synthesis of the fluorinated structure in chapter 3.1: First the ester was saponified, then the free carboxylic acid groups were coupled with H-Lys(Boc)-OMe·HCl and subsequently, the Boc protecting groups were removed. Finally, the target **56** was obtained after double cyclization to the DKP ring in gram scale and an overall yield of 10% over 9 steps. The fluorinated azobenzenes **22** and **69** where synthesized according to Knie *et al.*[61a] by oxidative dimerization of the respective anilines (Scheme 11).

Photophysical properties

Photophysical investigations were performed for the new LMWG **56**. The azobenzene unit was assumed to isomerize comparable to unsubstituted azobenzene: First the PSSs upon exposure to UV and visible light frequencies were determined by ^{1}H NMR spectroscopy measurements. The highest percentage of the *Z*-isomer was obtained after UV light irradiation ($PSS_{365\,nm}$ = 76% (*Z*)) and back isomerization was efficient under blue light irradiation ($PSS_{455\,nm}$ = 20% (*Z*)). Additionally, the photochromism of compound **56** in MeCN was determined by UV-Vis spectroscopy. The curves (Figure 14 a)) are in qualitative accordance with the measured PSS values and isosbestic points were determined at 238 nm, 287 nm, and 397 nm. Next, the thermal stability was assessed in water and AcOH (Figure 14 b), see also Figure 111 and Figure 112). The half-life of *Z*-**56** was 9 d in water at 20 °C and 158 min at 60 °C, which is comparable to literature values of unsubstituted azobenzenes.[61a] In AcOH, the thermal stability decreased significantly (30-fold) to 6.6 h at 20 °C. A possible explanation is that partial protonation at the DKP leads to enolization, which may withdraw electron density from the chromophore system and thus, reducing the isomerization energy barrier.

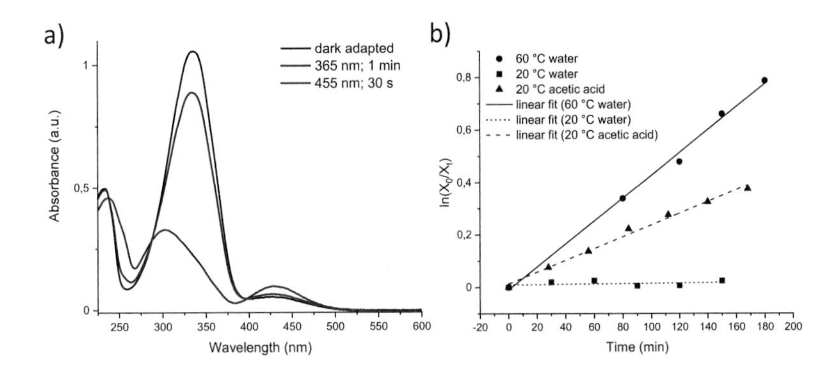

Figure 14: a) UV-Vis spectra of compound **56** after equilibration in the dark and at the PSS at 365 nm and 455 nm. The isosbestic point at 287 nm was used for quantification by HPLC analysis, for example to determine the thermal stability of the photoisomer. b) Linear fit of the decay of the *Z*-isomer of LMWG **56** for first-order kinetics. The thermal stability is significantly decreased in AcOH compared to water.

Biological compatibility

Due to the envisioned application of the smart materials in biological environment, biocompatibility is an important parameter to consider for **56**. Therefore, cytotoxicity was assessed for the hydrogelator **56** against mammalian cells (HeLa) using cell viability assays (MTT assays). Both the *E*-isomer (dark) and the with UV-light (365 nm) photoequilibrated mixture, containing predominantly *Z*-**56**, were of low toxicity with an IC_{50} at or above 1 mM (Table 21).

Hydrogelation of LMWG 56

First, the hydrogelation behavior of compound **56** without any additives was investigated (Table 16). In contrast to the fluorinated analogue (**55**, chapter3.2) the new LMWG **56** formed stable opaque gels in the range between 1.5 – 3.0 wt% in PBS buffer (pH 7.4). At lower concentration of 1.0 wt%, the gel was of low mechanical stability and collapsed upon shaking. In all successful gel formations, the thermal stability was high (>77 °C). In distilled water, 200 mM aqueous NaCl solution or Ringer's solution viscous, colored liquids were formed and thus, these solvents were deemed not suitable as the liquid component of the gel. Photoisomerization of the gelator molecule **56** at 1.5 wt% in PBS resulted in dissipation of the gel-to-sol (Figure 15 d)). Electron microscopy was performed to get a deeper insight into the macromolecular structure of the hydrogel material. TEM images of the compound **56** in PBS (1.5 wt%) revealed a distinct network in a sample which was equilibrated in the dark (Figure 15 b)). Upon irradiation, the network dissipated and weakened (Figure 15 c)). Coherent structural assemblies are visible in the SEM image (Figure 15 a)), however the network structure, which was visible in the TEM images, was not observed.

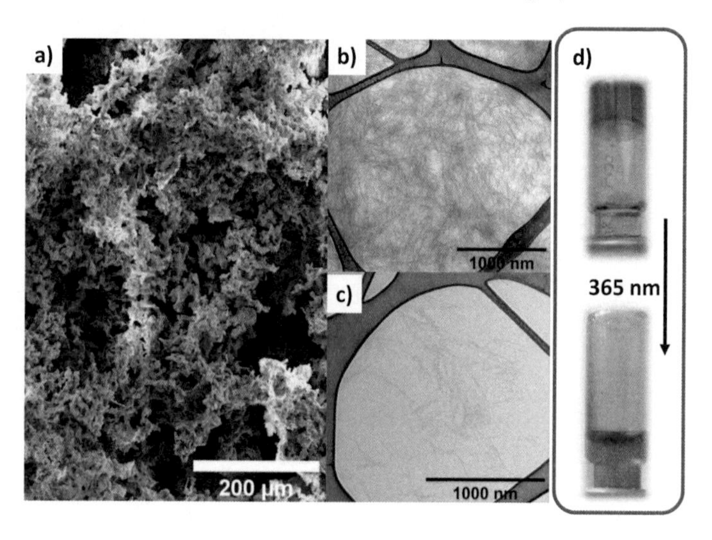

Figure 15: Electron microscopy of the supramolecular assemblies composed from **56** (1.5 wt%) under aqueous conditions (PBS). a) SEM analysis of the xerogel from the dark equilibrated sample; ordered structures are visible. b) TEM analysis of a dark equilibrated sample; a distinct network with entangled fibers is observed. c) TEM analysis after irradiation at 365 nm until liquefaction, the quantity of fibers decreased, and the network is partially disrupted. d) UV light irradiation (365 nm, 15 min) liquifies the opaque hydrogel of **56** (1.5 wt% in PBS).

Stabilization of hydrogelation by face-centered π-π stacking

Next it was investigated if the addition of electron-deficient aromatic systems such as F_6-AB **69** or F_{10}-AB **22** results in a stabilization of the fiber formation and thus an increased mechanical or thermal stability. The fluorinated azobenzenes were added in equimolar amount to the LMWG **56**, at a concentration of 1.0 – 2.0 wt% of the gelator in PBS buffer, resulting in inhomogeneous gels at

1.5 wt% and 2.0 wt%. The inhomogeneities were identified as insoluble F_6-AB **69** or F_{10}-AB **22** respectively (see Figure 16) and moreover, hydrogels were formed by LMWG **56** at these concentrations without additives, so no stabilizing effect was observed. Therefore, the LMWG **56** was combined with the higher soluble fluorinated LMWG **55**. However, at the interesting concentration of ~1.0 wt%, at which the model system **56** formed weak gels on its own, the addition of 1 eq fluorinated LMWG **55** resulted in destabilization to a viscous red liquid.

Figure 16: Gels composed of equimolar mixtures of F_6-AB **69** and **56** at 1.5 wt% (a); F_{10}-AB **22** and **56** at 1.5 wt% (b) and viscous liquid composed of 1.05 wt% **56** with 1 eq fluorinated LMWG **55** (c).

Stabilization of hydrogelation by ionic interactions

The charge complementary polymer alginate exposes multiple acid groups on single polymer chains and LMWG **56** formed stable hydrogels in PBS buffer, where multivalent ions are present. Therefore, inspired by the concept of double-network gelation, the combination of LMWG **56** and alginate to form composite hydrogels was investigated. First, equivalent (1:1 by weight) mixtures of **56** and alginate were prepared by a technique ("method A") where the compounds were suspended in distilled water (to avoid interference by salt ions) by sonication, followed by heating to the boiling point until complete dissolution and cooling to room temperature overnight. The most promising concentration was at 0.6 wt% **56** / 0.6 wt% alginate, at which an opaque orange gel with high thermal stability (T_m = 83 °C) was formed. By addition of alginate, the amount of LMWG **56** required to achieve gelation was reduced by half. This gel composition dissipates to a sol by UV light (365 nm, 15 min, Figure 17 d)) irradiation. This effect was reversible by following irradiation with blue light (455 nm). A mechanically stable gel was reconstituted after 15 min of irradiation, while a control sample, which equilibrated in the dark, remained a fluid. In darkness the fluid slowly (>3 h) solidified to a gel by thermal azobenzene *Z*→*E* isomerization.

Next, the alginate to **56** ratio was varied: Reducing the alginate percentage destabilizes the system. However, by increasing the ratio to 1:2 – 1:4, stable hydrogels (T_m = 76–83 °C) were formed. Additionally, the critical concentration of LMWG **56** decreased to 0.4 wt% by addition of 0.8 wt% of alginate, while retaining mechanical stability (T_m = 55 °C).

The morphology of the composite hydrogel (0.6 wt% **56** / 0.6 wt% alginate) was investigated by scanning- and transmission electron microscopy (Figure 17). Network-like structures built from fibers, which are twisted and entangled to thicker threads, were observed in the dark equilibrated samples (Figure 15 a) and b)). Irradiation at 365 nm to liquefaction resulted in a subtle network without the thicker threads (Figure 15 c)).

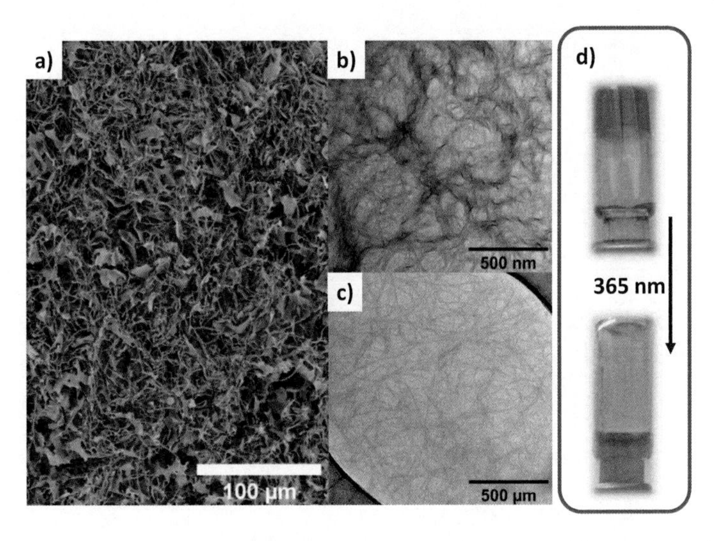

Figure 17: Electron microscopy images of the supramolecular assemblies composed from 1:1 by weight mixture of **56** (0.6 wt%) and alginate (0.6 wt%) in distilled water. a) SEM analysis of the xerogel from the dark equilibrated sample; a very pronounced network is visible. b) TEM analysis of a dark equilibrated sample; single fibers are twisted to entangled fibrous structures. c) TEM analysis after irradiation at 365 nm to liquefaction, the twisted and entangled fibers disappeared, a plain network remained. d) UV light irradiation (365 nm, 15 min) led to transition from the opaque hydrogel of **56** (0.6 wt%) and alginate (0.6 wt%) in distilled water to a sol.

Compared to the hydrogel formed without additives in PBS (1.5 wt%, Figure 15), the electron microscope images indicate a different morphology for the composite hydrogels with the alginate additive. The more complex twisted threads may originate from the ionic interactions between LMWG **56** and alginate.

To gain further insight into the gel strength of the composite hydrogels, rheological measurements were performed with a sample composed of 0.6 wt% **56** and 1.2 wt% alginate, which exhibits a clear and stable gel. First, both separate components were dissolved at their respective concentration in distilled water, where they remained liquid. After equally mixing them, the composite hydrogel was analyzed on a strain-controlled rotational rheometer. First, the LVE was determined by a strain sweep experiment at an angular frequency of

$\omega = 6.28$ rad s^{-1} and from the result (Figure 129) the strain $\gamma = 0.5\%$ was selected for the following frequency sweep in the range of 0.03-71.9 Hz. The resulting values for the storage modulus G' were higher than for the loss modulus G'' by approximately one order of magnitude with $\tan\delta = G''/G'$ in the range of 0.30 to 0.15, indicating a predominantly elastic response of the material. In addition, G' was nearly independent of the frequency.[187] The measurement results indicate the formation of a physical gel through ionic interactions between alginate and LMWG **56**, where **56** is the crosslinker of alginate polymer chains (Figure 18).

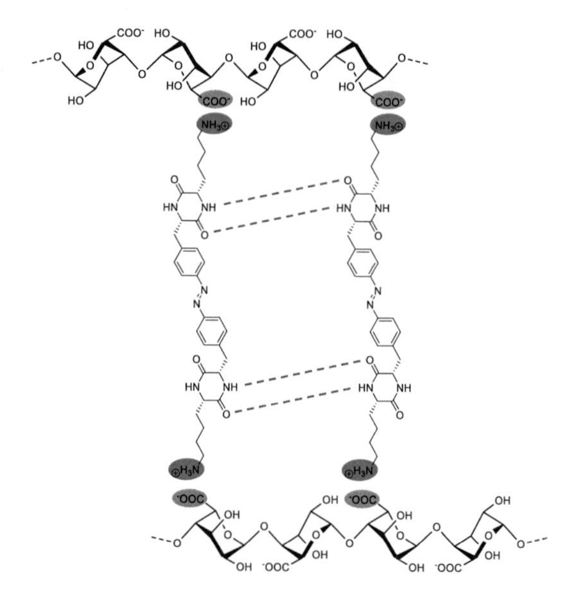

Figure 18: Hypothetical interaction between molecules of LMWG **56** and alginate polymer chains to form fibers.

For the interaction between alginate and LMWG **56** the lysine side chain acts as a Brønsted-Lowry-base. Therefore, a pH-dependency of the material was assumed and consequently the influence of pH on the gelation and gel stability was investigated. In the range of pH 4 to pH 8 mechanically stable hydrogels were formed. Further increase of the pH to 10 resulted in precipitation instead of gelation. This is explained by partial deprotonation of the lysine side chain

(Lysine: isoelectric point (pI) at 9.74 and pK_a for the alkylamine side chain at 10.53), which results in decreased aqueous solubility and weakened ionic interaction with the carboxylate groups of alginate.

PIANOWSKI group demonstrated physical encapsulation of cargo compounds (*e.g.* drugs or DNA) and release upon irradiation.[121b, 135b, 165] Drug delivery is a desired application for hydrogels, however, application is often limited by the preparation method to form cargo-loaded hydrogels, where heat is usually required. Preparation method A is also not compatible with heat-sensitive cargo. Thus, a new, mild method B was developed: Method B was accessible due to solubility of LMWG **56** in aqueous solutions at low concentration, still in the range where gel formation occurred in the composite alginate formulation. An aqueous solution of pure **56** was irradiated with UV light to enrich the fraction of *Z*-isomer, which hinders gel formation by combination with alginate solution. Separately, a solution of alginate was prepared, which can be doped with cargo compounds. Subsequently, both solutions can be mixed and solidified to a gel by irradiation with blue light, which reconstitutes the gelator's *E*-isomer. Equilibration overnight ensured homogenous gel formation, which was reflected in an increased melting temperature (T_m = 89 °C) compared to a gel with the same composition (0.6 wt% **56** and 0.6 wt% alginate; T_m = 83 °C) previously prepared with the method A. Furthermore, preparation method B gives access to a stable transparent yellow gel with the lowest critical gelation concentration observed for this material so far (0.3 wt% **56** and 0.3 wt% alginate, T_m = 80 °C).

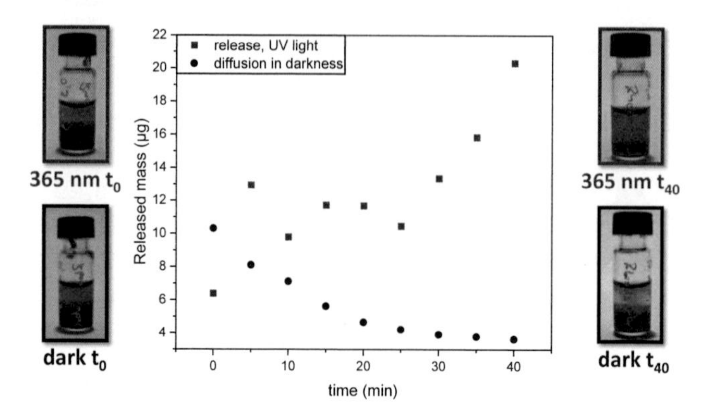

Figure 19: Light-induced release of fluorescent cargo (rhodamine B) *vs.* spontaneous diffusion in darkness from composite hydrogels (0.3 wt% **56** and 0.3 wt% alginate in water, method B). At t_0 the first buffer aliquot was placed on the gel. At t_{40} the 9th buffer aliquot was taken and quantified. The quantified amount of rhodamine B at these points in time is consistent with the color observed by eye.

Light-induced release of cargo was demonstrated and visualized by encapsulation of the fluorescent and colorful rhodamine B into a hydrogel formed from 0.3 wt% **56** and 0.3 wt% alginate in distilled water (250 µg cargo load into 500 µL of gel). The release was performed over a period of 40 min and was measured by covering the sample with an equal volume of PBS and replacing it periodically every 5 min (Figure 117). The procedure was performed for a sample under constant UV light (365 nm) irradiation and a sample incubated in the dark. The efficiency of the release was determined by quantification of the amount of rhodamine B in the buffer aliquots by a previously measured calibration curve (Figure 116). In total 113 µg (>45%) of rhodamine B were released in the irradiated sample, while the aliquots taken from the dark sample contained only 52 µg (<21%) of rhodamine B, which accumulated by diffusion from the surface gel layers. Furthermore, the irradiated gel turned into a sol over the period of the experiment. A detailed overview of the quantification is given in Figure 19. Leaking of rhodamine B in the dark is considerable, but it was already demonstrated that this factor depends on the cargo structure. Cargo, which can interact with the LMWG structure by stabilizing interactions (hydrogen bonds, π-π stacking, ionic interactions) shows significantly reduced leaking.[135b]

In a final experiment, the influence of Ca^{2+} ions on the composite LMWG **56** / alginate gels was assessed. Alginate is known to form hydrogels by addition of divalent cations, usually Ca^{2+}, to crosslink the polymer chains. Consequently, two gel samples (0.6 wt% **56** and 0.6 wt% alginate) were prepared and covered with equivalent volumes of 10 wt% aqueous $CaCl_2$ solution. The aliquot of $CaCl_2$ solution was exchanged every hour (4 times in total). During the process one gel sample was incubated in the darkness and the other gel sample was irradiated at 365 nm.

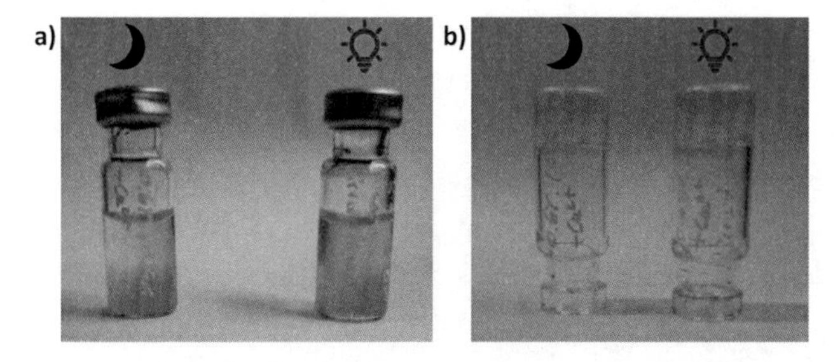

Figure 20: Hydrogels at 0.6 wt% gelator **56** and 0.6 wt% alginate prepared by Method A. a) 500 µL $CaCl_2$ solution (10 wt% in water) was added on top of the gels and equilibrated for 1 h. During this time, one sample was kept in the dark, while the other one was irradiated with UV light (365 nm). b) The remaining gel after 4 h total equilibration time with 4 aliquots of $CaCl_2$ solution.

During the first hour of incubation the color of the supernatant changed from colorless to yellow, which corresponds to the color of dissolved LMWG **56**. This effect was more pronounced for the irradiated sample. (Figure 20 a)). However, the irradiated sample has not been dissipated to a sol after 4 hours of irradiation (365 nm) and stable gels were observed for both samples (Figure 20 b)). Therefore, it was assumed that the LMWG **56** was slowly exchanged by the Ca^{2+} ions, which stabilize the alginate by strong salt bridges. Additionally, the polar *Z*-**56** which accumulated upon irradiation is better soluble in the aqueous environment and its migration through the gel network was facilitated. This experiment demonstrated that the gelator **56** / alginate composite gels can be transformed to Ca^{2+} / alginate gels and the gelator **56** is subsequently removable by irradiation. The remaining Ca^{2+} / alginate gels are completely biocompatible, as only digestible oligosaccharides and calcium ions are left, without the azobenzene-containing component.

Conclusion

A new photochromic LMWG **56** was developed and successfully synthesized to serve as a model system to assess the effects of π-π stacking and ionic interactions on gel formation and stability. The LMWG **56** formed hydrogels without additives under aqueous conditions (PBS). By combination with electron-deficient fluorinated azobenzenes it was observed that a higher aqueous solubility of the azobenzene additive is required to analyze stabilization of hydrogelation by face-centered π-π stacking. This may be achievable for example by attachment of polyethylene glycol chains or other polar functional groups (e.g. sulfonate groups)[188] *para* to the azo bond of the fluorinated azobenzenes. Furthermore, it was demonstrated that LMWG **56** formed composite supramolecular hydrogels with alginate below the critical gelation concentration of **56** alone and over a range between pH 4 to pH 8. Despite the stabilizing effect of the ionic interactions between alginate and the LMWG **56,** a complete gel-to-sol transition occurred under UV-light irradiation and reverse stabilization under blue light offered access to a new gel preparation method (B). This new method is particularly suitable for the encapsulation of heat sensitive cargo and its following release and offers new possibilities to apply hydrogels in biocompatible 3D photoprinting. Following these promising results, it is important to increase the biocompatibility by shifting the irradiation wavelength into the visible light region. Therefore, some alternative structures are presented in this thesis (*vide infra*).

3.4 Transfer of the composite alginate system to PAP-DKP-Lys

Motivation

In chapter 3.3 the stabilizing interaction between negatively charged alginate and bolaamphihiplic LMWG **56** was described. It was assumed that the double occurrence of lysine side chain amines in the symmetrical gelator molecule are key components for stabilized gelation. On the basis of these results, the interest in a comparison to the asymmetrical LMWG structures developed by J. Karcher arose. In this chapter, the LMWG PAP-DKP-Lys **52** (Figure 21) was investigated in composite gelation experiments.

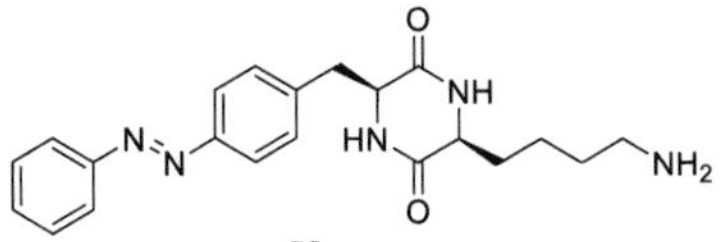

52

Figure 21: Structure of PAP-DKP-Lys **52**, which is the first Photoresponsive low-MW hydrogelator developed by Pianowski *et al.*[121b]

Photophysical properties

PAP-DKP-Lys **52** was thoroughly analyzed by J. Karcher regarding its photophysical behavior. However, the thermal stability had not been assessed yet, which was therefore determined in water and AcOH at 20 °C (see Figure 132) during the course of this work. The obtained values of 6 d in AcOH and 12 h in water are comparable to values of unsubstituted azobenzenes in the literature and in contrast to LMWG **56** of chapter 3.3, the stability in AcOH is not decreased.

Composite gelation with alginate

The hydrogelation properties of PAP-DKP-Lys **52** combined with various amounts of alginate were analyzed. Pure compound **52** was investigated by Johannes Karcher. Compound **52** formed stable gels in the range of 2.0 – 3.0 wt% in deionized water and can be stabilized by addition of NaCl or htDNA.[121b] At a lower concentration than 2.0 wt%, the gel was of low mechanical stability and sensitive upon shaking. Therefore, the behavior of the gel upon addition of sodium alginate was investigated. First, equivalent mixtures of **52** and alginate (1:1 by weight) were prepared in water starting at the lowest concentration, at which the LMWG **56** still formed gels (0.6 wt% each). Outstandingly stable and clear hydrogels were formed with a lower limit of 0.1 wt% **52** / 0.1 wt% alginate (>10-fold increased stability). Though at this concentration the hydrogels were of weak mechanical stability and thus, 0.2 wt% **52** / 0.2 wt% alginate is the most promising concentration (see Table 3). Combinations of alginate with **56** at 0.2 wt% remained liquid. Moreover, the gels which formed in combination with **52** have a high thermal stability (0.2 wt% **52** / 0.2 wt% alginate; T_m = 96 °C). Above this concentration the gels were stable in a boiling water bath. A possible explanation is based on the flexibility and packing density of the formed fibers. It is assumed that the asymmetric gelator can arrange to denser packed helical fibers, while exposing a multitude of lysine amines to the surrounding, which can interact with the negatively charged alginate.

Table 3: Hydrogelation experiments of composite alginate hydrogels formed with LMWG 56 or PAP-DKP-Lys **52**.

	Approx. conc.		Description	T_m °C
Gelator **52**	Gelator **56**	Alginate		
0.6 wt%		0.6 wt%	clear orange gel	>100
0.2 wt%		**0.2 wt%**	**clear yellow gel**	**96±2**
0.1 wt%		0.1 wt%	weak, clear yellow gel	48±5
1.5 wt%			weak gel	38
	0.6 wt%	0.6 wt%	opaque orange gel	83
	0.2 wt%	0.2 wt%	no gelation	-

Next, the alginate to **52** ratio was varied. Lowering the alginate proportion to 0.6 wt% **52** and 0.2 wt% alginate (3:1) resulted in clear stable hydrogels (T_m >100 °C). A further decrease of the alginate proportion led to opaque gels. The reversed ratio of 1:3 resulted in clear stable gels, which melted in a boiling water bath (see also Table 22).

The hydrogels were irradiated with UV light (365 nm). However, the gels did not dissipate to a fluid, even at the lowest critical gelation concentration (0.1 wt% **52** / 0.1 wt% alginate) and with an extended irradiation time (>2 hours). It was assumed that the high mechanical stability in combination with a light source, which requires high energy and thus generates a fair amount of heat in the irradiated sample and triggers thermal relaxation of the switch, thereby hinders the gel-to-sol transition. Then, the morphology of the composite hydrogel (0.6 wt% **52** / 0.6 wt% alginate) was investigated by electron microscopy (Figure 22 a) = SEM and b), c) = TEM; Figure 133 to Figure 135).

TEM images revealed few µm long fibers which accumulate to dense network bundles. UV light irradiation had no visible effect on the macromolecular structure. However, the conditions of samples under the electron microscope do not reflect the material under natural conditions and interpretation must be regarded with caution. The surface of the xerogel (SEM) is characterized by layered sheets of smooth texture. At the fracture edge frayed structures were visible.

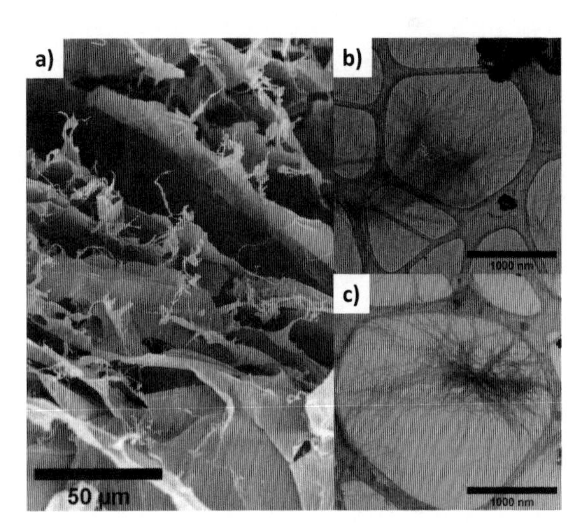

Figure 22: Electron microscopy of the supramolecular assemblies composed of 1:1 by weight mixture of **52** (0.6 wt%) and alginate (0.6 wt%) in distilled water. a) SEM analysis of the xerogel from the dark equilibrated sample; a layered sheet structure is visible. b) TEM analysis of a dark equilibrated sample; single fibers are

twisted to entangled condensed structures. c) TEM analysis after irradiation at 365 nm no liquefaction was observed. This is supported by morphological similarity to the dark sample.

Conclusion

PAP-DKP-Lys **52** formed composite supramolecular hydrogels with alginate below the critical gelation concentration of **52** alone and below the critical concentration of composite hydrogels with bolaamphiphilic gelator **56**. This behavior of strong stabilization was not expected, with regards to the asymmetrical structure of LMWG **52** with only one lysine side chain per molecule and the presumed necessity of two lysine amines per molecule to form crosslinks. However, the asymmetrical structure might enable a helical layout in the macromolecular fibers with a denser arrangement. In such a theoretical macromolecular structure the lysine side chains are presented on the outer side of the fibers and allow for ionic interactions and crosslinking with the alginate carboxylate groups. In this system, gel-to-sol transition by irradiation was not possible, but the superior mechanical and thermal stability of the hydrogels encourages further examination of composite hydrogels employing asymmetrical AB-DKP-Lys LMWGs.

3.5 Transfer of the composite alginate system to F_2-PAP-DKP-Lys

Motivation

F_2-PAP-DKP-Lys **57** (Figure 23) is readily synthetically accessible and has already been well investigated by J. Karcher.[165] The compound photoisomerizes under biocompatible green light irradiation and hydrogels formed by **57** in PBS buffer are quickly dissipated to a sol upon irradiation. Therefore, F_2-PAP-DKP-Lys **57** was selected as a next asymmetrical LMWG for investigation of composite hydrogels with the aim to obtain hydrogels with high stability and the ability to dissipate them into a fluid state under light irradiation. Furthermore, the biocompatibility of the LMWG should be expanded by development of a more suitable gelation system for first *in vitro* cell experiments.

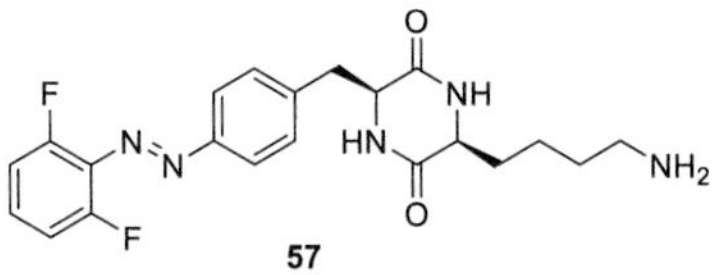

Figure 23: Structure of F_2-PAP-DKP-Lys **57**. The LMWG isomerizes under green light irradiation and was already applied for drug release.[165]

Composite gelation with alginate

The gelation properties of compound **57** have already been investigated by J. Karcher. In this thesis, the hydrogelation properties of F_2-PAP-DKP-Lys **57** combined with various amounts of alginate were analyzed. Compound **57** formed stable gels in the range between 3.0 – 7.0 wt% in PBS buffer, which dissipate to fluids under green light irradiation (30 min to 180 min required depending on the concentration and light intensity).

First, the general gelation behavior of composite alginate and **57** hydrogels was investigated. In equivalent (1:1 by weight) mixtures, the gelation of **57** was comparable to non-fluorinated **52**. Stable clear hydrogels (Figure 24) with high thermal stability (T_m >97 °C) were formed between 0.6 wt%/0.6 wt% and 0.2 wt%/0.2 wt%. A further decrease of concentration in equivalent mixtures resulted in clear fluids. Gels of these mixtures were irradiated with green light.

The gel of 0.2 wt% **57** / 0.2 wt% alginate dissipated to fluid after 120 minutes of irradiation $PSS_{523\,nm}$ = 66 %. Reversal of this effect by irradiation with violet light (410 nm) did not occur. At the $PSS_{410\,nm}$ of LMWG **57** 20% *Z*-isomer is present and thermal stability is high (half-life in PBS at 22 °C is at 109 d) therefore, gel reconstitution is hindered.

Figure 24: Clear hydrogel with a diameter of 25 mm (0.6 wt% of F_2-PAP-DKP-Lys (**57**); 0.6 wt% alginate). Due to high mechanical stability the shape remained intact upon handling with a spatula.

Based on this, the alginate to **57** ratio was varied. The composition was designed to contain a higher amount of LMWG to have a higher fraction of phototriggered component. Starting from the equivalent mixture 0.2 wt% **57** / 0.2 wt% alginate, where photodissipation was possible, the alginate concentration was lowered to 0.1 wt%. This mixture formed a clear yellow gel and gel-to-sol transition occurred after irradiation with green light. However, gel reconstitution was not possible, since the overall gelator and alginate concentration was too low. Next, the fraction of gelator was increased to obtain a mixture of 0.3 wt% **57** / 0.1 wt% alginate, which formed clear yellow gels with T_m = 100 °C. Irradiation with green light (523 nm, 135 min) turned the gel into sol. Here, irradiation with violet light (410 nm, 30 min) resulted in reversible gel formation. The reconstituted gel had a decreased thermal stability (T_m = 91 °C), which correlates with the $PSS_{410\,nm}$ of LMWG **57** 20% *Z*-isomer. These findings supported the viability of the milder preparation method B, which was introduced in chapter 3.2 and was also successfully transferred to composite alginate / asymmetrical AB-DKP-Lys LMWG hydrogels. A summary of the gelation experiments is given in Table 23.

To gain a deeper insight into gel strength of composite alginate and **57** hydrogels, rheological measurements were performed with samples consisting of 0.6 wt% **57** and 0.6 wt% alginate or of 0.3 wt% **57** and 0.1 wt% alginate. The LVE was determined by strain sweep experiments (Figure 148 and Figure 151) at an angular frequency of ω = 6.28 rad s^{-1}.

From the strain sweep experiments γ = 0.1 % was selected for the frequency sweep experiments (Figure 25). The values for the storage modulus *G'* were higher than for the loss modulus *G''* by approximately one order of magnitude, resulting in a tan δ = *G''*/*G* = 0.05 at 1 Hz for both compositions. The values for both *G'* and *G''* of the sample with higher concentrations were increased compared to the lower concentrated sample which is consistent with the behavior under irradiation or melting. Compared to the bolaamphiphilic model system in chapter3.3, the storage and loss modulus of composite hydrogels formed by alginate / asymmetrical **57** were one order of magnitude higher. This is reflected in the overall lower critical gelation concentration and higher melting temperatures at the same time.

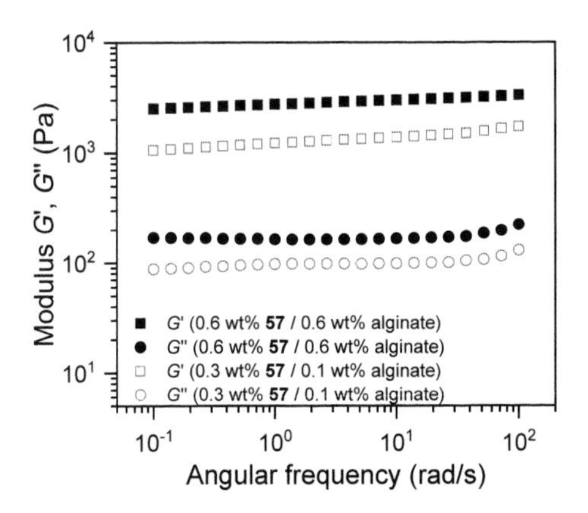

Figure 25: Frequency sweep of a sample consisting of 0.6 wt% hydrogelator **57** and 0.6 wt% alginate compared to 0.3 wt% hydrogelator **57** and 0.1 wt% alginate.

Next, the rheological behavior of the optimal mixture of 0.3 wt% **57** / 0.1 wt% alginate after gel-to-sol transition and following gel reconstitution was investigated. In previous experiments green light irradiation of this composition led to gel-to-sol transition, for the frequency sweep the sample was not transformed to a sol, but to a highly viscous material by shorter irradiation time. This was due to flat borderless lower rheometer plate, which was not suitable for liquids.

The frequency sweep of the viscous material (Figure 26) revealed that *G'* is higher than *G''* at frequencies above 20 rad/s, which corresponds to a viscoelastic solid. At 20 rad/s storage and loss modulus are equal, while at lower frequencies *G''* is higher than *G'*. This intersection can be interpreted as a transition from liquid-like to solid-like.[189] This behavior corresponds to the observations during the gelation experiments, where at higher concentrations (0.3 wt% **57** / 0.3 wt% alginate) gel-to-sol transition under irradiation was hindered due to a significant proportion of the *E*-isomer at PSS_{523} being present (~34%),[165] maintaining the gel structure.

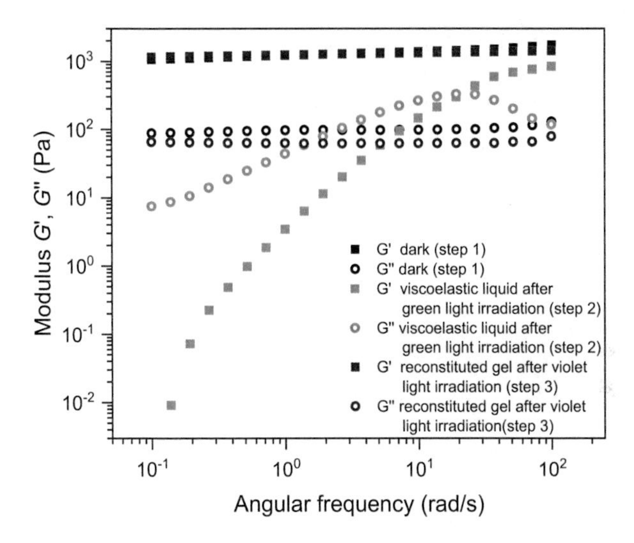

Figure 26: Frequency sweep of a sample consisting of 0.3 wt% hydrogelator **57** and 0.1 wt% alginate. First, the dark state was measured (step 1), then the sample was irradiated directly on the lower plate. The gel was first irradiated with green light (523 nm) for 30 min (Step 2). Following irradiation with violet light (410 nm) for 160 seconds reconstituted the gel after additional 20 minutes of equilibration (step3).

Consequently, green light irradiation led to weakening of gel structure and, depending on the concentration, the weakened structure collapsed to a sol. At this point, violet light irradiation was applied to isomerize the gelator molecule **57** ($PSS_{410\,nm}$ = 20% *Z*-isomer)[165] and reverse the gel dissipation.

Direct measurement after violet light irradiation (Figure 27) revealed a constant *G'* comparable to *G'* in the dark state. However, the loss modulus *G''* was frequency dependent, being in the same range (>4 rad/s) as in the measurement after green light irradiation. This suggests that the material partially behaved as a viscoelastic liquid and network formation was not completed. Therefore, another measurement was performed with additional equilibration time after violet light irradiation and parallel behavior of *G'* and *G''* throughout the entire frequency range of gel-like material was reconstituted.

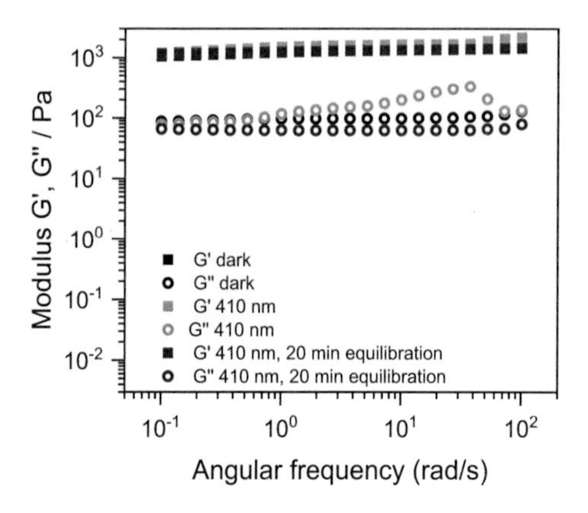

Figure 27: Frequency sweep of a sample consisting of 0.3 wt% hydrogelator **57** and 0.1 wt% alginate. The gel was first irradiated with green light (523 nm) for 30 min to reach the viscoelastic liquid state. The following irradiation with violet light (410 nm) for 160 seconds reconstituted the gel. Additional 20 minutes of equilibration were necessary to obtain the same frequency independent behavior as in the dark state.

These new rheological insights were correlated with microscopy images of the composite hydrogels. Samples of various concentrations were prepared according to method A (Table 26) and investigated using SEM (Figure 140 to Figure 145). An overview of the effects of green light irradiation on the structure under electron microscopic conditions is shown in Figure 28. In a) the dark state of a xerogel composed of 0.3 wt% **57** and 0.1 wt% alginate had a layered sheet macromolecular structure. These sheets had a rather smooth surface with spherical protrusions which are regularly distributed. Overall, the microscope images of the dark states gave the impression that the sheets were built from single, compact aligned fibers. The material after green light irradiation in b) showed similar sheets, which were intact in patches, but the structure had mainly degraded to a frayed network.

This network conveyed an impression of fragility which is consistent with the observations during rheology investigations of gelation, which showed that green light irradiation does not dissipate the entire network. More precisely, the macromolecular structure was weakened to a point at which small mechanical forces were sufficient to collapse the structure.

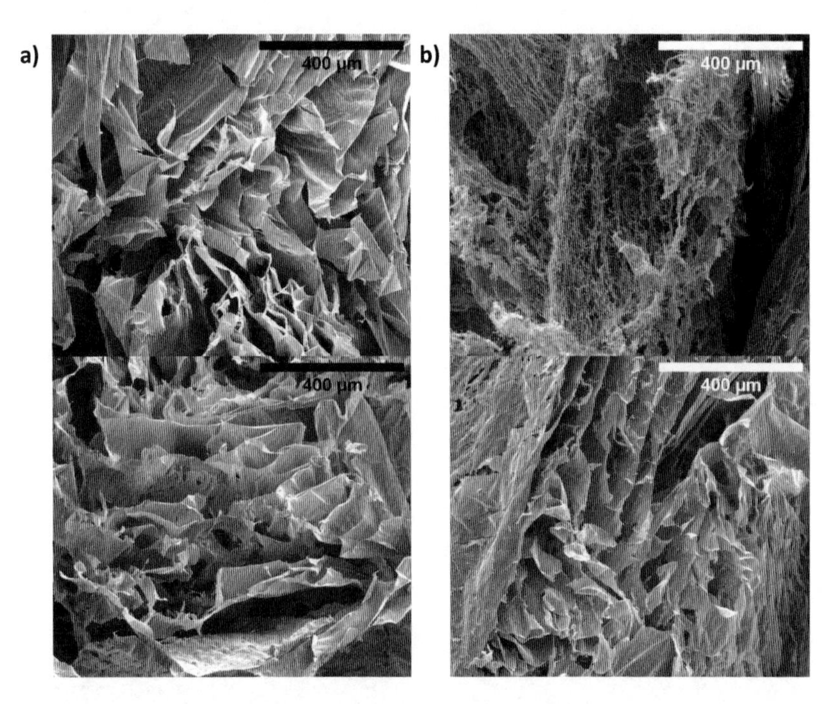

Figure 28: Scanning electron microscopy of the xerogel prepared from **57** (0.3 wt%) and 0.1 wt% alginate in water to investigate supramolecular assemblies. a) The sample was equilibrated in the dark. Layered sheets were visible. b) The sample was irradiated with green light (523 nm) until a sol was formed, then the sample was frozen and dried. While layered sheets were still visible, areas where the smooth sheets were transformed to a frayed network dominated.

One potential application of the composite hydrogels is tissue engineering or mediation of cell growth. For this purpose, a biocompatible formulation is required. The biocompatibility of the natural polysaccharide alginate was proven by several applications[141, 190] and the cytotoxicity (EC_{50} against HeLa cells) of pure **57** was previously determined to exceed 0.5 mM[165]. However, distilled water is not compatible with cell growth and maintenance. Therefore, optimal gelation conditions were screened for gels formed in Dulbecco's Modified Eagle Medium (DMEM). The critical gelation concentration in DMEM increased compared to gels formed in distilled water to 0.6 wt% **57** and 1.5 wt% alginate. It is assumed that the ionic composition, which also comprises divalent cations, and additives in DMEM had a negative effect on the molecular assembly.

Nevertheless, the critical gelation concentration of the composite hydrogel in DMEM is lower than for pure **57** in PBS buffer, so alginate had a stabilizing effect even in DMEM. The thermal stability was at T_m = 75±11 °C and thus compatible with typical conditions for cell growth. Next, the mechanical strength of the new biocompatible hydrogel was analyzed by rheological measurements. A frequency sweep within the LVE regime (Figure 152) revealed values of storage modulus G' (275 Pa at 16 Hz) higher than loss modulus G'' (75 Pa at 16 Hz). The respective values simultaneously decreased towards lower frequencies. The measurement indicated a predominant elastic response and the formation of gel-like material with $\tan\delta = G''/G = 0.21$ at 1.14 Hz. These results are in good agreement with observations made under the electron microscope. The xerogel of **57** (0.6 wt%) and 1.5 wt% alginate in DMEM (Figure 29) showed a porous network. Compared to composite hydrogels in distilled water the pores of composite hydrogels in DMEM were larger and the structure was not ordered in a layered sheet manner.

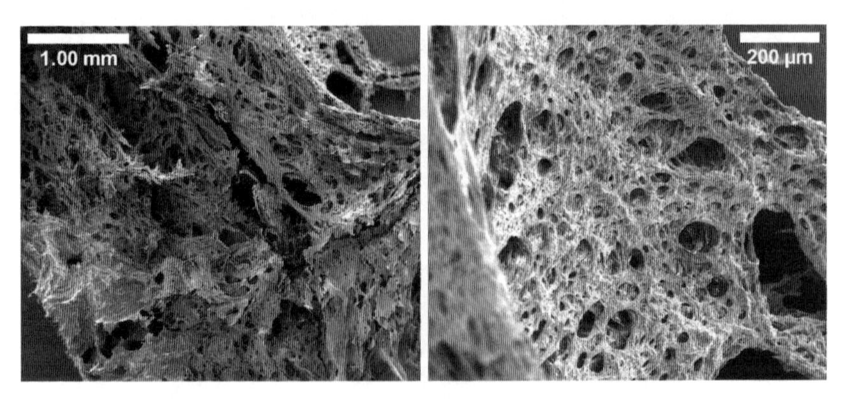

Figure 29: Scanning electron microscopy of the xerogel prepared from **57** (0.6 wt%) and 1.5 wt% alginate in DMEM to investigate supramolecular assemblies. The sample was equilibrated in the dark.

One possible application of the composite hydrogels formed from fluorinated **57** and alginate is in the field of biology due to the biocompatibility of the materials and the isomerization wavelengths within the visible region. Therefore, the material was sent as part of a collaboration to the group of Prof.

Véronique Orian-Rousseau with the idea to grow organoids inside of the hydrogel material. Green light irradiation of grown organoids encapsulated in composite hydrogel was envisioned to be a mild release method. However, more adjustments to gel preparation and compositions are required. One issue was that the organoids inside the gel need to be supplemented with nutrients by aqueous medium, which must be frequently replaced. This excess of supplemented medium slowly dissolved the gel, as the contact surface of gel and media is too large.

Excursus: Pure F_2-PAP-DKP-Lys 57 gelation in DMEM

At this point another collaboration was made with the group of Prof. Stefan Zahler in Munich. A material with modulable stiffness was required for cell growth on the surface of the hydrogel. For this purpose, a specific medium (M199 10x, ThermoFisher cat. No. 11825015) was used as the liquid gel component and pure compound **57** as LMWG. Without bovine serum the M199 medium showed sufficient heat tolerance to form gels with LMWG **57** by Method A, which was suitable for cell growth on top of the gel, because the cells were added after gel formation. Opaque gel-like material was formed in the range of 2-4 wt% (Table 25). Gel-to-sol transition by light irradiation was not observed at this range of concentration, but rather softening of the surface was detected by touching the gel surface with a spatula. Quantification of this observation was attempted by rheological measurements.

First, the LVE was determined by a strain sweep experiment at an angular frequency of ω = 6.28 rad s^{-1} and from the result (5.7.6, Figure 154) the strain γ = 0.19% was selected for the following frequency sweep (Figure 30). In the dark state G' was nearly independent of the frequency and approximately one order of magnitude higher than G'' with $\tan\delta = G''/G = 0.09$ at 1.14 Hz. These values indicate a gel-like material. The material was then irradiated with green light (523 nm) for 60 minutes and over this period of time the gel was covered with a glass petri dish to avoid evaporation. In the subsequent frequency sweep it was revealed that both the values of G' and G'' decreased by almost one order of magnitude respectively. With these insights the material was sent to

Munich. So far, the cells adhere to the gel surface, but do not spread, which implies that improvements of the system on either biological side or gel composition are necessary.

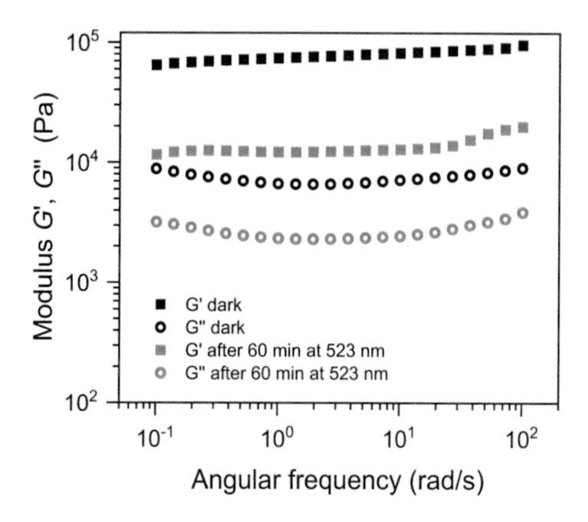

Figure 30: Frequency sweep of a sample consisting of 2.0 wt% hydrogelator **57** in medium199. The sample was irradiated directly on the lower plate. The gel was first measured in dark state and then irradiated with green light (523 nm) for 60 min.

Influence of divalent Ca^{2+} cations on the composite hydrogels

Due to previous results (see Stabilization of hydrogelation by ionic interactions in 3.2), the question arose how the composite alginate / asymmetrical AB-DKP-Lys LMWG hydrogels behave upon addition of divalent cations in form of Ca^{2+}. For this experiment, hydrogel samples (500 µL, 0.3 wt% **57** and 0.1 wt% alginate, Method A) were prepared and covered with 500 µL of a 10% aqueous $CaCl_2$ solution. The aliquot of $CaCl_2$ solution was exchanged every hour (3 times in total) and the sample was kept in darkness during the process (Figure 138b)). The supernatant was less colored after each step and after the last one, a clear solution and a colorless gel - which shrank approximately to 1/8 of its former size - was left (Scheme 23).

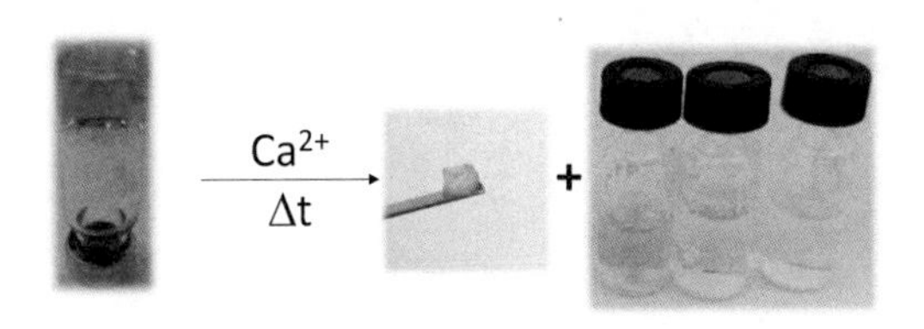

Scheme 23: A hydrogel sample (0.3 wt% **57** and 0.1 wt% alginate) was incubated with 10 wt% aqueous $CaCl_2$ solution. Colorful supernatant was collected each hour until a clear, shrunken gel and an off-white solution were left.

This experiment demonstrated that a stable alginate-Ca^{2+} hydrogel can be formed from a composite hydrogel and the LMWG **57** is removable for the most part. Moreover, the shape of the gel was maintained. Such a system can be applied to fixate the structure of a composite hydrogel, which can be formed by Method B mediated by light irradiation, with no heating being required during the whole process. The remaining alginate-Ca^{2+} hydrogel is composed of material of completely natural origin. The prospective application of this system is 3D printing of gel material.[191] The idea is to form hydrogel structures precisely under irradiation, while the material in the dark remains liquid and can be removed by washing. In the following, the light-printed structure can be incubated with $CaCl_2$ solution to fixate the structure and remove the non-natural photochromic gelator.

Excursus: Interaction of host F_2-PAP-DKP-Lys 57 with cargo guest molecules

PIANOWSKI group demonstrated the encapsulation and release of various additives ranging from antibiotics, anticancer and anti-inflammatory drugs to DNA oligomers.[135b, 165] For these cargo molecules, variable rates of diffusion were observed. Furthermore, the released anti-cancer agent plinabulin elicited unusually high solubility when encapsulated into the hydrogel. This behavior suggests specific interactions between the LMWG and the cargo molecules.

To investigate the influence of supramolecular interactions, 4 wt% hydrogels of **57** were prepared in deuterated PBS and selected additives were incorporated, see Table 4. Then, various NMR spectroscopic measurements (1D selective NOESY, 2D NOESY and diffusion ordered spectroscopy DOSY) were performed by Fabian Hoffmann (group of Luy and Muhle-Goll).

Table 4: Hydrogels were prepared in deuterated PBS buffer with or without additives and transferred to NMR tubes.

Hydrogelator **57**	Additive
20 mg (4 wt%)	-
20 mg (4 wt%)	Ciprofloxacin HCl, 8 mg
20 mg (4 wt%)	Glucose, 3.7 mg
20 mg (4 wt%)	NaCholate, 9.5 mg
20 mg (4 wt%)	Ciprofloxacin HCl, 92 µg

The results of these measurements are discussed in detail in the submitted publication (Hoffmann, Leistner, *et al.*, 2023, *submitted*). The key message of the experiment however, is that the guest molecules are encapsulated by specific supramolecular interactions of LMWG and cargo molecule and only to minor extend physically restricted in their movements. Derived from the observations, an optimal cargo structure is functionalized with acidic groups and aromatic residues. Based on that, the design of a hydrogelator tailored to basic cargo molecules encompasses the functionalization of the LMWG with acidic side chains such as glutamic or aspartic acid instead of lysine.

Conclusion

F_2-PAP-DKP-Lys **57** formed composite supramolecular hydrogels with various favorable properties, e.g. the mixed system with alginate having a critical gelation concentration below the one of **57** alone and below the critical concentration of composite hydrogels with bolaamphiphilic gelator **56**. Furthermore, gel-to-sol transition occurred under green light irradiation, which was not possible for non-fluorinated PAP-DKP-Lys **52** in composite hydrogels with alginate. Reconstitution of the gel could be performed by irradiation with violet light, which enabled the use of the mild preparation method B. The macroscopic observations after green light irradiation were supported by electron microscopy images and rheological measurements. It was demonstrated that composite hydrogels with alginate are suitable candidates for biological experiments with living cells, even though some optimizations are necessary either from biological side or hydrogel composition. Moreover, LMWG **57** was replaced by Ca^{2+} ions by equilibration with a $CaCl_2$ solution. In this process, the LMWG diffused to the supernatant and consequently, only oligosaccharides of natural origin remained in the reformulated hydrogel. Based on these results, new opportunities for 3D-printing are enabled including photo-controlled printing and subsequent replacement of the non-natural gelator.

Additionally, the interaction of host F_2-PAP-DKP-Lys **57** with cargo guest molecules for drug delivery application was investigated in closer detail by NMR spectroscopy in collaboration with Dr. Claudia Muhle-Goll and Prof. Burkhard Luy.

3.6 A new red light triggered LMWG – Cl_4-PAP-DKP-Lys

Motivation

The previous chapters revealed that asymmetrical AB-DKP-Lys LMWGs were proven superior to the symmetric versions with respect to stable gel formation and critical gelation concentration. However, so far only the symmetrical LMWG **58** with the conjugated TFAB chromophore core is addressable with red light. Therefore, it is desirable to synthesize asymmetrical AB-DKP-Lys LMWGs, which photoisomerize under red light illumination. This goal had already been pursued during the Bachelor Thesis of Anna-Lena Leistner and was continued during the PhD work. The aim was to synthesize a LMWG linked to a tetra-*ortho* chloro AB (see Figure 31). This photoswitch design is of sufficient structural similarity to the tetra-*ortho* fluoro ABs, which had already been successfully applied in LMWGs, so a related gelation behavior was assumed. In previous attempts, the implementation of the tetra-*ortho* chloro AB into an amino acid scaffold was not successful. Finding a solution for the synthetic problems to form the tetra-*ortho* chloro AB amino acid would open up many potential applications in biology, since the chromophore isomerizes within the therapeutic window (600-900 nm).

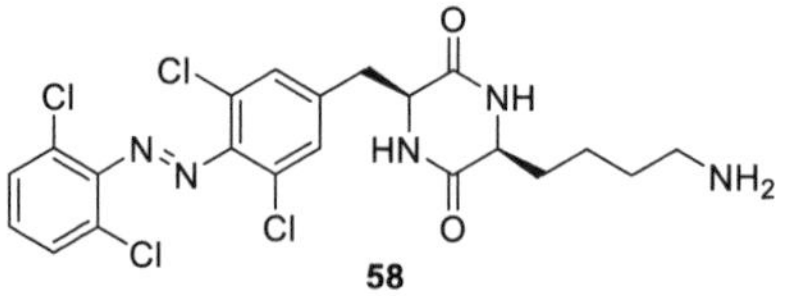

Figure 31: Target structure Cl_4-PAP-DKP-Lys **58**. The Tetra-*ortho*-chlorinated azobenzene photoisomerizes under red light irradiation.

Synthesis

The synthetic access to asymmetrical tetra-*ortho* chloro ABs is limited, compared to the fluorinated derivatives. Mills reaction provides low yields and tedious purification processes, while classical azo coupling is disabled by steric hindrance of the chlorine substituents. Azo bond formation with sterically demanding *ortho* substituents by addition of lithium organyls to diazonium salts was not successful in early attempts during the Bachelor Thesis in 2017. Consequently, a different strategy – the late-stage functionalization – developed by TRAUNER group[192] was pursued in this thesis. The palladium-catalyzed *ortho*-chlorination of azobenzene was first reported in 1970 by Fahey *et al.*, though the requirement of chlorine gas discouraged the applicability.[193] However, after the report of *ortho*-iodination in combination with $Pd(OAc)_2$ as catalyst, the palladium(II)-catalyzed C-H activation encouraged new attempts at chlorinations.[194] Consequently, the TRAUNER group performed multiple tetra-*ortho*-chlorinations on various azobenzenes.[192] As the late-stage chlorination requires an *ortho* unsubstituted azobenzene, the synthesis towards the LMWG target structure **58** started by development of the azobenzene amino acid. An overview of the synthetic strategy is given in Scheme 24.

The design of this amino acid considers that the chlorination requires harsh reaction conditions, which were not compatible with the Boc protecting group employed in the syntheses towards earlier AB-DKP-Lys LMWGs. In more detail, the natural amino acid phenylalanine (Phe, **70**) was first nitrated in *para* position in multigram scale.

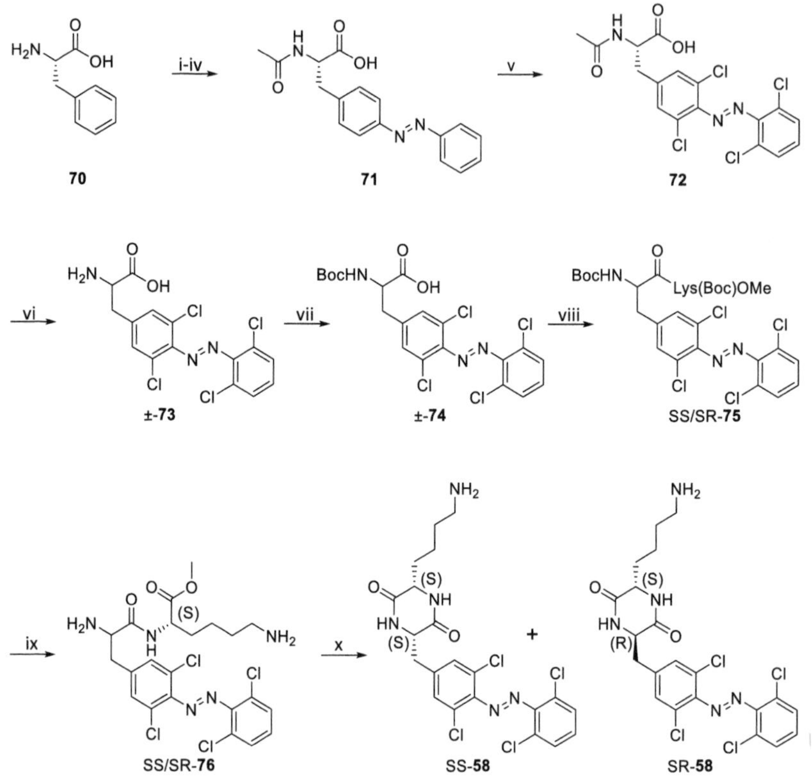

Scheme 24: Synthetic strategy towards target compound **58**. **i**) H_2SO_4, HNO_3, 0 °C, 3 h; **ii**) Ac_2O, MeOH, reflux, 6 h; **iii**) H_2, Pd/C, MeOH, 20 °C, 18 h; **iv**) Nitrosobenzene, AcOH, 20 °C, 24 h; **v**) NCS, $Pd(OAc)_2$, AcOH, 140 °C, 2.5 h; **vi**) 6 M HCl, 110 °C, 24 h; **vii**) Boc_2O, $NaHCO_3$, dioxane/water, 20 °C, 20 h; **viii**) HBTU, DIPEA, H-Lys(Boc)-OMe·HCl, DMF, 20 °C, 2 h; **ix**) TFA, DCM, 20 °C, 1 h; **x**) AcOH, 4-methylmorpholine, DIPEA, butan-2-ol, 120 °C, 2 h.

To avoid side reactions in later steps, such as polymerization during chlorination, which was observed during attempts with and without Boc protection, the amine was protected by an acetyl protection group. The reaction was done on multigram scale, and the product **110** was obtained with fair yield of 69%. Next, the nitro group was reduced quantitatively to a primary amine by hydrogen gas in a heterogenous catalysis (Pd/C). The primary amine was required for the following Mills reaction with commercially available nitrosobenzene to form the azobenzene **71** in good yield of 75%.

The subsequent key step of the synthetic strategy was the tetra-*ortho* chlorination of the azobenzene amino acid. The directing effect of this reaction is based on coordination of the palladium (II) catalyst to the azo bond. A possible mechanism is demonstrated in the catalytic cycle, shown in Scheme 25.

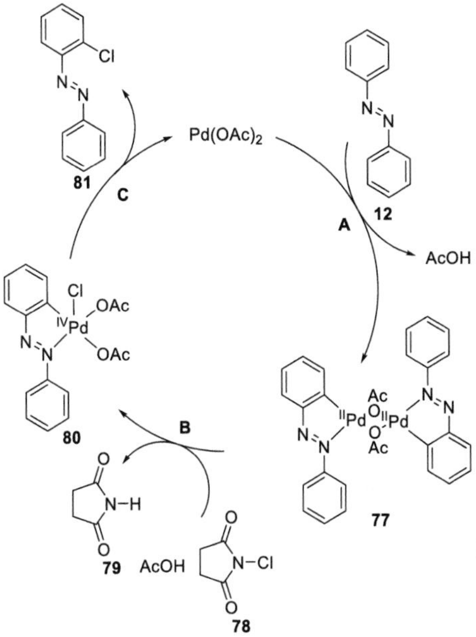

Scheme 25: Possible catalytic cycle for the azo-bond-directed C-H bond activation based on the mechanistic proposal of palladium(II)-catalyzed alkoxylation.[195] A: the azobenzene (**12**) coordinates to $Pd(OAc)_2$ and forms intermediate **77** by cyclopalladation. B: oxidative addition of NCS (**78**) and acetic acid (AcOH) give Pd(IV) complex **80**. C: the product **81** is released by reductive elimination.

To obtain the tetra-substituted product, the azobenzene passes the catalytic cycle four times. Excess of NCS was necessary to avoid mono-, di- and tri-substituted side products. The reaction only proceeded in sealable reaction tubes approximately 20 °C above the boiling temperature of acetic acid. These harsh conditions were not suitable for the non-protected or Boc protected azobenzene amino acid, as decomposition and polymerization were observed for such derivatives. The reaction conditions for tetra-*ortho* chlorination of acetylated azobenzene **71** were optimized, so that fair yields of 54% were obtained on gram scale.

In the following, the acetamide was deprotected. This reaction can either proceed by the reaction with hydrazine, hydroxide bases or in the presence of strong acids at elevated temperatures.[196] Due to the stability of **72** during the harsh conditions of the chlorination reaction, the acidic conditions were chosen for acetamide deprotection. Compound **72** was suspended in 6 M HCl and heated to reflux for 24 h (Scheme 26). The product **73** precipitated in the course of the reaction, which facilitated the purification process. However, heating of amino acids in strongly acidic or basic solutions is known to result in racemization. Therefore, chiral RP-HPLC chromatography (Figure 169) was performed and revealed total racemization of the compound **73** during deprotection.

6 M HCl
110 °C, 24 h
78%

72

73-(*S*), 50% + **73**-(*R*), 50%

Scheme 26: Deprotection of the acetamide under strongly acidic conditions. The product is obtained as racemate.

It was decided to proceed with the racemate, since a chiral amino acid should be coupled to the racemic mixture in a later stage, resulting in diastereomers which are easier to separate. For this peptide coupling, the primary amine of the azobenzene amino acid **73** requires protection. Boc-protection was chosen, as the side chain of the later used lysine building block also carried a Boc protection group, enabling a simultaneous deprotection. Consequently, a Boc protecting group was introduced to compound **73** in very good yield (**74**, 83%). This building block is particularly interesting as such modified amino acids can be applied in (solid phase) peptide synthesis and provide access to photoswitchable peptides or proteins.[197]

The following peptide coupling of **74** with H-Lys(Boc)-OMe·HCl proceeded nearly quantitatively and afforded 3.00 g of the product **75** (98%). Then, the Boc protecting groups were removed quantitatively, and in a final cyclization reaction the (3*S*,6*S*)-3-(4-aminobutyl)-6-(Cl_4-PAP)piperazine-2,5-dione **58** and (3*S*,6*R*)-3-(4-aminobutyl)-6-(Cl_4-PAP)piperazine-2,5-dione **58** were obtained. For one diastereomer the cyclization reaction proceeded faster and after two hours reaction time a diastereomer ratio of 7:3 was formed (Figure 32). For analytics, the isomers were separated by preparative reversed phase HPLC. The more polar product which eluted first and was formed during the reaction in excess will be further referred to as **58-P1** and the less polar compound as **58-P2**.

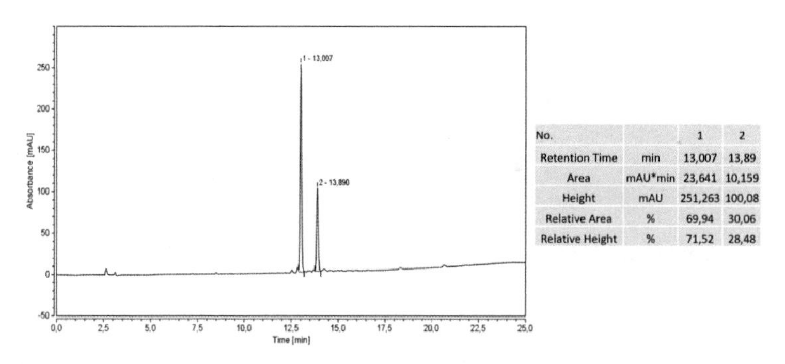

No.		1	2
Retention Time	min	13,007	13,89
Area	mAU*min	23,641	10,159
Height	mAU	251,263	100,08
Relative Area	%	69,94	30,06
Relative Height	%	71,52	28,48

Figure 32: Chromatogram at 265 nm after a HPLC analysis (20 min gradient of 5-95% MeCN in H_2O, 0.1% TFA) of the compound **58** with dr 7:3 after the reaction.

HPLC purification usually affords TFA salts of the products. pH strongly influences later gel formation and therefore, an ion exchange was performed with Amberlite® IRA-900 giving the hydrochloride salts. The isolated isomers were characterized separately, especially the hydrogen at the decisive chiral center showed a significant difference in the chemical shift in ^{1}H-NMR spectroscopy (**58-P1** (3.81 ppm) and **58-P2** (3.65 ppm)). Besides, the C=O stretching vibration of the conjugated ketones in the DKP ring differs between the isomers (Figure 33).

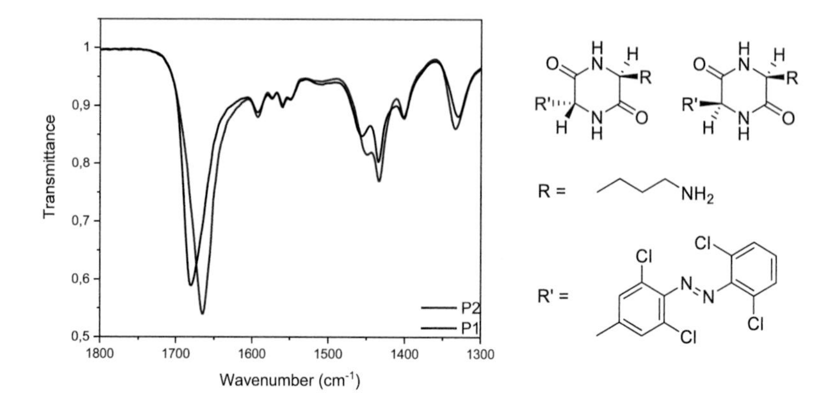

Figure 33: IR spectrum of **58-P1** and **58-P2**. The C=O stretching vibration is shifted from 1680 cm^{-1} (**58-P1**) towards smaller wavenumbers 1664 cm^{-1} (**58-P2**).

Unequivocal assignment of **58-P1** and **58-P2** to the correct chemical structure was not possible so far. The overall low solubility in fast evaporating solvents hampered the approaches to grow crystals and slow evaporation of solutions in mixtures of water and acetonitrile resulted in amorphous material.

Photophysical properties

Photophysical investigation of diastereomers **58-P1** and **58-P2** was performed separately. The azobenzene unit was assumed to isomerize comparable to the tetra-*ortho* chloro-azobenzenes reported in the literature[92]. First, the isosbestic points were determined by UV Vis spectroscopy and light irradiation (Figure 34) of irradiated samples of **58-P1** and **58-P2** in 10% H_2O in MeCN. Even though the molar extinction coefficient $\varepsilon_{630\ nm}$ is below 10 $M^{-1}\ cm^{-1}$ qualitative observation of the spectra shows that red light (660 nm) irradiation resulted in the strongest shift of the molar absorptivity originating from the dark state. Both isomers show distinct isosbestic points at 272 nm and 379 nm. The third point deviated for both compounds (**58-P1**: 479 nm; **58-P2**: 475 nm). PSSs were determined by HPLC supported quantification at the isosbestic point at 272 nm of solutions irradiated with visible light frequencies.

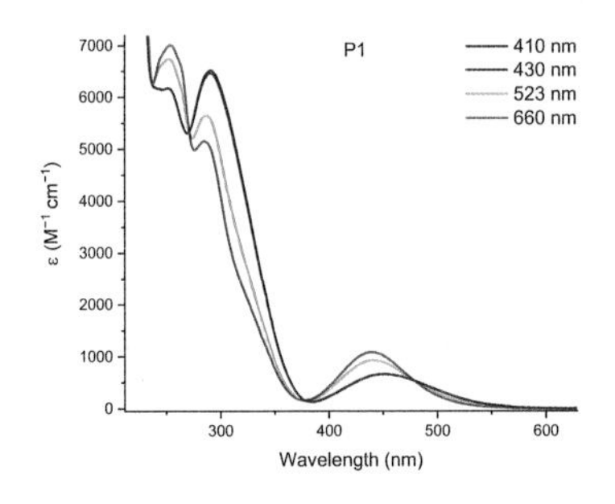

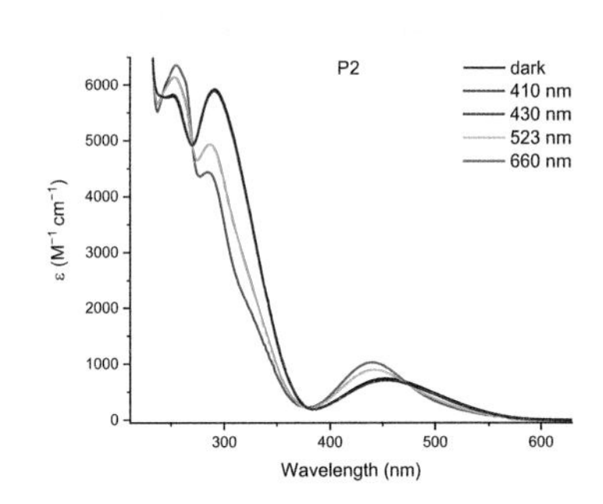

Figure 34: Molar absorptivity of compound **58 P1** and **58 P2** (1.50 mM, 10% H_2O in MeCN) after irradiation. The shown curves correspond to the photostationary states; absorption >600 nm is visible.

The highest percentage of the *Z*-isomer was obtained after red light irradiation ($PSS_{660\,nm}$ = 85% (*Z*)) and back isomerization was efficient under violet light irradiation ($PSS_{410\,nm}$ = 12% (*Z*)). According to the literature[92], green light irradiation resulted in lower percentage of the *Z*-isomer. This is supported by the observations during UV Vis spectroscopy (Figure 34, Figure 155 to Figure 158).

Table 5: PSS of **58** determined by isosbestic point at 272 nm in a 500 μM concentrated solution in DMSO by HPLC. Values agree with NMR analysis and are depicted as % of *Z*-isomer. * a 630 nm cut off filter was used.

Compound	410 nm	430 nm	455 nm	523 nm	660 nm*
P1	12	14	19	53	85
P2	12	14	19	55	85

Next, thermal stability was assessed in AcOH and 1:1 mixtures of water and MeCN for both isomers. The half-life of *Z*- **58-P1** was slightly decreased compared to *Z*- **58-P2** under all tested conditions (Figure 159 to Figure 161). The half-life at 60 °C in AcOH is 4.39 h for **58-P1** and 5.07 h for **58-P2**. Compared to similar azobenzenes, the described half-lives are lower than TFAB derivatives but higher than the new red shifted TFAB **36** reported in chapter 3.1.

Hydrogelation of LMWG 58

The next step was to verify whether the new LMWG **58-P1** and **58-P2** form gels under aqueous conditions. It was assumed that hydrogelation is possible due to structural similarity to the previously investigated LMWGs. However, the 3D conformation of the *E*-azobenzene unit is changed from coplanar (*E*-TFAB[198]) to a twisted structure[92] due to the bulky chlorine substituents. Therefore, a different gelation behavior was expected, compared to the TFAB-based LMWGs.

First, the diastereomer mixture obtained directly from the final reaction was tested upon its gelation behavior. Compound **58** was suspended in PBS buffer (Table 28) or Ringer's solution (Table 29) and boiled for a short period of time. Compound **58** formed homogenous hydrogels in aqueous solutions under physiological conditions (PBS buffer, pH 7.4) at the concentration of 0.1 wt%, which were stable at the inverted vial test. This value of the critical concentration is over one order of magnitude lower compared to the fluorinated analogue (F_4-PAP-DKP-Lys[135b]). Further decrease of concentration to 0.05 wt% resulted in mechanically weak gels. Hydrogels formed in Ringer's solution (0.1 wt%–0.3 wt%) were less stable and therefore, the focus was placed on gelation in PBS buffer. In PBS buffer, hydrogels at a concentration of 0.3 wt% and higher were stable in a boiling water bath, while at lower concentration an unusual behavior was observed: Instead of a collapse to a sol, the gel shrank and released slightly colored liquid (see Figure 35). This process was reversible by heating over the boiling point in a closed vessel.

Figure 35: Hydrogels of compound **58** (dr 7:3, P1:P2 directly obtained from reaction) were formed in the range of 0.1 wt% to 2.0 wt% under physiological conditions (PBS buffer, pH 7.4). Instead of melting in form of gel-to-sol transition, the gel was shrinking and releasing liquid.

Then, the question arose, whether the separated diastereomers **58-P1** and **58-P2** would independently form gels or if their combination resulted in the superior gelation behavior. In a first experiment the separated isomers were recombined to a final ratio of 7:3 (**P1**:**P2**) to validate that the purification and ion exchange did not change the gelation properties. At a concentration of 0.1 wt% stable gel formation with similar shrinking behavior was observed, while at higher concentration of 0.2 wt% an increased stability was detected. However, all gels formed from HPLC purified material showed no complete dissolution of the LMWG. An excerpt of the results is shown in Table 6. In PBS, pure **58-P1** formed gels at 0.2 wt% and shrinking was observed starting from 90 °C. To obtain a material with comparable behavior, 0.3 wt% of pure **58-P2** were necessary. Equivalent mixtures of **P1** and **P2** also formed stable gels at 0.2 wt%. Irradiation of the gel samples with red light resulted in shrinking similar to the reaction upon exposure to heat (Figure 38 a)). Irradiation with blue light of 455 nm did not reconstitute the original gel. A possible explanation of the shrinking behavior is based on the twisted structure of the *E*-tetra-*ortho*-chloro azobenzene where π–π stacking is not possible in the same fashion as for fluorinated or non-substituted AB-DKP-Lys LMWGs. Trauner group reported three conformations for the *E*-tetra-*o*-AB in one single crystal and suggested twisted, but highly flexible structures.[92] Consequently, at least two stable three-dimensional arrangements are possible for the *E*-isomer and isomerization to the *Z*-isomer enables further arrangements. Further investigations are necessary to be able to make well-founded statements regarding the shrinking effect.

Table 6: Gelation properties of compound **58**.

Composition (Ratio **P1**:**P2**, gel in 500 µL PBS)		Approx. concentration	Appearance after 5 min cooling	T_m (°C)
Mixture after reaction		0.2 wt%	Clear gel	92[2]
After HPLC separation				
P1	**P2**			
7	3	0.2 wt%	Clear gel[1]	>99[2]
1	1	0.2 wt%	Clear gel[1]	91[2]
1	0	0.2 wt%	Clear gel[1]	90[2]
0	1	0.3 wt%	Clear gel[1]	90[2]

[1] no complete dissolution, an example is given in the picture; [2]shrinking and releasing liquid instead of melting, reversible by heating >100 °C

The morphology of hydrogel formed from Cl_4-PAP-DKP-Lys (**58**; 1.0 wt% Figure 166, 0.5 wt% Figure 165, 0.1 wt% Figure 164 in PBS) was investigated at the microscale level with SEM. The freeze-dried xerogel samples were handled carefully as they were fragile at this low level of material. A porous network was revealed for LMWG **58** in PBS buffer (see Figure 36). Though at lower concentration only rudimentary network structures were visible under the electron microscope. That indicates that stabilizing interactions mainly occured in the aqueous environment presented by the fluid component of the hydrogel. This implicates that strong solvent dependent interactions like hydrogen-bonds dominate in hydrogel formation with LMWG **58** and supports the previously stated hypothesis about hindered π–π stacking and shrinking effects.

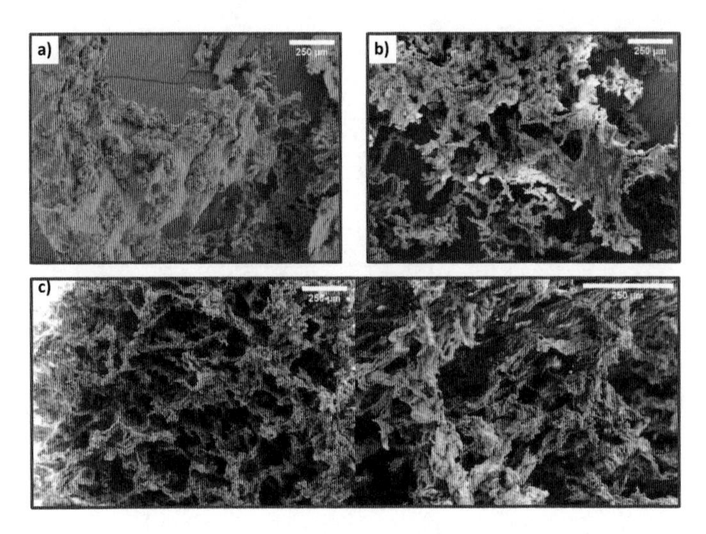

Figure 36: Electron microscopy of the supramolecular assemblies composed from **58** (dr: 7:3 P1:P2) under aqueous conditions in PBS buffer. At low concentration of a) 0.1 wt% and b) 0.5 wt% the rudimentary network is visible. c) At 1.0 wt% a better overview of structural entanglement is given.

Next, the mechanical stability of the hydrogels formed from LMWG **58** (7:3 isomer mixture after reaction) was assessed by rheological measurements. A hydrogel sample consisting of 1.0 wt% **58** in PBS buffer was analyzed on a strain-controlled rotational rheometer. First the LVE was determined, followed by a frequency sweep at 0.01% strain (Figure 163), where $\tan\delta = G''/G' = 0.10$ at 1.14 Hz was measured. The resulting values for the storage modulus G' are higher than for the loss modulus G'' comparable to the fluorinated analogue at 1.7 wt%[135b]. Additionally, a composition at lower concentration (0.3 wt%) was analyzed. Again, the LVE was determined, followed by a frequency sweep at 0.01% strain (Figure 162). The resulting values for the storage modulus G' are higher than for the loss modulus G'' with $\tan\delta = G''/G' = 0.14$ at 1.14 Hz but overall, both parameters decreased compared to the higher concentrated (1.0 wt%) sample. Another measurement was performed at the lower level of concentration. A temperature dependent frequency sweep (Figure 37) at constant angular frequency $\omega = 6.28$ rad s^{-1} supported the results obtained for melting temperatures (Table 28). The sample consisting of 0.3 wt% **58** directly from the reaction (7:3) is stable at least upon heating to 80 °C. Higher temperatures during rheology were not possible due to water evaporation. Consequently, gelator **58** formed gels (without additives) at lower critical gelation concentrations and with higher heat stability than all other gel variants developed by the PIANOWSKI group.[121b, 135b, 165]

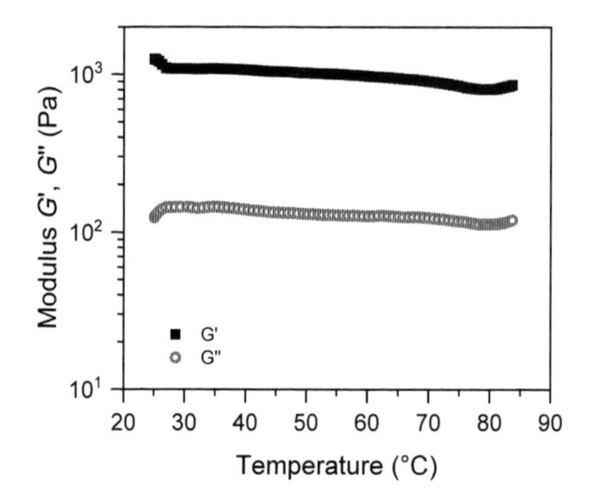

Figure 37: Temperature dependent frequency sweep of a sample consisting of 0.3 wt% hydrogelator **58** in PBS at constant angular frequency ω = 6.28 rad s^{-1}.

Drug delivery is a desired application for hydrogels and was demonstrated for various hydrogel compositions by the PIANOWSKI group (see also chapter 3.2). So far, during release upon light irradiation a complete gel-to-sol transition took place. In the following experiment (Figure 38 b)) it was proven that cargo release is also possible by shrinking hydrogels formed from LMWG **58**. Light-induced release of cargo was demonstrated and visualized by encapsulation of the fluorescent dye 5(6)-carboxyfluorescein into a hydrogel formed from 0.2 wt% **58**. The vial was inverted, so that the gel was on top and irradiated at 660 nm over a period of 1.5 h. At the end of the irradiation period, a fluorescent liquid accumulated at the bottom of the vial.

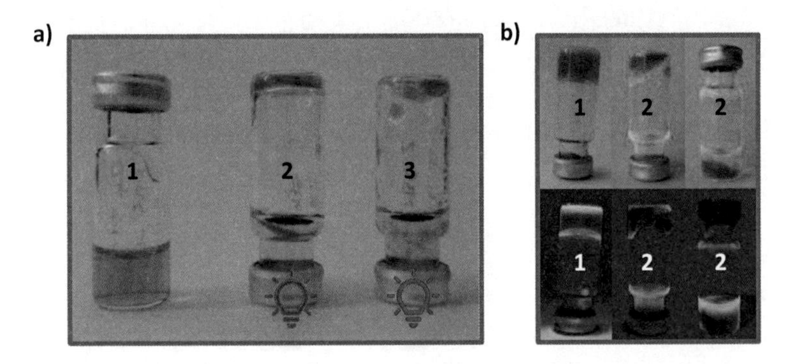

Figure 38: a) Hydrogel (0.1 wt% **58** in PBS) in a crimp vial. *1* = equilibration in darkness; *2* = status after red light (660 nm) irradiation; *3* = status after blue light (455 nm) irradiation. b) The fluorescent dye 5(6)-carboxyfluorescein was added during formation of a hydrogel (0.2 wt% **58** in PBS) and pictures were taken under ambient light or under an UV flashlight. *1* = equilibration in darkness; *2* = status after red light (660 nm) irradiation; the fluorescent cargo was released from the gel.

Biological compatibility

Compatibility of the smart materials with the biological environment in which they are supposed to be employed is an important parameter. Cytotoxicity should be excluded, especially for potential drug delivery systems. Therefore, this parameter was assessed for the hydrogelator **58** using cell viability assays (MTT assays). A human cancerous (HeLa) cell line was treated with increasing concentrations of the gelator **58** (**P1** and **P2**) in order to determine the range of IC_{50} values. Under potential therapeutic circumstances, the material may occur in two forms – either as pure *E*-isomer (the thermally stable form) that forms hydrogels in aqueous solutions, or as a mixture obtained upon irradiation of gels composed of *Z*-**58** with red light. The latter mixture was formed by irradiation of a stock solution of *E*-**58** with red light (660 nm), until the photostationary state was reached (60 min). The experimental protocols and results are summarized in chapter 5.8.6. The IC_{50} value of the irradiated mixture (containing both the *E*-**58** and *Z*-**58**) against HeLa cells can be classified as moderate cytotoxic activity. More precisely, **P1** exhibited an IC_{50} between 100 µM and 10 µM, while **P2** was slightly more toxic at 10 µM. When HeLa cells were exposed to the pure (non-irradiated) solution of *E*-**58**, the IC_{50} value increased to ca. 1 mM (**P1**), which is classified as inactive, while in case of **P2** there was no difference between dark and irradiated mixture, so the IC_{50} remains at 10 µM.[199]

Conclusion

A novel asymmetric AB-DKP-Lys LMWG with *ortho*-chlorine substitution at the azobenzene unit was developed and synthesized. In this context, a synthetic strategy towards a tetra-*o*-chloro azobenzene amino acid was provided. These new compounds photoisomerize upon red light irradiation inside the "therapeutic window" and thus are particularly interesting in a biological context. The new LMWG **58** formed mechanical stable hydrogels at extraordinary low concentrations under physiological conditions in PBS (pH 7.4). The LMWG **58** was obtained as a mixture of diastereomers, which demonstrated hydrogelation as a mixture and after separation. Instead of gel-to-sol transition, shrinking and release of liquid was observed under irradiation. This effect was successfully exploited to release fluorescent cargo as proof of principle. The cytotoxic effects of both diastereomers were determined as moderate to low, which further supports the possible application for drug release in a biological context. Small differences in the investigated properties were observed between the diastereomers. Consequently, more attention will be given in the future to the idea of synthesizing analogous LMWGs with D-lysine to obtain all possible isomers.

3.7 A new red light triggered LMWG – Cl_2-F_2-PAP-DKP-Lys

Motivation

The design of Cl_4-PAP-DKP-Lys **58** provided access to a red light triggered LMWG with low critical gelation concentration. However, the shrinking behavior upon heating or light irradiation differs from previous AB-DKP-Lys LMWGs. Therefore, to obtain an AB-DKP-Lys LMWG reacting on red light and behaving comparable to the fluorinated derivatives, the dfdc-AB design (0, Tetra-*ortho* substituted azobenzenes) was implemented into the gelator structure. The dfdc structural motif was reported to isomerize fast and with high $PSS_{E\rightarrow Z}$ under green light irradiation, and slow with high $PSS_{E\rightarrow Z}$ under red light irradiation. Compared to the *E*-tetra-*o*-chloro AB the dfdc-AB analogue is characterized by a decreased distortion (see Figure 39). Here, both aryl rings are oriented in parallel, slightly displaced planes.[92] According to the rather planar phenyl rings it was expected that the dfdc-PAP-DKP-Lys **59** displays comparable gelation properties to the fluorinated derivatives.

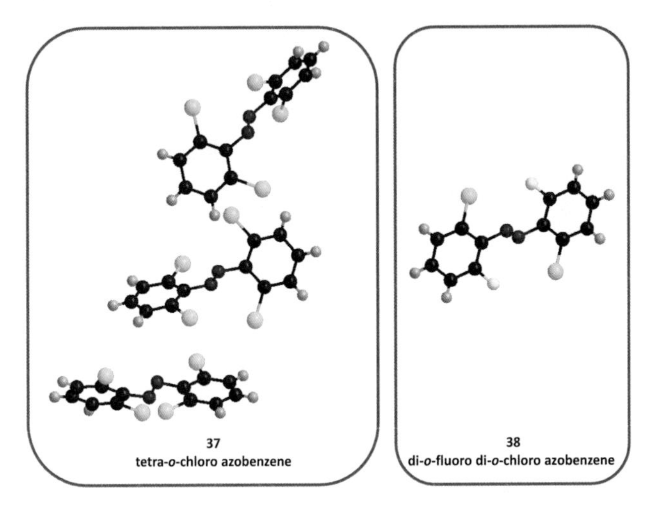

Figure 39: X-ray structures of *E*-tetra-*o*-chloro azobenzene **37** (CCDC 1954066) and *E*-di-*ortho*-fluoro di-*ortho*-chloro azobenzene **38** (CCDC 1954065).[92]

Synthesis

Synthesis towards **59** started from the natural amino acid L-serine and was performed by Mario Michael Most during his Bachelor Thesis under supervision of Anna-Lena Leistner.[200] The synthetic strategy is based on the preliminary work to synthesize the tetra-fluoro derivative[135b] with Negishi-Cross-Coupling as the key step to generate the unnatural amino acid. This reaction required an adequately protected iodoalanine (**85**, Scheme 27) and protected bromoaniline **112**. Starting from L-serine, the methyl ester was formed as a first step in excellent yield, followed by *N*-acetylation in very good yield. Then, the alkyl iodide was formed by an Appel reaction, which was challenging due to low stability of the product. Best results were obtained when reaction and purification were performed on the same day with subsequent storage of the product in the freezer. Nevertheless, purity was only improved to 78% with triphenylphosphine oxide and decomposition product methyl 2-acetamidoacrylate (**86**) as contaminants.

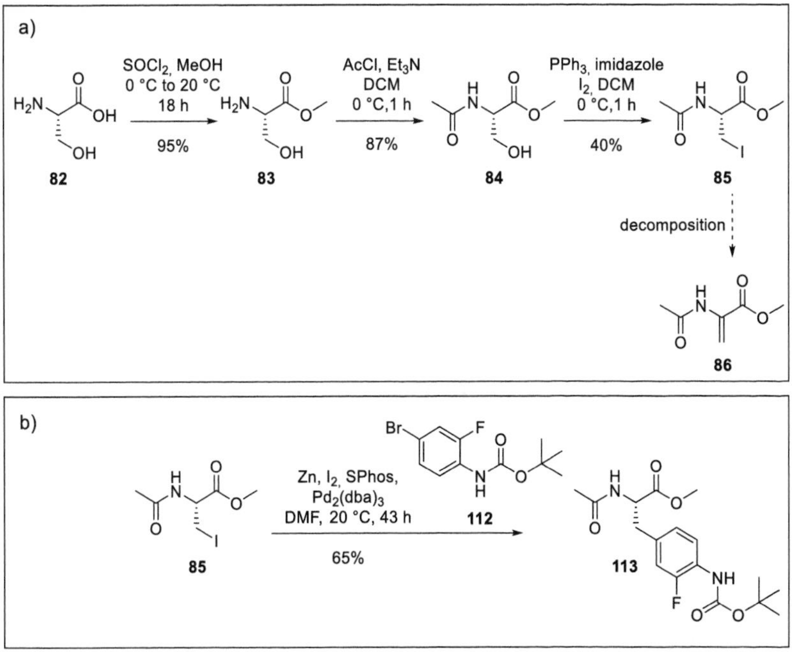

a)

$SOCl_2$, MeOH, 0 °C to 20 °C, 18 h, 95%; AcCl, Et_3N, DCM, 0 °C, 1 h, 87%; PPh_3, imidazole, I_2, DCM, 0 °C, 1 h, 40%; decomposition

82 83 84 85 86

b)

Zn, I_2, SPhos, $Pd_2(dba)_3$, DMF, 20 °C, 43 h, 65%

85 112 113

Scheme 27: a) Synthesis towards the protected iodoalanine **85** and its decomposition to methyl 2-acetamidoacrylate (**86**). b) Negishi coupling between Boc protected **112** and protected iodoalanine **85**.

Next, 4-Bromo-2-fluoroaniline was Boc protected to avoid side reactions and coordination of the Pd catalyst to the amine during Negishi coupling. The *tert*-butyl (4-bromo-2-fluorophenyl)-carbamate (**112**) was obtained in fair yield (52%). Subsequent Negishi coupling between the synthesized building blocks **85** and **112** proceeded with fair yield on gram scale (**113**, 1.17 g, 65%). Previous attempts without protection did not produce the desired product. Moreover, earliest attempts of Negishi coupling between **87** and **88** resulted in an interesting, unexpected side product (**89**, Scheme 28 and Figure 40).

Br
F Cl
NH_2
88
Zn, I_2, SPhos,
$Pd_2(dba)_3$
DMF, 20 °C, 43 h
44%
87
89

Scheme 28: Earliest attempts at Negishi coupling as the key synthetic step towards Cl_2-F_2-PAP-DKP-Lys **59**. An unexpected product was formed during the reaction.

The starting material **88** was commercially available, but so far, no single crystal and x-ray structure were obtained to verify the correct constitution. Thus, it was not possible to draw conclusions whether a halogen exchange took place or if the wrong starting material was sold from the supplier. However, additionally to the Negishi cross coupling an intramolecular cyclization resulted in a side product, the structure of which was elucidated by x-ray analysis (Figure 40). The formed product **89** was not suitable for further reactions towards the desired novel hydrogelator, so no further analysis of the reaction was performed in this thesis.

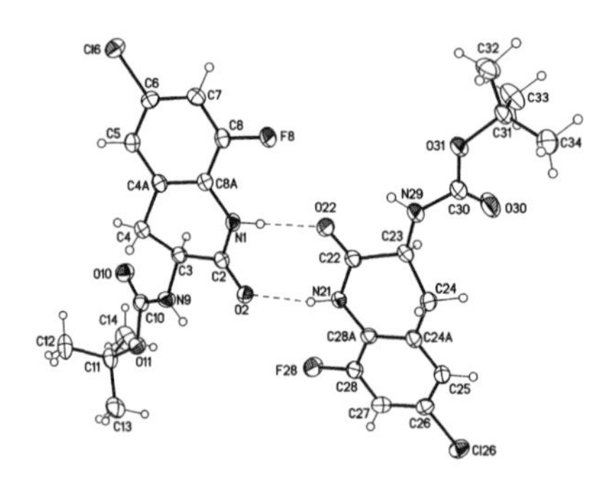

Figure 40: X-ray structure of the unexpected product **89** of the Negishi coupling described in Scheme 28.

The synthesis towards Cl_2-F_2-PAP-DKP-Lys (**59**) was pursued with compound **113** by quantitative deprotection of the amine to form **90**. For the following Mills reaction to form an azobenzene, the second building block 1-

Fluoro-2-nitrosobenzene **91** was synthesized by oxidation of the respective aniline with oxone®. The nitroso compound was prone to decomposition and procedures from the literature did not give satisfying purity. Therefore, the purification was optimized and sublimation at 50 °C and 3 mbar resulted in high purity, though storage in darkness and at −18 °C is advised. Mills reaction between **90** and **91** formed the azobenzene with one fluorine atom on each phenyl ring in *ortho* position (**92**, Scheme 29).

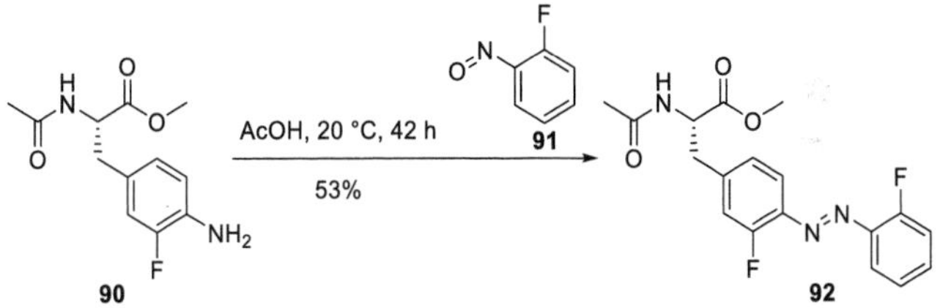

AcOH, 20 °C, 42 h
91
53%
90
92

Scheme 29: Mills reaction between 1-fluoro-2-nitrosobenzene **91** and the aniline **90** towards the azobenzene **92**.

In the following step, the late-stage chlorination at the unsubstituted *ortho* positions was performed to obtain the dfdc-azobenzene **114** (57%). Next, the acetamide was deprotected, but the harsh acidic conditions simultaneously catalyzed the ester cleavage, and the free amino acid **115** was obtained. Chiral RP-HPLC chromatography was performed and in contrast to the racemic tetrachloro analogue **73** the dfdc-AB amino acid **115** had an enantiomeric excess of 80% (Figure 170). The next steps were performed according to the procedures developed for the tetra-choro derivative. The amine was protected, using a Boc protecting group to enable regiospecific peptide coupling with H-Lys(Boc)-OMe·HCl and obtain the dipeptide **117** with a yield of 62% over two steps. A following Boc-deprotection proceeded quantitatively. Final cyclization to form the DKP ring was performed on small scale and preparative HPLC purification with subsequent ion exchange gave 29 mg of the final product **59** as orange solid (16%). As the amino acid **115** was used in 80% ee, the more polar diastereomer was formed in excess and cyclization gave one main product. Due to low amount, it was not possible to isolate the less polar diastereomer. The formation of the product was verified by NMR-spectroscopy (Figure 41) and HRMS (*m/z* calcd. for $C_{21}H_{22}N_5O_2F_2Cl_2$ [M+H] = 484.1113 Da, found 484.1111 Da.).

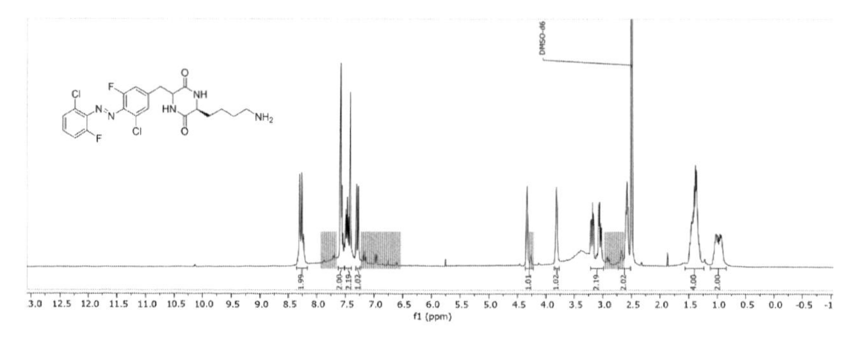

Figure 41: Structure and ^{1}H-NMR spectrum of Cl_2-F_2-PAP-DKP-Lys **59**. ^{1}H NMR (400 MHz, DMSO) δ = 8.28 (q, J = 12.3 Hz, 2H), 7.65 – 7.53 (m, 2H), 7.51 – 7.43 (m, 1H), 7.42 (s, 1H), 7.34 – 7.25 (m, 1H), 4.33 (d, J = 5.1 Hz, 1H), 3.81 (d, J = 5.6 Hz, 1H), 3.19 (dd, J = 13.6, 4.8 Hz, 1H), 3.05 (dd, J = 13.6, 5.1 Hz, 1H), 2.58 (t, J = 7.7 Hz, 2H), 1.40 (pq, J = 10.3, 5.4 Hz, 4H), 1.10 – 0.85 (m, 2H) ppm. The product was partially isomerized to the *Z*-isomer (blue highlighted parts), verified by HPLC analysis (Figure 171).

Conclusion

A novel asymmetrical AB-DKP-Lys LMWG with di-*ortho*-fluoro di-*ortho*-chloro substitution at the azobenzene was developed and synthesized. So far, no further investigation was done, due to limited time and amount of material. However, future progress in this project will be facilitated by the presented synthetic pathway.

3.8 New AIE-gen – Naphthalimide-(F_2)-PAP-DKP-Lys conjugate

Motivation

AIE-gens are promising candidates for application in bioimaging, chemosensing, optoelectronics and stimuli-responsive systems.[168] In examples from the literature the well-known fluorophore naphthalimide was linked with hydrogen bond-promoting peptidic structures to form AIE-gens.[178-179] As described in the previous chapters (3.1–3.6), the cyclic dipeptide DKP repeatedly demonstrated its positive qualities in the assembly of supramolecular structures. Furthermore, examples from the literature report strong hydrogen bonding under aqueous conditions, which can be disrupted by addition of DMSO.[179] (F_2)-PAP-DKP-Lys-naphthalimide **93/94** (Figure 42) was designed based on similar DKP containing structures, which are known to act as aggregation-induced emission fluorophoric systems.[179] Here, the DKP structure induces spontaneous self-assembly by intermolecular hydrogen bonding. Therefore, samples in the literature were first dissolved in DMSO, which interferes with intermolecular hydrogen bonding, followed by increasing the water fraction in the system to reenable aggregation. PAP-DKP-Lys-naphthalimide **93** contains an azobenzene photoswitch, which exhibits photomodulable polarity.[40] Consequently, it was assumed that the compound is better soluble as its *Z*-isomer and thereby photoisomerization may influence aggregation.

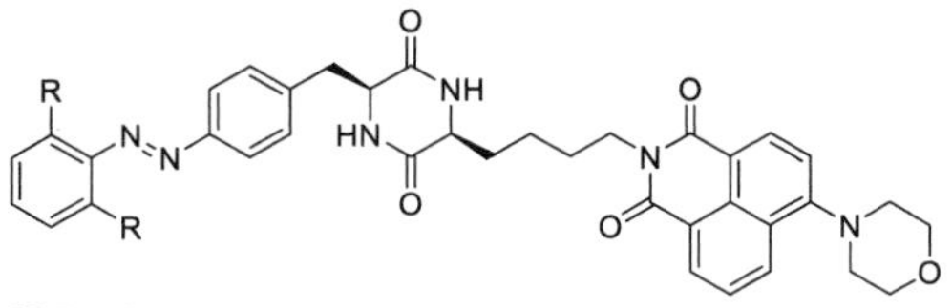

R
N
N
R
O
NH
HN
O
O
N
O
N
O

93 R = H
94 R = F

Figure 42: Structural design of (F_2)-PAP-DKP-Lys-naphthalimide **93/94** – a proposed AIE-gen, based on well-investigated LMWG structures.

Synthesis

The 4-morpholino substituted naphthalic anhydride was selected as fluorophore due to promising properties in DKP-NI conjugates by Balachandra *et al.*[179] and fluorescence emission which is hardly overlapping with the azobenzene absorption. The compound **119** was synthesized according to the literature[201] in very good yield (80%). In the subsequent reaction, the naphthalic anhydride reacted in a Gabriel synthesis with the primary amine of LMWG **52** or **57**. The new target structures **93** and **94** were obtained after preparative HPLC purification and verified by NMR spectroscopy and HRMS (chapter 5.10.1 and 5.10.6).

Fluorescence properties

Fluorescence spectra were recorded for compounds **93** and **94** in various solvents (Figure 43). Both compounds exhibit strong solvatochromism. Detailed emission and excitation spectra are shown in experimental part 5.10.3 and 5.10.7. Non-fluorinated **93** demonstrated low solubility in MeOH, Acetone, ^{i}PrOH and MeCN, whereas **94** was completely soluble at a concentration of 50 µM. In polar solvents such as DMF the fluorescence emission was bathochromically shifted (**93**: $\lambda_{max\ em.}$ = 527 nm; **94**: $\lambda_{max\ em.}$ = 530 nm), compared to less polar THF (**93**: $\lambda_{max\ em.}$ = 501 nm) or $CHCl_3$ (**94**: $\lambda_{max\ em.}$ = 503 nm) respectively. However, the furthest bathochromic shift of emission was observed in the polar protic solvent hexafluoroisopropanol (HFIP) for both compounds (**93**: $\lambda_{max\ em.}$ = 541 nm; **94**: $\lambda_{max\ em.}$ = 544 nm).

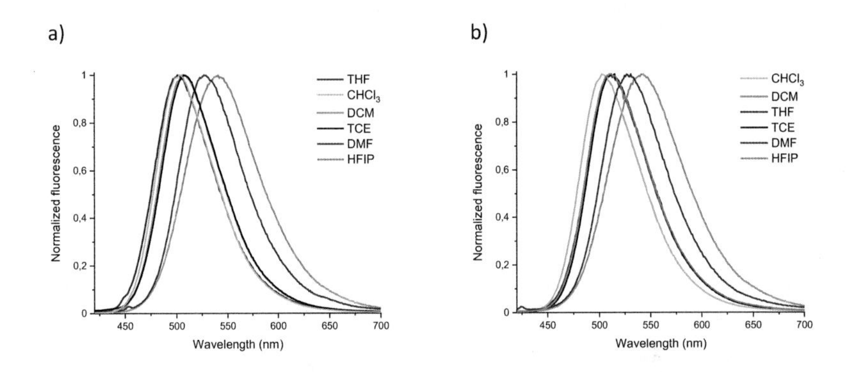

Figure 43: Fluorescence emission spectra (excitation at 400 nm) were measured of 50 µM solutions in various solvents and normalized. a) PAP-DKP-Lys-naphthalimide (**93**) and b) F_2-PAP-DKP-Lys-naphthalimide (**94**) were both probed for solvatochromic behavior.

In order to investigate the behavior upon isomerization, solutions of PAP-DKP-Lys-naphthalimide **93** were irradiated, and the fluorescence was measured. The experiments were performed in triplicates in the solvent 1,1,2,2-tetrachlorethane (TCE). First, the emission spectrum upon excitation at 400 nm was recorded for solutions which were previously equilibrated in the dark, followed by a sequence of irradiation: Starting with irradiation at 365 nm for 10 min, followed by 10 min irradiation at 455 nm and subsequently at 365 nm for 10 min (Figure 176 to Figure 178).

In all samples only minor changes of the fluorescence spectrum were observed after the first irradiation at 365 nm (PSS_{365nm} = 36% *Z*-isomer, Figure 172). Consequently, there was no isomerization-dependent fluorescence behavior observed. Surprisingly, the fluorescence intensity increased slightly (factor 1.1 to 1.2) upon irradiation with light of 455 nm (PSS_{455nm} = 30% *Z*-isomer). In the following, irradiation with a 365 nm LED was repeated to verify if the same fluorescence level as after the initial irradiation (365 nm) could be restored. As illustrated in Figure 44, the fluorescence was again slightly increasing, furthermore a 1.6 to 2.2-fold increase of fluorescence intensity was observed upon 1 h equilibration in the dark after the irradiation.

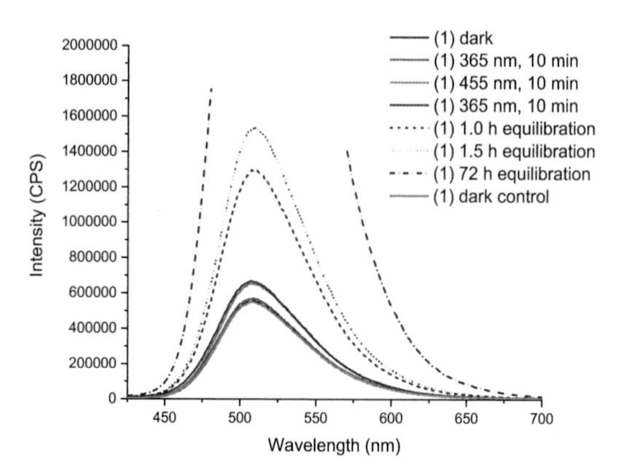

Figure 44: Emission spectra (excitation at 400 nm) of compound **93** in TCE (measurement number 1). Upon irradiation the fluorescence was increasing over time. after 72 h the spectrum was recorded from 425 nm to 480 nm and was continued from 570 nm to 700 nm to avoid damage to the fluorescence spectrometer when keeping the slits at the same value.

Longer equilibration time led to further enhancement of fluorescence intensity; after 72 h the spectrum was recorded from 425 nm to 480 nm and was continued from 570 nm to 700 nm to avoid damage to the fluorescence spectrometer when keeping the slits at the same value. Nevertheless, the spectrum after 72 h demonstrated a significant increase in the fluorescence level (over 3 times higher).

Based on the low solubility of **93**, ultrasonic treatment was applied to the solutions during the investigation. Here a new phenomenon – a hypsochromic shift of the fluorescence emission – was observed. This effect followed a distinct pattern and was occurring after ultrasonic treatment of solutions in TCE (with an increase in fluorescence emission). Even more interesting is the fact that another hypsochromic shift was induced by irradiation of the solution with light at 365 nm ($Em_{max, dark}$ = 509 nm; $Em_{max, treated}$ = 480 nm). The effect was demonstrated in Figure 45, where the experiment was done in independent triplicates (Figure 179 to Figure 181) by each time dissolving fresh compound **93** in TCE to obtain a 50 µM solution.

Furthermore, a control experiment where the samples were kept in the dark was performed (Figure 182). Three independent samples were dissolved in TCE at a concentration of 50 µM and stored in the dark. Here, no hypsochromic shift was observed and additional ultrasonic treatment after a week of equilibration in the dark did also not affect the emission maximum.

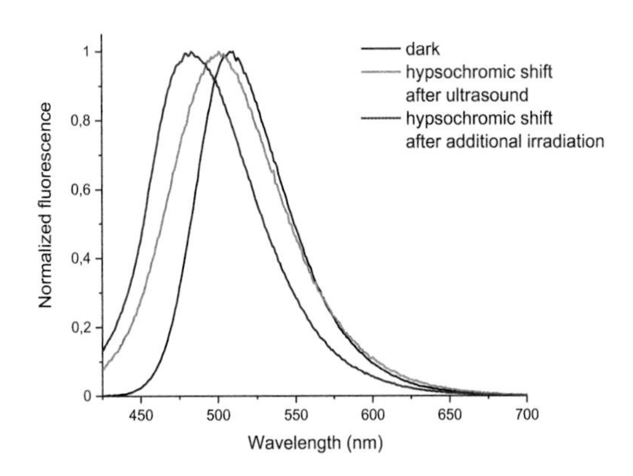

Figure 45: Demonstration of the hypsochromic shift described above. Normalized fluorescence emission: Dark state ($\lambda_{max\ em.}$ = 509 nm), ultrasonic treatment ($\lambda_{max\ em.}$ = 501 nm), and irradiation at 365 nm to the endpoint ($\lambda_{max\ em.}$ = 480 nm).

The influence of the solvent on the hypsochromically shifted fluorescence emission phenomenon was investigated. Compound **93** was dissolved at a concentration of 50 µM in HFIP, DMF, DMSO, THF, DCM, $CHCl_3$ and again in TCE as a control. The solutions were freshly prepared, and fluorescence was measured after each step (Figure 46). After assessment of the dark state, the solution was treated directly for 30 min with ultrasonic waves. In the next step the sample was equilibrated for 60 min in darkness. In the following the solution was again treated with ultrasonic waves and finally, the solution was irradiated with light at 365 nm. For selected samples further equilibration and ultrasound periods with subsequent fluorescence measurements were performed (see Figure 183).

The results of this experiment demonstrated that in highly polar solvents (dielectric constant ε >15) such as HFIP (protic), DMF (aprotic) and DMSO (aprotic) the fluorescence emission remained rather constant. In solvents of lower polarity, the fluorescence emission increased gradually over the course of the described experimental sequence. The hypsochromically shifted fluorescence emission phenomenon was only observed in TCE. It was assumed that formation of aggregates was the origin of abnormal fluorescence behavior, as the different irradiation and ultrasonic sequences may influence the aggregation process in various ways.

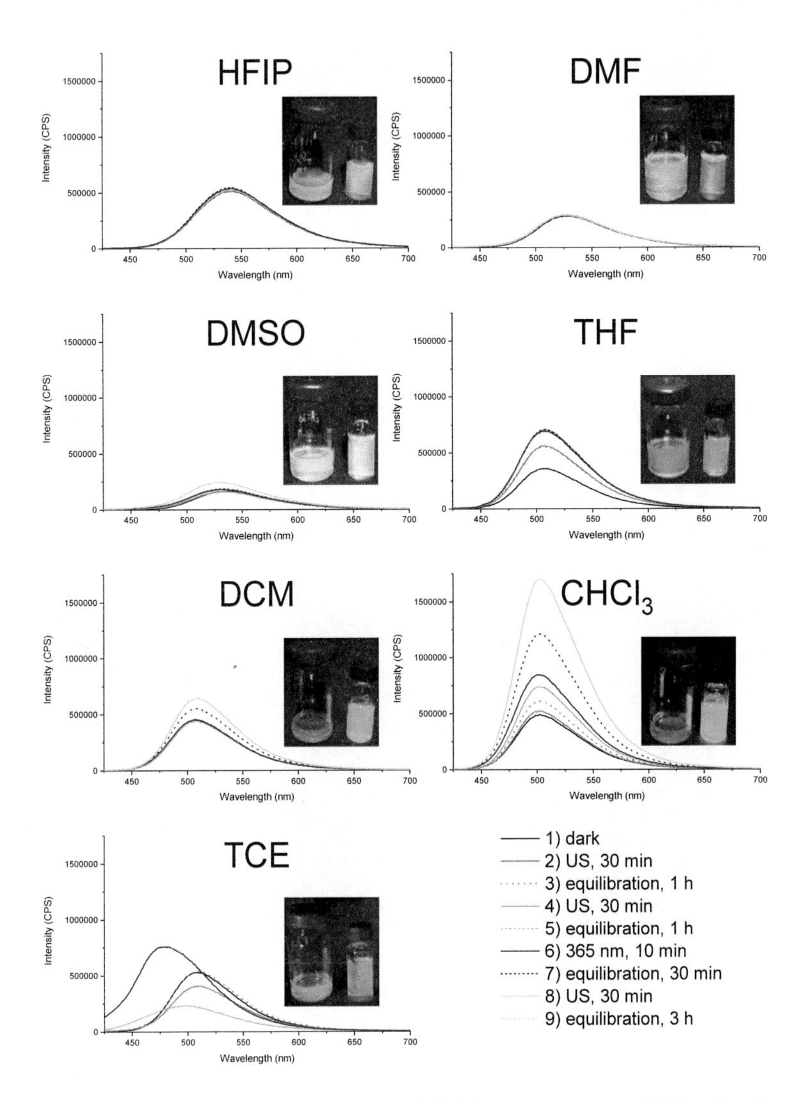

Figure 46: Transfer of the protocol for the hypsochromic shift of the emission maximum (TCE) to other solvents. The solvents are sorted by polarity from polar (top) to less polar (bottom). The inserted pictures show each a bigger vial containing the dark equilibrated solution and the solution after the last applied exposure respectively. US = ultrasound.

The effect of hypsochromically shifted emission was only observed in the solvent TCE. To correlate the effects to macromolecular structures, SEM images were recorded. For this experiment, samples at the four decisive stages observed in TCE were prepared and applied on silicon wafers (Figure 47). Sample **a)** was taken directly after dissolution of the compound **93 (dark)** Due to the drying period of 1 h, sample **a)** had additional equilibration time, in which subtle network formation started. Sample **b)** was taken after ultrasonic treatment and 3 h of equilibration **(fluorescence increase)**. Here, a distinct network-like structure was formed on the wafer**.** Sample **c)** was taken after following additional ultrasonic treatment **(slight hypsochromic shift)**. In this sample, small particles, which adhered to the network, were visible. Sample **d)** was irradiated at 365 nm for 10 min after the previous steps **(final hypsochromic shift)**. The microscopic images revealed that the irradiation disrupted the network and released the adhered particles. In all samples large structural assemblies were observed, which were most pronounced in sample **b)** (for more details see Figure 185).

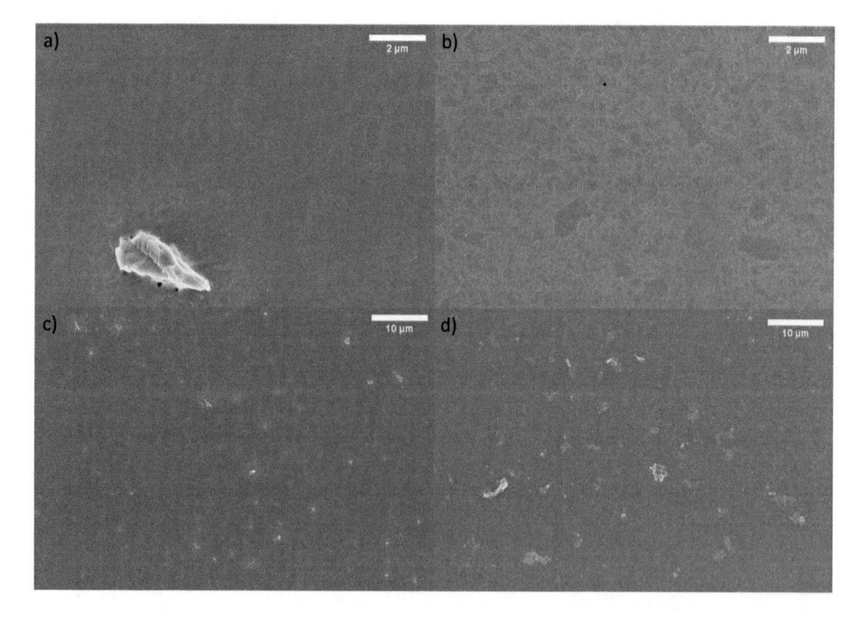

Figure 47: SEM images of sample **93** in TCE (50 µM). a) A network started to form during wafer preparation; b) a dense network was detected on the wafer, some small particles are enclosed; c) the network is still present and distinct particles are visible; d) the network disintegrates upon irradiation and small particles remain.

Another sample **e)** was prepared after equilibration of the hypsochromically shifted fluorescence sample **d)** overnight, with the respective microscope pictures being depicted in Figure 48. The fragmented network and the particles observed in sample **d)** assembled to spheric nanoparticles with submicrometer size.

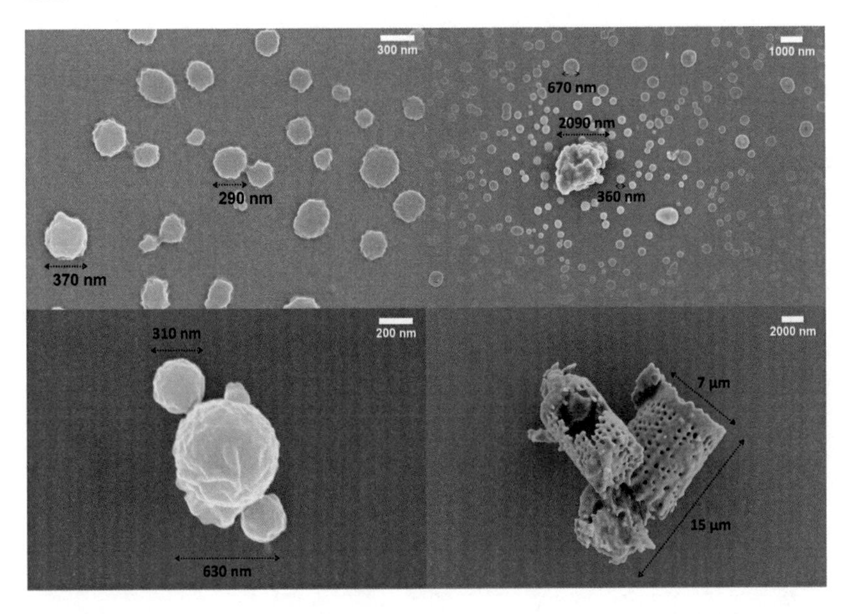

Figure 48: SEM images of sample e). The fragments and small particles (visible in Figure 185 d)) assembled to spheric nanoparticles, which tend to agglomerate.

To verify the observed particles in the microscope images, DLS measurements were performed. Three samples prepared the same way as sample **e)** were analyzed for their particle size (Figure 49). The results varied slightly in between the replicates, though the values were all in the same magnitude between 400 nm and 700 nm. These particle sizes correlated with the structures visualized by electron microscopy.

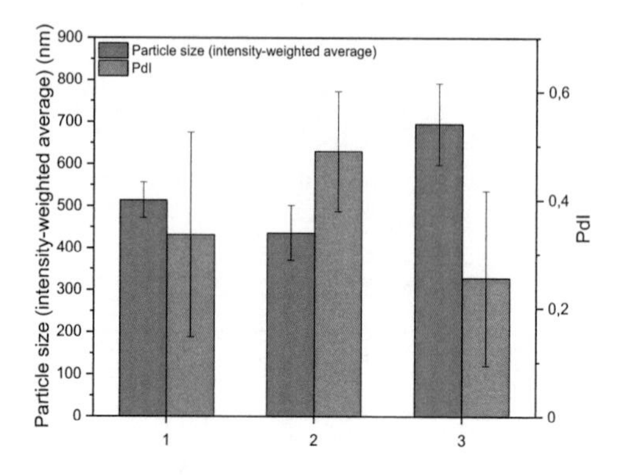

Figure 49: Results of the DLS measurement. Three independent measurements (number of technical replicates n≥4 respectively) were performed. The particle size is in good agreement with the microscopy-observations.

Conclusion

Novel potent AIE-gens were synthesized and investigated. It was demonstrated that aggregation occurred in the least polar solvent, TCE, for compound **93**. The effect was expressed by fluorescence emission increase, due to formation of large agglomerates and network-like structures. In TCE, an exclusive hypsochromic shift was observed following certain patterns: suspensions of aggregates in TCE were treated by ultrasound and UV-light irradiation to trigger the hypsochromic shift. The effect was related to the formation of nanoparticles (400 nm to 700 nm in diameter), as proven by DLS and microscope analysis. Further experiments are necessary to explain the absence of the hypsochromically shifted fluorescence emission phenomenon and thereby linked nanoparticle formation in THF, DCM and $CHCl_3$. It is assumed that the process and the possible geometries during the formation of aggregates and networks is complex, in addition, light irradiation produces mixtures of photoisomers, which further complicates the process. More detailed analysis is necessary to gain a deeper understanding about the relation between photoirradiation, solubility, aggregate formation, and fluorescence behavior. Moreover, the better soluble and more polar compound **94** will be analyzed in terms of fluorescence behavior upon sonication or light irradiation.

4 Summary and Outlook

Hydrogels are soft materials, which often find application in biomedical context. They classify by their self-healing properties, thixotropy, on-demand reversibility and stimuli responsiveness, for example to light. The photoswitches inside photochromic hydrogels react in spatiotemporal precision to light irradiation and are excellent candidates for photo-controlled drug release. In the scope of this thesis, novel photochromic supramolecular hydrogelators (**52**, **54-59**, Figure 50) were developed, synthesized, and characterized. Furthermore, the hydrogen bond promoting peptide structure, which can be found in the DKP motif of all discussed LMWGs, was linked with the fluorophore naphthalimide to obtain AIE-gens **93** and **94**.

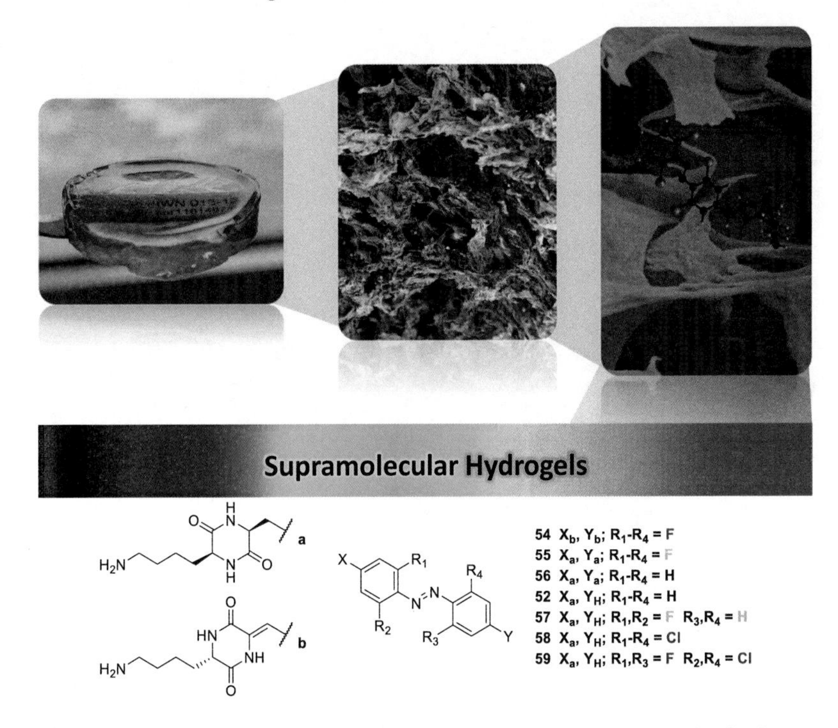

Figure 50: Photochromic supramolecular hydrogelators, which were developed or optimized in this thesis.

The versatility of azobenzene containing DKP-Lys hydrogelators was demonstrated in this thesis. The distinct stability enhancement in composite hydrogels with alginate, as well as the high stability of red-light-triggered system **58** are remarkable.

Table 7: Important parameters of the examined photochromic supramolecular hydrogelators.

LMWG	Critical gelation conc.	T_m °C	Light-triggered liquefaction	Light-triggered gelation
54	1 wt%[1]	84 ± 14	660 nm	no
55	No gelation <2 wt%[1,2,3]	[4]	[4]	[4]
56	1.5 wt%[2]	77	[4]	
56 + alginate	0.6 wt% + 0.6 wt%[2]	83	365 nm	455 nm
52	2.0 wt%[3]	51 ±4	365 nm	[4]
52 + alginate	0.2 wt% + 0.2 wt%[3]	96±2	no	[4]
57	3.0 wt%[2]	67 ± 23	523 nm	[4]
57 + alginate	0.3 wt% + 0.1 wt%[3]	100	523 nm	410 nm
58	0.1 wt%[2]	74[5]	660[5]	no

[1]Ringer's solution; [2]PBS buffer; [3]water; [4]not determined; [5]shrinking

4.1 Fluorinated azobenzenes switchable with red light

In this thesis, a new approach to design bathochromically shifted azobenzenes was demonstrated by conjugation of azobenzenes with unsaturated substituents. The new chromophore was photophysically examined and can be incorporated into larger structures. The bis-*para*-substituted aldehyde **36** was coupled by aldol condensation to two DKP-Lys units, forming a symmetric, red light triggered smart material **54** (Figure 50, Table 7). In aqueous media under physiological conditions **54** formed viscous materials, which can be liquified to a non-viscous fluid by red-light irradiation. However, they were of microcrystalline rather than of hydrogel nature. Consequently, the conjugated TFAB chromophore is a valuable building block to form photochromic biomaterials and photopharmacology agents. Overall, the investigation of material formed by **54** in aqueous media indicated that a flexible structure is essential for efficient self-assembly of fibrous networks and, consequently, proper hydrogels.

4.2 Symmetrical chiral TFAB gelator

The non-conjugated version **55** of the LMWG **54** (Figure 50) was synthesized and the gelation behavior was investigated and compared to the previously obtained smart material. The reduced π-π stacking ability of the hydrophobic core leads to increased influence of the hydrophilic lysine side chains and results in pH-sensitive material without gel formation under the tested conditions (Table 7).

4.3 Symmetrical chiral gelator – model system for the addition of dopants

A model system photochromic LMWG **56** (Figure 50) was developed and the effects of π-π stacking and ionic interactions on gel formation and stability were determined. The designed system was not suitable to assess the effects of π-π stacking. By combination and attempted gelation with electron-deficient fluorinated azobenzenes, it was observed that a higher aqueous solubility of the azobenzene additive is required to analyze stabilization of hydrogelation by face-centered π-π stacking. This may be achievable, for example, by attachment of polyethylene glycol chains or other polar functional groups like sulfonates[188] in *para* position to the azo bond of the fluorinated azobenzenes. However, it was revealed that the acidic polymer alginate reduced the critical gelation concentration, improved the stability (Table 7), and enhanced the biocompatibility. The LMWG **56** formed composite supramolecular hydrogels with alginate below the critical gelation concentration of **56** alone and over a range between pH 4 to pH 8. Despite the stabilizing effect of the ionic interactions between alginate and LMWG **56**, a complete gel-to-sol transition occurred under UV-light irradiation and reverse stabilization under blue light offered access to a new gel preparation method (B). This new method is particularly suitable for the encapsulation of heat-sensitive cargo and its following release.

4.4 Transfer of the composite alginate system to (F_2)-PAP-DKP-Lys

The composite system was extended to existing LMWGs (**52** and **57,** developed by PIANOWSKI group.[121b, 165], Figure 51). Both compounds formed composite supramolecular hydrogels with alginate below the critical gelation concentration of the respective LMWG alone and below the critical concentration of composite hydrogels with bolaamphiphilic gelator **56**.

Figure 51: Clear hydrogel with a diameter of 25 mm (0.6 wt% of F_2-PAP-DKP-Lys (**57**); 0.6 wt% alginate). Due to high mechanical stability the shape remained intact upon handling with a spatula.

This behavior of strong stabilization was not expected with regards to the asymmetrical structure of LMWG **52** and **57** with only one lysine side chain per molecule and the presumed necessity of two lysine amines per molecule to form crosslinks. However, the asymmetrical structure might enable a helical layout in the macromolecular fibers with a denser arrangement. In such a theoretical macromolecular structure the lysine side chains are presented on the outer side of the fibers and allow for ionic interactions and crosslinking with the alginate carboxylate groups. Gel-to-sol transition occurred for **57** under green light irradiation, which was not possible for non-fluorinated **52** in composite hydrogels with alginate. Reconstitution of the gel could be performed by irradiation with violet light, which enabled the use of mild preparation method B for LMWG **57**.

In addition, it was demonstrated that composite hydrogels with alginate are suitable candidates for biological experiments with living cells, even though some optimizations are necessary either from biological side or hydrogel composition. By equilibration with a $CaCl_2$ solution the photochromic LMWG **57** is replaced by Ca^{2+} ions. In this process, the LMWG diffused to the supernatant and consequently, only natural materials remained in the reformulated, stable hydrogel. Based on these results new opportunities for 3D-printing are enabled, including photo-controlled printing and subsequent replacement of the non-natural gelator. Appropriate experiments are to be designed and carried out. Additionally, the interaction of host F_2-PAP-DKP-Lys **57** with cargo guest molecules for drug delivery application was investigated in closer detail by NMR spectroscopy in collaboration with Dr. Claudia Muhle-Goll and Prof. Burkhard Luy (*submitted*).

4.5 New red light triggered LMWG – Cl_4/Cl_2-F_2-PAP-DKP-Lys

A synthesis route for an *ortho* chloro (**73**) and difluoro-dichloro (**115**) azobenzene amino acid was established. Those compounds were further functionalized towards LMWGs (**58** and **59,** Figure 50, Table 7), which operate with light inside the therapeutic window (600-900 nm) and are potential materials for photo-controlled drug release. These new compounds photoisomerize upon red light irradiation and thus are particularly interesting in biological context. The new LMWG **58** formed mechanical stable hydrogels at extraordinary low concentrations under physiological conditions in PBS (pH 7.4). The stabilizing effect strongly depends on the twisted structure of *ortho*-chloro azobenzene, which determines its pliability and hydrophobic stacking, and thereby affects the macromolecular arrangement in fibers.

The LMWG **58** was obtained as a mixture of diastereomers, which also demonstrated hydrogelation after separation. Instead of gel-to-sol transition, shrinking and release of liquid was observed under irradiation. This effect was successfully exploited to release fluorescent cargo as proof of principle. The cytotoxic effects of both diastereomers were determined as moderate to low, which further supports the application for drug release in biological context. Small differences in the investigated properties were observed between the diastereomers. Consequently, more attention will be given in the future to the idea of synthesizing analogous LMWGs with D-lysine to obtain all possible isomers. Further experiments are being conducted to extend the release system to biological active cargo or drugs. In addition, the characterization of the hydrogelation behavior of **59** is still pending.

4.6 New AIE-gens - Solvent dependent aggregation

Finally, AIE-gens **93** and **94** were synthesized and investigated (Figure 52). Solvent-dependent fluorescence emission maxima were observed for both compounds, and in solvents of lower polarity aggregation occurred for compound **93**. The effect was expressed by fluorescence emission increase due to formation of large agglomerates and network-like structures. In TCE, an exclusive hypsochromic shift in fluorescence emission was observed, depending on the respective treatment of the solution (irradiation, sonication, equilibration in the dark. The effect was correlated to the formation of nanoparticles (400 nm to 700 nm in diameter), as proven by DLS and SEM analysis. Further experiments are necessary to explain the absence of the hypsochromically shifted fluorescence emission phenomenon and thereby linked nanoparticle formation in other solvents of low polarity (THF, DCM and $CHCl_3$). Moreover, the better soluble (more polar) compound **94** will be analyzed with respect to fluorescence behavior upon sonication or light irradiation.

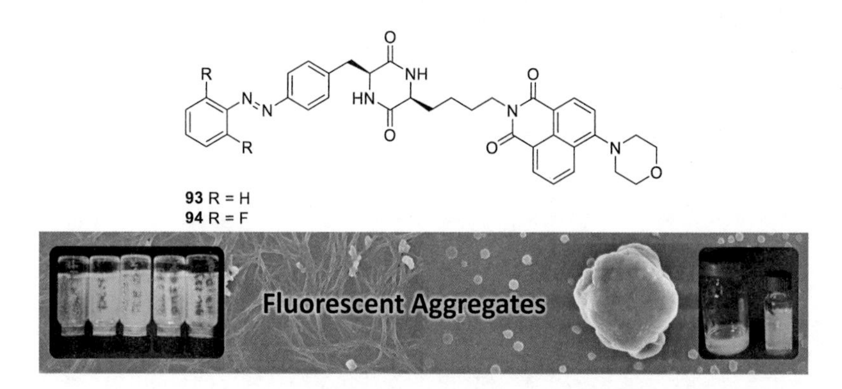

Figure 52: Structure of the novel AIE-gens **93** and **94**. Fluorescence solvatochromism and aggregation were observed.

5 Experimental part

5.1 General remarks

All reagents and starting materials are commercially available (Sigma-Aldrich, Fluorochem, chemPur, Alfa Aesar or BLDpharm) and were used as supplied unless otherwise indicated. Solvents of technical quality were distilled prior to use.

In the experiments where physiological conditions were required, Dulbecco's Phosphate-Buffered Saline (DPBS) buffer pH 7.4, (-/-): no calcium, no magnesium, Gibco™ from Thermofisher, cat.#: 14190136, abbreviated below and in the thesis simply as "PBS buffer", was used.

All experiments with photoisomerizable molecules in the visible light spectrum were performed in absence of sunlight, for example with brown glassware, or colourless glassware wrapped with aluminium foil, working in a room with dimmed light. All experiments were conducted in air unless noted otherwise and all reactions containing air- and moisture-sensitive compounds were performed under an argon atmosphere using oven-dried glassware applying common Schlenk-techniques. Liquids were added *via* steel cannulas and solids were added directly in powdered shape. For aqueous extraction, for all analytic samples and for the HPLC, deionized water (diH_2O) from Millipore was used unless noted otherwise. For certain reactions, flat-bottom crimp neck vials from CHROMAGLOBE with aluminum crimp caps were used. Reactions at low temperatures were cooled using flat dewars produced by Isotherm (Karlsruhe) with water/ice or isopropanol/dry ice mixtures.

In general, solvents were removed at preferably low temperatures (<60 °C) under reduced pressure with a rotary evaporator. Solvents for solvent mixtures were separately volumetrically measured. If not indicated otherwise, solutions of inorganic salts are aqueous, saturated solutions.

All chemical structures were drawn in Chemdraw Professional 20.1.0.110 (PerkinElmer Informatics, Inc.).

5.2 Analytics and devices

Analytical Balance

Used devices: Sartorius Basic (PreZion: 0.001 g), Sartorius Secura26-1S (0.002 mg) and Sartorius M2P Micro Balance (0.001 mg).

Solvents and reagents

Solvents of technical quality were distilled prior to use. Solvents of p.a. (per analysis) quality were commercially purchased (ACROS ORGANICS, FISHER SCIENTIFIC, SIGMA ALDRICH) and used without further purification. Absolute solvents (DCM, Tetrahydrofurane (THF)) were prepared by purification by MB SPS 5 from M. BRAUN and stored under an argon atmosphere.

Reagents were commercially purchased from industry companies (ABCR, ACROS, ALFA AESAR, FISHER SCIENTIFIC, MERCK) and used without further purification if not stated otherwise.

Other reagents and solvents were abbreviated as follows: cyclohexane (cH), triethylamine (Et_3N), ethyl acetate (EtOAc).

Reaction control

Analytical TLC was carried out using silica coated aluminium plates (silica 60, F_{254}, layer thickness: 0.25 mm) with fluorescence indicator by MERCK. The spots were detected by fluorescence quenching of UV-light at $\lambda = 254$ nm or $\lambda = 366$ nm and subsequent staining with phosphomolybdic acid solution (5% phosphomolybdic acid in ethanol, dip solution); potassium permanganate solution (1.00 g potassium permanganate, 2.00 g AcOH, 5.00 g sodium bicarbonate in 100 mL water, dip solution); Mostain solution (5% $(NH_4)6Mo_7O_{24}$ in 10% sulfuric acid with 0.03% $Ce(SO_4)$-2) or β-naphthol solution (10% β-naphthol, 10% AcOH and 80% ethanol, dip solution) followed by heating in a hot air stream. For amino acids, the Seebach reagent (2.5% phosphomolybdic acid, 1.0% cerium(IV) sulfate tetrahydrate, 6.0% conc. sulfuric acid, 90.5% water; dipping solution) was used with subsequent heating in a hot-air flow. Amines were detected by dipping the TLC-plate in ninhydrin solution (0.2% ninhydrin in 0.1% AcOH in EtOH, dip solution) and subsequently heating in a hot-air stream. Alcohols were detected by dipping the TLC-plate in $KMnO_4$-solution (1.5 g of $KMnO_4$, 10 g K_2CO_3 and 10 mL 10% NaOH in 200 mL of water) and subsequently heating in a hot air stream.[202]

Chromatography

Crude products were purified according to literature procedure by column chromatography *via* flash-chromatography.[203] Silica gel 60 (0.063×0.200 mm, 70–230 mesh ASTM) (MERCK), Geduran® Silica gel 60 (0.040×0.063 mm, 230–400 mesh ASTM) (MERCK) or Celite® (FLUKA) and sea sand (calcined, purified with hydrochloric acid, RIEDEL–DE HAËN) were used as stationary phase. The flow of eluants was regulated with a pressure valve to typical 1.3 bar with pore size of 2 (40 – 100 μM).

Analytic and Preparative HPLC

Analytic HPLC was performed with the Thermofisher UltiMate 3000 system containing a degasser, pump, autosampler, column compartment and diode array detector. There, the flow rate was 1 mL/min at 25 °C on a stationary PerfectSil Target (MZ-Analytik) C18 column (3-5 μm, 4.0 mm × 250 mm). For the elution, if not noted otherwise, a linear gradient (20 min) was used from 5% to 95% MeCN (0.1% TFA) in diH_2O (0.1% TFA). Chromeleon 7 software was used for data extraction.

Preparative HPLC separation was performed with a PURIFLASH® 4125 from INTERCHIM, equipped with InterSoft V5.1.08 software and a UV diode array detector. The stationary phase was a VDSpher column with C_{18}-M-SE, 250×20 mm and 10 μm from VDSOPTILAB. For the elution, if not noted otherwise, a linear gradient was used from 5% to 95% MeCN (0.1% TFA) in diH_2O (0.1% TFA) and with a flow rate of 10 mL/min at 25 °C.

Lyophilization

Substances which were purified by HPLC were dried by lyophilization. The device used was a CHRIST LDC-1 ALPHA 2-4 system.

Nuclear Magnetic Resonance spectroscopy (NMR)

NMR spectra were recorded using the following device: BRUKER Avance 400 (400 MHz), BRUKER Avance 500 DRX 500 (500 MHz), ^{13}C-NMR: BRUKER AM 400 (100 MHz), BRUKER Avance 500 DRX 500 (125 MHz), ^{19}F-NMR: BRUKER Avance 400 (376 MHz). All measurements were carried out at room temperature. The chemical shift of ^{19}F-NMR-spectra is calculated by the device without reference. The following solvents from EURISOTOP were used: chloroform-d_1, acetonitrile-d_3, methanol-d_4, water-d_2 and DMSO-d_6. Chemical shifts δ were expressed in parts per million (ppm) and referenced to chloroform-d_1 (^{1}H: δ=7.26 ppm, ^{13}C: δ=77.0 ppm), acetonitrile-d_3 (^{1}H: δ=1.94 ppm, ^{13}C: δ=1.32, 118.26 ppm), methanol-d_4 (^{1}H: δ=7.26 ppm, ^{13}C: δ=49.0 ppm), water-d_2 (^{1}H: δ=4.79 ppm) and DMSO-d_6 (^{1}H: δ=2.50 ppm, ^{13}C: δ=39.43 ppm).[204] The signal structure is described as follows: s=singlet, d=doublet, t=triplet, q=quartet, quin=quintet, b=broad singlet, m=multiplet, kb=complex area, dt=doublet of triplets. The spectra were analyzed according to the first order. All coupling constants are absolute values and expressed in Hertz (Hz). NMR spectra were analyzed in MestReNova v14.1.2-25024 © 2020 Mestrelab Research S. L.

Infrared spectroscopy (IR)

The infrared spectra were recorded with a Bruker, Alpha P instrument. All samples were measured by attenuated total reflection (ATR). The positions of the absorption bands are given in wavenumbers $\tilde{\nu}$ in cm^{-1} and were measured in the range from 3600 cm^{-1} to 500 cm^{-1}.

Characterization of the absorption bands was done in dependence of the absorption strength with the following abbreviations: vs (very strong, 0–9%), s (strong, 10–39%), m (medium, 40–69%), w (weak, 70–89%), vw (very weak, 90–100%).

Mass spectrometry (EI-MS, FAB-MS and ESI)

Mass spectra were recorded on a FINNIGAN MAT 95 mass spectrometer using electron ionization-mass spectrometry (EI-MS) or fast atom bombardment-mass spectroscopy (FAB-MS). For FAB measurements, *m*-nitrobenzyl alcohol (3-NBA) was used as the matrix. The software of FAB and EI adds the mass of one electron. The molecular ion is abbreviated [M+] for EI-MS, the protonated molecular ion is abbreviated [M+H] for FAB-MS. Electrospray ionization-mass spectrometry (ESI-MS) spectra were recorded on a THERMO FISHER SCIENTIFIC Q EXACTIVE mass spectrometer. Calibration was carried out using premixed calibration solutions (THERMO FISHER SCIENTIFIC). The molecular fragments are stated as ratio of mass per charge *m/z*.

UV/VIS-spectroscopy

The UV-Vis spectra were measured with the Lambda 750 from PERKINELMER at 20 °C, slit=2 nm and an UV/Vis/NIR spectrometer Cary 500 (Varian), referenced against pure solvent. The absorbance was measured from 200°nm to 800 nm. The spectra were measured in Quartz cuvettes of 10 mm optical path length at 20 °C. Baseline correction was performed manually to correct for solvent composition.

Fluorescence spectroscopy

Fluorescence spectra were recorded on a Fluoromax-4 (Jobin Yvon - HORIBA) equipped with a Haake AC200 thermostat from Thermo Scientific at 20 °C. The spectra were recorded using FluorEssence v3.5. Graphs were plotted or fitted in OriginPro 2020b 9.7.5.184 (Academic) Copyright © 1991-2020 OriginLab Corporation.

Ultrasonic bath

An analogous high power (35 kHz) ultrasonic bath with timer and heater, SUPER RK 52 H from Bandolin, was used to sonicate samples.

Rheological measurements

Rheological measurements were performed using ARES-G2 Rheometer (TA Instruments) at room temperature. Graphs were plotted or fitted in OriginPro 2020b 9.7.5.184 (Academic) Copyright © 1991-2020 OriginLab Corporation.

DLS measurements

Hydrodynamic radii were determined with a ZETASIZER Nano-S ZEN1600 DLS spectrometer (Malvern Instruments Limited, Worcestershire, UK). Evaluation was done with Zetasizer Software 8.02 © 2002-2021 Malvern Panalytical and the analysis model was based on multiple narrow modes (high resolution).

LED and isomerization

Sample irradiation (for photoisomerization and characterization of photophysical data) was performed using LED diodes with following emission maxima: 10 W LED diodes: 365, 405, 460, 523, 590, 623 and 660 nm – from LED ENGIN or 3 W LED diodes: 365, 380, 400, 410, 420, 430, 450, 470, 490, 530, 590 and 605 nm – from LEDS-GLOBAL. For the time of irradiation, samples were maintained at constant temperature (22 ± 2 °C) using a metal cooling block unless otherwise noted. Graphs were plotted or fitted in OriginPro 2020b 9.7.5.184 (Academic) Copyright © 1991-2020 OriginLab Corporation.
Irradiation intensities of the respective LEDs were determined using the PowerMax USB (type PS19Q) sensor device (Coherent®) in five independent measurements. The detector (diameter 19 mm) was located at a distance of 55 mm from the light source, identical to the position of irradiated samples.

Table 8: The LEDs applied to irradiate the photoswitchable compounds were characterized using the PowerMax USB (type PS19Q) sensor device (Coherent®).

λ_{max} of the LED diode [nm]	Max Power [mW/cm^2]	Max mean power measured [W]	Min Power [mW/cm^2]	Min mean power measured [W]
365 nm	-	-	5.61E-01	1.59E-03
380 nm	8.67E-02	2.46E-04	-	-
407 nm	9.39E+01	2.66E-01	8.59E+01	2.43E-01
430 nm	1.85E+01	5.23E-02	1.19E+01	3.38E-02
450 nm	1.13E+00	3.21E-03	8.86E-01	2.51E-03
470 nm	1.69E+01	4.80E-02	1.59E+01	4.52E-02
490 nm	7.64E+00	2.17E-02	4.98E+00	1.41E-02
523 nm	7.08E+00	2.01E-02	6.56E+00	1.86E-02

5.3 Fluorinated azobenzenes switchable with red light

5.3.1 Synthesis

A.-L. Leistner, S. Kirchner, J. Karcher, T. Bantle, M. L. Schulte, P. Gödtel, C. Fengler, Z. L. Pianowski, *Chem. Eur. J.* **2021**, *27*, 8094.

DOI: 10.1002/chem.202005486

Fluorinated Azobenzenes Switchable with Red Light

Author contributions

The synthesis and characterization was done by the first author, with contributions by J.Karcher, T. Bantle. and M. L. Schulte. All calculations were performed by S. Kirchner with the GAUSSIAN09 program package,[180] employing the B3LYP/6311G* level of theory.[181]

Compounds **35**,**36** and **60** to **64** were synthesized by Anna-Lena Leistner during her Master thesis[182], therefore only the synthesis of new compound **54** and the improved synthesis towards **36** is discussed (see below).

Fmoc-Lys(Boc)-Gly-OMe (95)

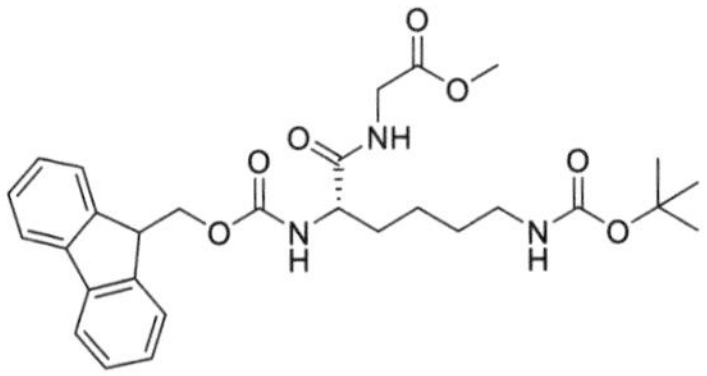

Fmoc-Lys(Boc)-OH (15.0 g, 32.0 mmol, 1.00 eq), HOBt H_2O (5.93 g, 38.7 mmol, 1.21 eq) and EDC HCl (7.43 g, 38.7 mmol, 1.21 eq) were dissolved in in DCM (600 mL) and cooled to 0 °C. Methyl 2-aminoacetate HCl (4.02 g, 32.0 mmol, 1.00 eq) and DIPEA (13.4 g, 18.5 mL, 104 mmol, 3.24 eq) were added and the reaction was stirred at 0 °C for 1 h. The reaction mixture was warmed up to room temperature and stirred overnight. Half of the solvent was evaporated, the residue was washed with water and dried over Na_2SO_4. The crude product was purified by flash column chromatography (99:1 to 95:5 DCM/MeOH) and the pure product was obtained as colorless foam (9.23 g, 17.1 mmol, 53%).[205]

^{1}H NMR (400 MHz, CDCl$_3$): δ = 7.76 (d, J = 7.5 Hz, 2H), 7.59 (d, J = 7.5 Hz, 2H), 7.40 (t, J = 7.4 Hz, 2H), 7.31 (td, J = 7.5, 1.2 Hz, 2H), 6.64 (s, 1H), 5.53 (s, 1H), 4.66 (s, 1H), 4.41 (d, J = 7.5 Hz, 2H), 4.21 (t, J = 6.8 Hz, 2H), 4.08 – 3.99 (m, 2H), 3.74 (s, 3H), 3.14 – 3.06 (m, 2H), 1.92 – 1.85 (m, 1H), 1.73 (s, 1H), 1.73 – 1.66 (m, 1H), 1.49 (s, 3H), 1.43 (s, 9H). **^{13}C NMR (101 MHz, CDCl$_3$)**: δ = 172.1, 170.2, 156.4, 143.9, 141.4, 127.9, 127.2, 125.2, 120.1, 79.3, 67.2, 54.8, 52.5, 47.3, 41.3, 40.0, 32.1, 29.7, 28.6, 22.5 ppm. **TLC:** R_F = 0.30 developed in 95:5 DCM/MeOH. **HRMS (FAB):** *m/z* calcd. for $C_{29}H_{37}N_3O_7$ $[M+H]^+$ = 540.2710 Da, found 540.2709 Da (Δ = -0.17 ppm). **IR (ATR, $\tilde{\nu}$)** = 3299 (m), 3065 (vw), 3041 (vw), 2973 (w), 2938 (w), 2861 (w), 1740 (w), 1681 (vs), 1650 (vs), 1527 (vs), 1477 (w), 1446 (s), 1391 (m), 1366 (s), 1343 (w), 1298 (m), 1268 (vs), 1247 (vs), 1232 (vs), 1210 (vs), 1167 (vs), 1102 (s), 1088 (s), 1033 (s), 1018 (s), 1009 (m), 984 (m), 936 (w), 902 (w), 894 (w), 866 (w), 854 (w), 778 (w), 756 (s), 735 (vs), 656 (vs), 645 (vs), 620 (s), 596 (s), 586 (s), 562 (m), 547 (s), 524 (m), 514 (m), 503 (m), 490 (m), 467 (w), 462 (w), 445 (w), 426 (m), 402 (w), 382 (m) cm^{-1}.

Cyclo(Lys(Boc)-Gly) (96)

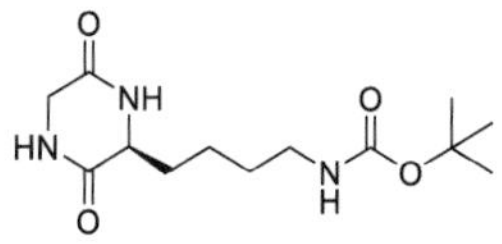

The linear dipeptide Fmoc-L-Lys(Boc)-Gly-OMe (**95**, 2.00 g, 3.71 mmol, 1.00 eq) was dissolved in DCM-piperidine (v/v: 80/20; 18.5 mL) and stirred at room temperature for 14 h in total. Formation of the product was observed by gelation. The gel was filtered and the filtrate was further stirred overnight. The product was filtrated, combined with the first filtrate, washed with DCM followed by water and dried. To eliminate trapped impurities, the solid was grinded with a mortar and washed again with DCM followed by H_2O and dried. The product was obtained as colorless solid (650 mg, 2.28 mmol, 61%).[205]

^{1}H NMR (400 MHz, DMSO-*d*$_6$): δ = 8.15 (s, 1H), 7.98 (s, 1H), 6.77 (t, *J* = 5.7 Hz, 1H), 3.84 – 3.59 (m, 3H), 2.89 (q, *J* = 6.4 Hz, 2H), 1.64 (qd, *J* = 9.1, 8.6, 4.7 Hz, 2H), 1.46 (s, 1H), 1.37 (s, 9H), 1.31 – 1.18 (m, 3H). **^{13}C NMR (101 MHz, DMSO-*d*$_6$)**: δ = 168.0, 166.1, 155.6, 77.3, 54.1, 44.3, 43.8, 32.5, 29.2, 28.3, 22.4, 21.8, 21.4 ppm. **HRMS (EI):** *m/z* calcd. for $C_{13}H_{23}N_3O_4$ [M^+] = 286.1767 Da, found 286.1768 Da (Δ = 0.35 ppm). **IR (ATR, ṽ)** = 3366 (w), 3197 (w), 3166 (w), 3078 (w), 3053 (w), 3013 (vw), 2968 (w), 2955 (w), 2932 (w), 2868 (w), 1677 (vs), 1560 (w), 1521 (vs), 1466 (s), 1449 (m), 1388 (w), 1366 (m), 1334 (s), 1290 (m), 1248 (s), 1230 (w), 1169 (vs), 1142 (m), 1111 (w), 1082 (m), 1054 (w), 1041 (w), 1004 (m), 952 (w), 926 (w), 897 (w), 868 (w), 810 (s), 762 (m), 734 (w), 673 (w), 650 (w), 640 (w), 578 (m), 564 (m), 501 (w), 467 (w), 452 (s), 445 (s), 419 (w), 405 (w), 387 (m) cm^{-1}.

Di-Boc-cyclo(Lys(Boc)-Gly) (67)

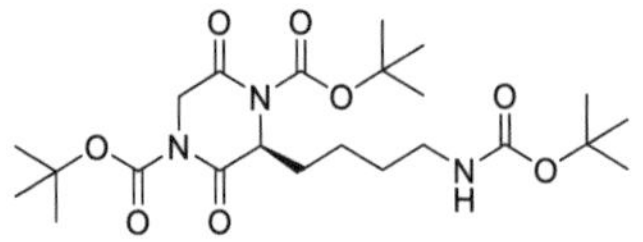

To a mixture of cyclo(Lys(Boc)-Gly) **96** (640 mg, 2.24 mmol, 1.00 eq) and Boc_2O (1.03 g, 4.71 mmol, 2.10 eq) in DMF (2.6 mL) was added *N,N*-dimethylpyridin-4-amine (578 mg, 4.73 mmol, 2.11 eq) and the reaction mixture was stirred at room temperature for 2 h. Then, the reaction mixture was diluted with EtOAc, washed with aqueous $KHSO_4$, and dried over Na_2SO_4. After removal of the solvent, the crude product was purified by column chromatography (3:7 EtOAc/cH). The pure product was obtained as colorless solid (704 mg, 1.45 mmol, 65%).

^{1}H NMR (400 MHz, $CDCl_3$): δ = 4.88 – 4.66 (m, 2H), 4.54 (s, 1H), 4.14 (d, *J* = 18.4 Hz, 1H), 3.11 (q, *J* = 6.4 Hz, 2H), 1.95 – 1.77 (m, 2H), 1.54 (s, 9H), 1.53 (s,

9H), 1.43 (s, 9H). **^{13}C NMR (101 MHz, $CDCl_3$)**: δ = 166.4, 164.6, 156.1, 150.1, 149.9, 85.2, 85.1, 60.1, 48.9, 40.1, 32.4, 29.6, 28.6, 28.1, 28.0, 23.1 ppm. **TLC:** R_F = 0.40 developed in 3:7 EtOAc/cH. **HRMS (EI):** *m/z* calcd. for $C_{23}H_{40}N_3O_8$ [M^+] = 486.2810 Da, found 486.2808 Da (Δ = -0.47 ppm). **IR (ATR, ṽ)** = 3394 (vw), 2979 (w), 2934 (w), 2868 (vw), 1779 (m), 1720 (vs), 1513 (w), 1476 (w), 1456 (w), 1391 (w), 1367 (s), 1282 (vs), 1247 (vs), 1140 (vs), 1004 (w), 973 (w), 936 (w), 928 (w), 870 (w), 846 (m), 775 (m), 739 (w), 633 (w), 605 (w), 589 (w), 565 (w), 523 (w), 511 (w), 487 (w), 477 (w), 462 (w), 439 (w), 412 (w), 404 (w), 397 (w), 382 (w), 375 (w) cm^{-1}.

4-Amino-3,5-difluorobenzaldehyde (66)

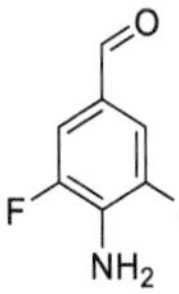

A mixture of 2,6-difluoroaniline (**65**, 4.00 g, 31.0 mmol, 1.00 eq) dissolved in DMSO (620 mL), conc. aqueous hydrochloric acid (37.2 g, 25.0 mL, 1.02 mol, 32.9 eq) and $CuCl_2$ (8.33 g, 62.0 mmol, 2.00 eq) was stirred at 90 °C for 19 h. Then the mixture was cooled down to 0 °C in an ice water bath. The pH was adjusted to pH = 8, which was accompanied by a color change to green. In the following, the aqueous phase was extracted with DCM. The combined organic layers were washed with water and dried over Na_2SO_4. The crude product was purified by column chromatography (DCM/Et_3N/MeOH 98:1:1) and the pure product was obtained as an off white solid (3.58 g, 22.8 mmol, 74%).

^{1}H NMR (400 MHz, $CDCl_3$): δ = 9.73 (t, *J* = 1.9 Hz, 1H), 7.42 – 7.35 (m, 2H), 4.34 (s, 2H) ppm. **^{13}C NMR (101 MHz, $CDCl_3$)**: δ = 188.9 (t, *J* = 2.6 Hz), 152.2 (d, *J* = 7.9 Hz), 149.8 (d, *J* = 7.9 Hz), 130.7 (t, *J* = 16.3 Hz), 125.3 (t, *J* = 6.2 Hz), 112.7 - 112.5 (q, *J* = 7.4 Hz) ppm. **^{19}F NMR (376 MHz, $CDCl_3$):** δ = -132.18 ppm. **TLC:** R_F = 0.68 developed in DCM/Et_3N/MeOH 98:1:1. **HRMS (EI+):** *m/z* calculated for $C_7H_5O_1N_1F_2$ [M^+] = 157.0339 Da, found = 157.0339 Da (Δ = –0.44 ppm).

IR (ATR, ṽ) = 3434 (s), 3332 (s), 3216 (vs), 3094 (w), 3044 (w), 3034 (w), 2851 (w), 2788 (w), 2744 (w), 1674 (s), 1629 (vs), 1599 (vs), 1572 (vs), 1531 (vs), 1463 (vs), 1408 (w), 1363 (s), 1341 (vs), 1269 (vs), 1145 (m), 1119 (vs), 1016 (m), 994 (m), 952 (vs), 882 (s), 851 (s), 820 (m), 707 (vs), 579 (s), 482 (s), 455 (m), 441 (m), 432 (m), 425 (m), 395 (m) cm^{-1}.

(*E*)-4,4'-(diazene-1,2-diyl)bis(3,5-difluorobenzaldehyde) (36)

4-Amino-3,5-difluorobenzaldehyde (**66**, 1.00 g, 6.32 mmol, 1.00 eq) was dissolved in 100 mL of dry DCM and 1.90 mL DBU (1.94 g, 12.7 mmol, 2.00 eq) was added. The solution was stirred for 5 min at room temperature and was then cooled down to −78 °C. NCS (1.70 g, 12.7 mmol, 2.00 eq) was added in small portions and the solution was stirred for 10 min. The mixture was quenched with 100 mL aqueous $NaHCO_3$ solution. The organic layer was separated, washed with water (50 mL) and 1 M aqueous HCl (50 mL), and then dried over Na_2SO_4. The drying agent was filtered off and the solvent was removed under reduced pressure. The crude product was first filtered through silica with DCM as eluent and evaporated to dryness. The resulting violet crystals where carefully washed with MeOH. The product was obtained as violet crystals (211 mg, 680 µmol, 21%).

^{1}H NMR (400 MHz, DMSO-*d*$_6$): δ = 10.04 (s, 2H), 7.94 – 7.91 (m, 4H) ppm. **^{13}C NMR (101 MHz, DMSO-*d*$_6$):** δ = 190.5, 156.1 (d, *J* = 3.6 Hz), 153.5 (d, *J* = 3.6 Hz), 139.0 (t, *J* = 7.9 Hz), 133.8 (t, *J* = 10.3 Hz), 113.8 (dd, *J* = 20.9, 3.3 Hz) ppm. **^{19}F NMR (376 MHz, DMSO-*d*$_6$):** δ = −119.29 ppm. **HRMS (FAB+):** *m/z* calcd for $C_{14}H_6O_2N_2F_4$: 310.0365 Da [M+H], found: 310.0363 Da (Δ = -0.65 ppm). **IR (ATR, ṽ)** = 3074 (w), 2873 (w), 1688 (s), 1617 (w), 1569 (s), 1443 (s), 1388 (m), 1323 (m), 1307 (m), 1204 (m), 1117 (m), 1053 (s), 1000 (m), 988 (s), 895 (w), 869 (s), 788 (m), 755 (w), 714 (m), 613 (s), 588 (m), 532 (m), 489 (m), 456 (m), 456 (w) cm^{-1}. **UV-Vis (MeCN):** λ_{max} = 196, 314, 475 nm.

Bis-(Boc-cyclo(Lys(Boc))-(2,6-difluoro-4-vinyl-azobenzene) (68)

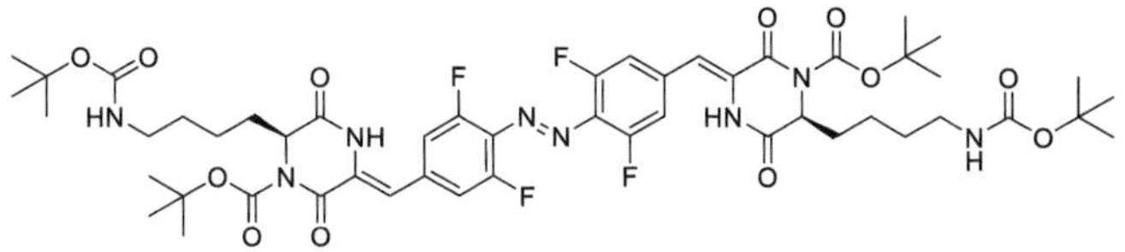

A mixture of (*E*)-4,4'-(diazene-1,2-diyl)bis(3,5-difluorobenzaldehyde) (**36**, 120 mg, 387 µmol, 1.00 eq) and Di-Boc-cyclo(Lys(Boc)-Gly) (**67**, 413 mg, 851 µmol, 2.20 eq) in THF (3.5 mL) was cooled to 0 °C in an ice water bath. KO*t*Bu (96.4 mg, 859 µmol, 2.22 eq) was added and the solution was stirred for 2 h at room temperature. Then the reaction mixture was partitioned between EtOAc and NH_4Cl. The crude product was purified by column chromatography (3:7 to 4:6 EtOAc/cH). The pure product was obtained as dark red solid (206 mg, 0.196 mmol, 51%).

^{1}H NMR (400 MHz, DMSO-*d*$_6$): δ = 10.76 (s, 2H), 7.78 – 7.39 (m, 4H), 6.95 (s, 2H), 6.76 (t, *J* = 5.7 Hz, 2H), 4.62 – 4.38 (m, 2H), 2.89 (q, *J* = 6.3 Hz, 4H), 1.81 (dp, *J* = 20.4, 6.5, 6.1 Hz, 4H), 1.50 (s, 18H), 1.45 – 1.37 (m, 8H), 1.34 (s, 18H). **^{13}C NMR (101 MHz, DMSO-*d*$_6$)**: δ = 166.6, 159.3, 156.0 (d, *J* = 5.2 Hz), 155.6, 153.4 (d, *J* = 5.3 Hz), 150.3, 138.1 (t, *J* = 11.1 Hz), 130.1 (t, *J* = 10.0 Hz), 129.3, 115.6, 114.2 (d, *J* = 22.1 Hz), 83.7, 77.4, 58.7, 33.6, 28.9, 28.3, 27.6, 22.3 ppm. **^{19}F NMR (376 MHz, DMSO-*d*$_6$):** δ = -121.36 ppm. **TLC:** R_F = 0.23 developed in 4:6 EtOAc/cH. **HRMS (FAB):** *m/z* calcd. for $C_{50}H_{65}N_8O_{12}F_4$ [M+H] = 1045.4653 Da, found 1045.4655 Da (Δ = 0.26 ppm). **IR (ATR, ṽ)** = 3369 (vw), 2976 (w), 2932 (w), 2866 (w), 1772 (w), 1691 (vs), 1633 (m), 1616 (m), 1567 (w), 1510 (w), 1477 (w), 1451 (m), 1366 (vs), 1283 (s), 1232 (vs), 1143 (vs), 1047 (s), 1011 (m), 890 (w), 847 (m), 806 (w), 773 (m), 745 (m), 694 (w), 662 (w), 642 (m), 632 (m), 606 (m), 595 (m), 575 (w), 552 (w), 541 (w), 514 (w), 493 (w), 472 (m), 460 (m), 450 (m), 443 (m), 433 (m), 425 (m), 415 (w), 392 (w), 380 (w) cm^{-1}.

Bis-(cyclo(Lys)-(2,6-difluoro-4-vinyl-azobenzene) (54)

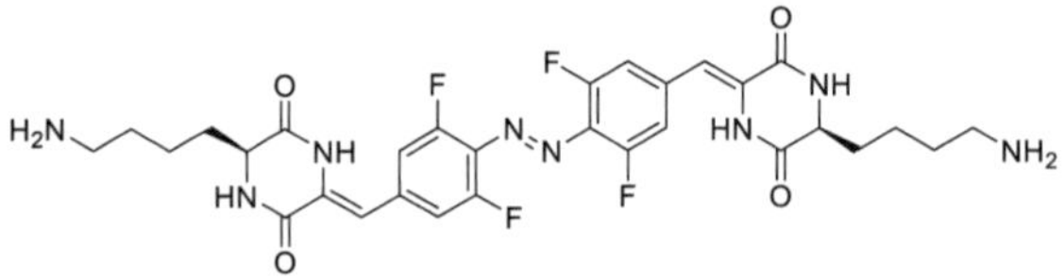

Bis-(Boc-cyclo(Lys(Boc))-(2,6-difluoro-4-vinyl-azobenzene) (**68**, 295 mg, 266 µmol, 1.00 eq) was dissolved in DCM (3.70 mL) and 2,2,2-trifluoroacetic acid (3.70 mL) was added. The mixture was stirred for 2 h at room temperature. The solvent was removed under reduced pressure. Excess TFA was removed by azetropic distillation using toluene. No further purification was necessary and the TFA-salt of the product Bis-(cyclo(Lys)-(2,6-difluoro-4-vinyl-azobenzene) was obtained as an orange solid (202 mg, 0.266 mmol, quant.).

^{1}H NMR (400 MHz, DMSO-*d*$_6$): δ = 10.48 (s, 2H), 8.66 (d, *J* = 2.5 Hz, 2H), 7.80 (s, 6H), 7.46 (d, *J* = 11.4 Hz, 4H), 6.71 (s, 2H), 4.13 (td, *J* = 5.4, 2.2 Hz, 2H), 3.46 (s, 8H), 2.79 (q, *J* = 6.6 Hz, 4H), 1.78 (tt, *J* = 14.6, 7.6 Hz, 4H), 1.56 (p, *J* = 7.6 Hz, 4H), 1.49 – 1.30 (m, 4H). **^{13}C NMR (101 MHz, DMSO-*d*$_6$)**: δ = 167.2, 159.3, 156.0 (d, *J* = 5.5 Hz), 153.5 (d, *J* = 5.6 Hz), 138.7 (t, *J* = 11.0 Hz), 130.1, 129.6 (t, *J* = 9.9 Hz), 114.5 – 112.3 (m), 110.6, 54.6, 38.6, 33.1, 26.7, 20.8 ppm. **^{19}F NMR (376 MHz, DMSO-*d*$_6$):** δ = -121.62 ppm. **HPLC:** R_t = 8.9 min; 5-95% MeCN in H_2O with 0.1% TFA. **HRMS (ESI):** *m/z* calcd. for $C_{30}H_{32}N_8O_4F_4$ [M+H] = 645.2555 Da, found 645.2544 Da (Δ = -1.70 ppm). **IR (ATR, $\tilde{\nu}$)** = 3187 (w), 3170 (w), 3152 (w), 3064 (w), 3058 (w), 3048 (w), 3033 (w), 2956 (w), 2928 (w), 2874 (w), 2817 (w), 2749 (w), 1679 (vs), 1619 (vs), 1562 (m), 1543 (m), 1534 (m), 1475 (w), 1458 (m), 1425 (vs), 1387 (s), 1351 (s), 1327 (m), 1299 (w), 1200 (vs), 1179 (vs), 1130 (vs), 1050 (vs), 1003 (w), 982 (w), 948 (w), 905 (w), 885 (w), 834 (vs), 798 (vs), 755 (s), 721 (vs), 680 (m), 645 (m), 635 (m), 589 (m), 554 (w), 534 (m), 520 (m), 499 (s), 479 (m), 459 (s), 438 (s), 416 (m), 407 (m), 385 (m) cm^{-1}.

5.3.2 Photophysical properties

Photostationary states

Table 9: Absorption maxima of the compounds **36**, **54**, **60**, **61** and **63** (for each photoisomer separately).

λ_{max} (nm)	**60**	**61**	**63**	**36**	**54**
E-isomer	312, 452	318, 457	311, 466	314, 475	376
Z-isomer	286, 418	292, 417	275, 421	277, 424	304, 433

Photostationary states were determined by ^{1}H NMR measurements for samples (10 mg/mL; in some cases – e.g. **36** – when the solubility in d_3-MeCN was lower, the saturated supernatant was taken for the NMR measurements) equilibrated for 8 hours under the indicated light wavelength (λ_{max} of the respective LED light diode; in case of the 660 nm LED additionally equipped with a 630 nm cut-off filter SCHOTT RG-630 to eliminate the <630 nm component of the emitted light). ^{1}H NMR measurements for **sample 36** were prepared in DMSO and irradiated for 3 h (660 nm) or 1 h (470 nm). Unsubstituted **TFAB 35** was synthesized according to the literature.[89]

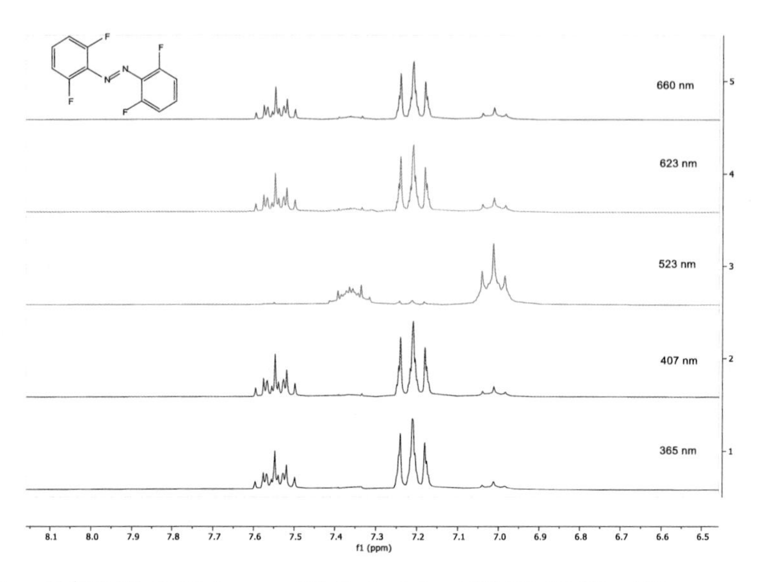

Figure 53: ^{1}H NMR in CD_3CN of compound **35** after irradiation at 660, 623, 523, 407 and 365 nm.

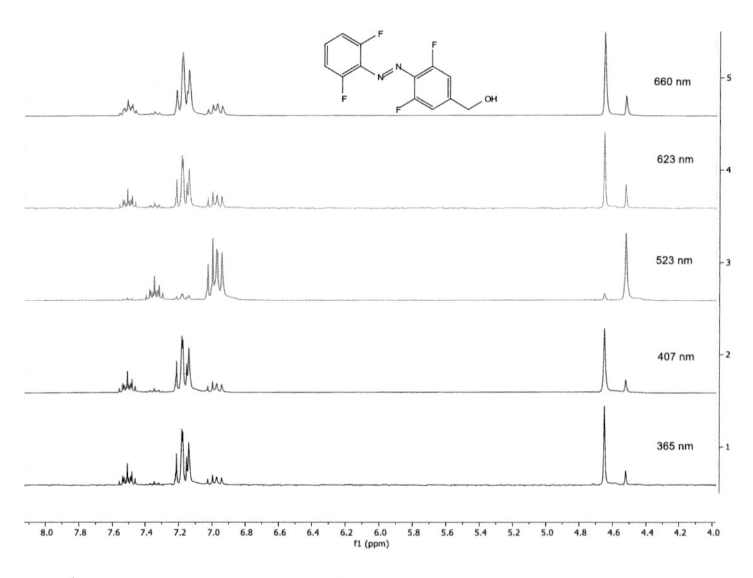

Figure 54: ^{1}H NMR in CD_3CN of compound **60** after irradiation at 660, 623, 523, 407 and 365 nm.

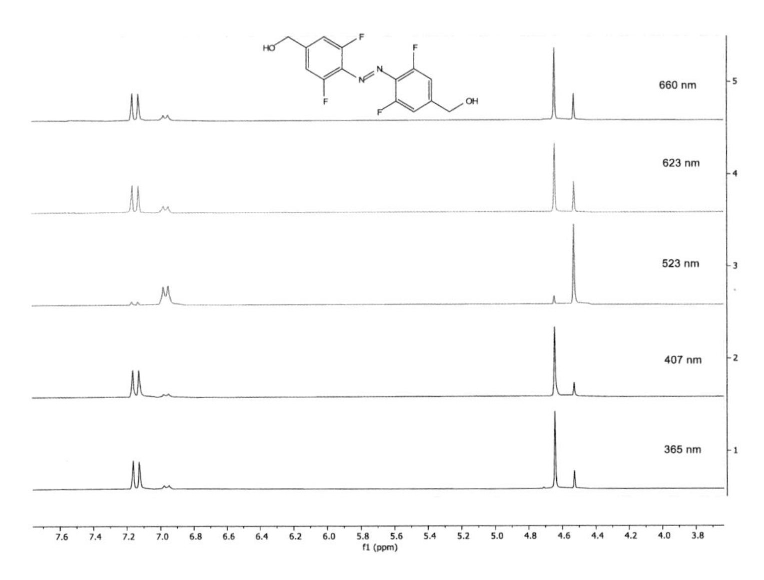

Figure 55: ^{1}H NMR in CD_3CN of compound **61** after irradiation at 660, 623, 523, 407 and 365 nm.

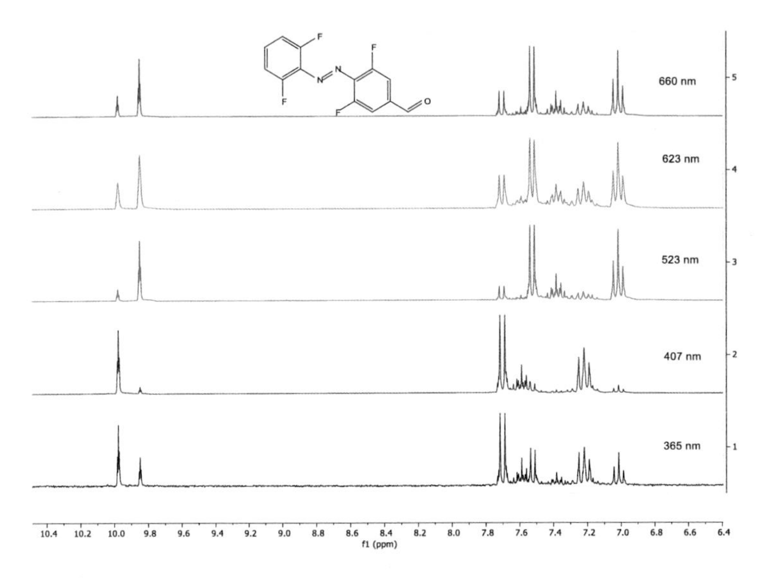

Figure 56: ^{1}H NMR in CD_3CN of compound **63** after irradiation at 660, 623, 523, 407 and 365 nm.

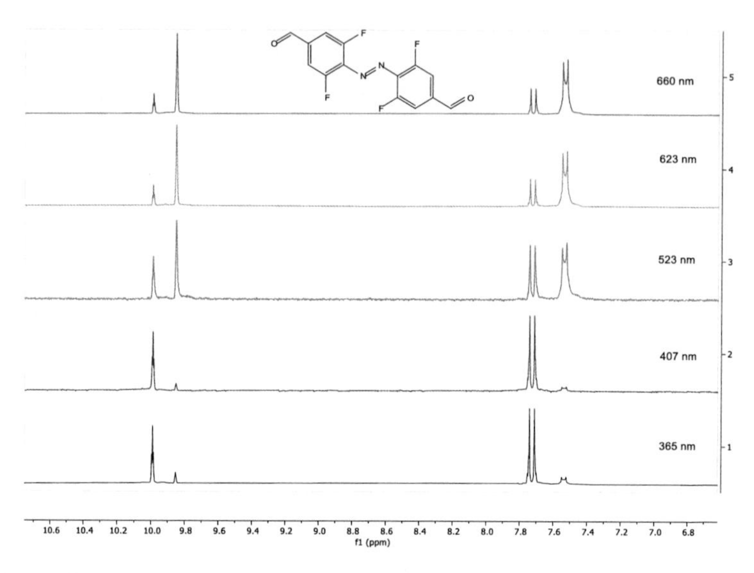

Figure 57: ^{1}H NMR in CD_3CN of compound **36** after irradiation at 660, 623, 523, 407 and 365 nm.

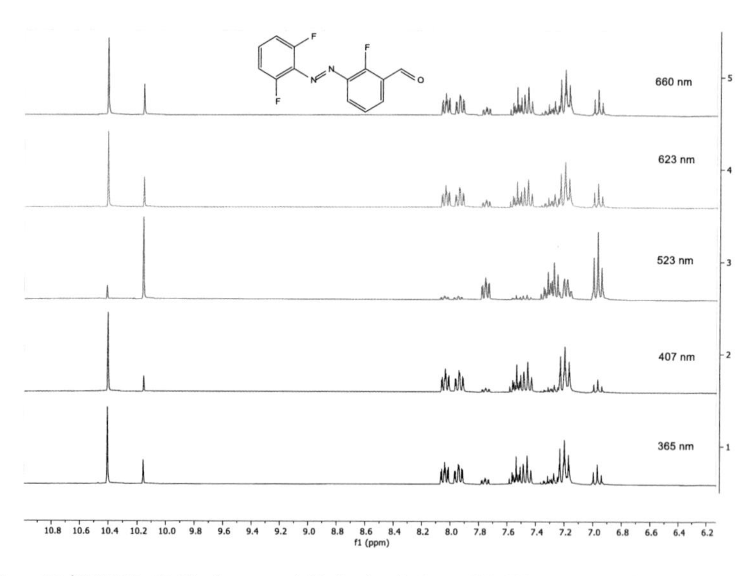

Figure 58: ^{1}H NMR in CD_3CN of compound **64** after irradiation at 660, 623, 523, 407 and 365 nm.

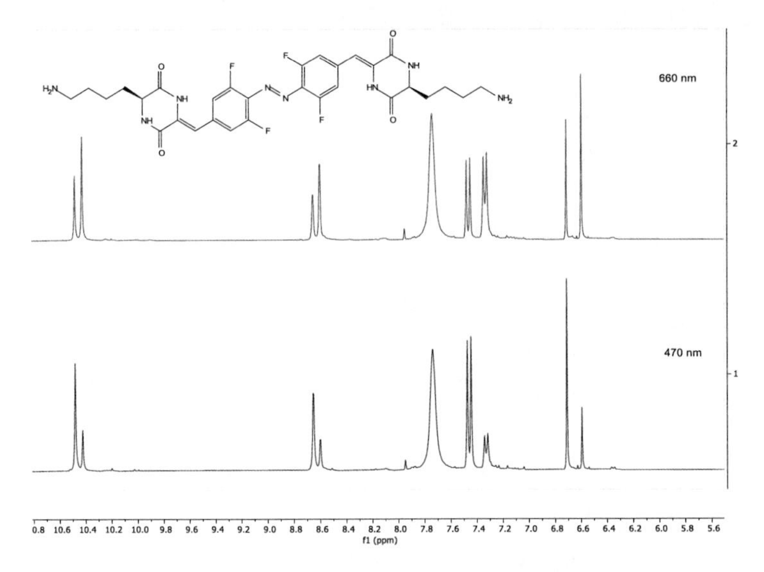

Figure 59: ^{1}H NMR in DMSO-d_6 of compound **54** after irradiation at 660 nm for 3 h and at 470 nm for 1 h.

Photostability - Switching cycles of compound 54

The photostability of compound **54** was determined by cycling a 1 mM solution in DMSO between the PSS at wavelengths 470 nm and 660 nm. In the experiment, the solution was irradiated for 20 min with light at the respective wavelength, followed by an analytical HPLC measurement. The chromatograms of the initial dark state and the last cycle at 470 nm are shown in Figure 61 and Figure 62 to prove that no degradation occurred.

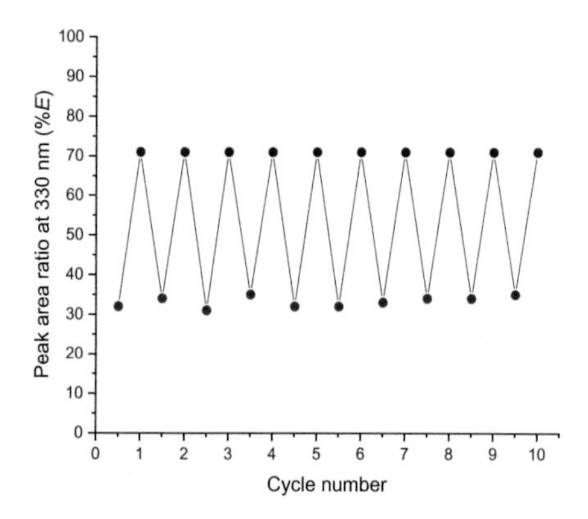

Figure 60: Switching cycles of compound **54**. Fluctuations after red light irradiation appear in consequence of the low thermal stability and heat generation of the 660 nm LED.

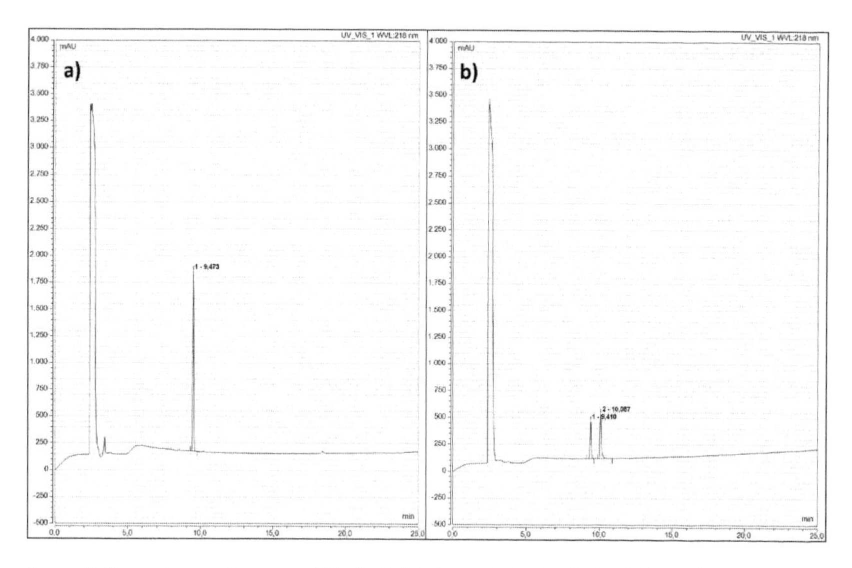

Figure 61: Chromatogram of compound **54**. a) initial dark state before irradiation. b) last cycle after irradiation at 660 nm.

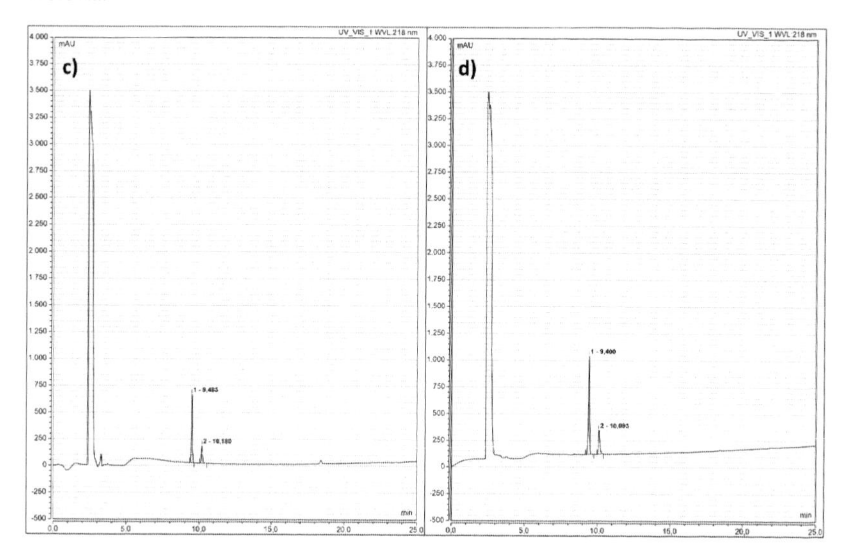

Figure 62: Chromatogram of compound **54**. c) irradiation at 470 nm (cycle 1). d) irradiation at 470 nm (cycle 10).

Thermal stability

The thermal stability was determined for compounds **63, 36** and **54.** Samples were prepared in MeCN (**63, 36**) or DMSO (**54**) and irradiated at 623 nm to yield a high amount of *Z*-isomer. The solution was kept at 60 °C or additionally at 20 °C for DMSO (**54**) and the isomer ratio was determined by HPLC in intervals. The collected samples of **54** were kept below 0 °C and respectively thawed directly before HPLC injection. The obtained data was processed by calculating the $\ln(X_0/X_t)$, where X is the percentage of the respective *Z*-isomer and linear fitting (Equation (1)) of the obtained values. The calculated slope corresponds to the isomerization rate constant k which is used to calculate the half-life $t_{1/2}$.

(1) $$x_t = x_0 \cdot e^{-k \cdot t} \leftrightarrow ln\left(\frac{x_0}{x_t}\right) = k \cdot t$$

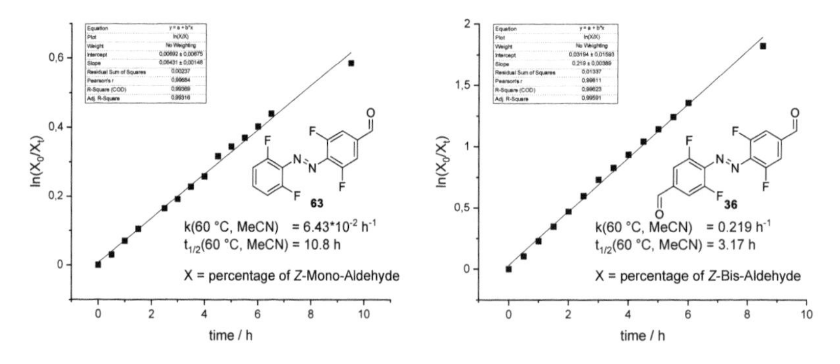

Figure 63: Linear fit of the decay of the *Z*-isomer of compounds **63** and **36** at 60 °C in MeCN for first-order kinetics.

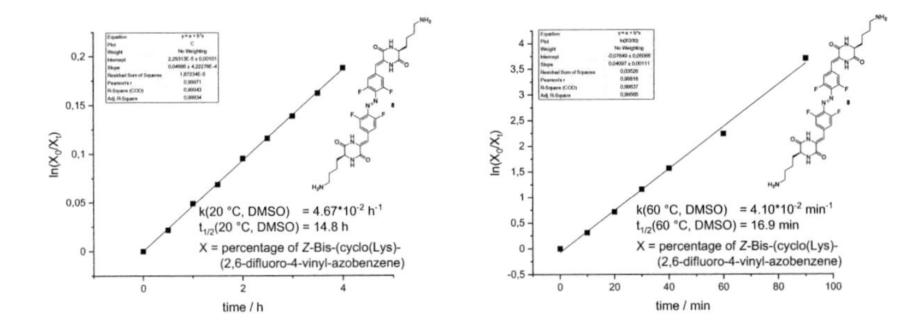

Figure 64: Linear fit of the decay of the Z-isomer of compound **54** at 20 °C and 60 °C in DMSO for first-order kinetics.

The half-life of *Z*-**63** (10.8 h) is comparable to other TFAB derivatives, whereas the half-life of *Z*-**36** (3.2 h) is generally lower and similar to the value of the unsubstituted azobenzene.[61a]

Computations

All calculations were performed by Susanne Kirchner with the GAUSSIAN09 program package,[180] employing the PBE0-D3/def2-TZVP level of theory.[181] This level performed well in studies on basic properties for similar systems.[21-22] We optimized geometries of *Z*- and *E*-isomers and performed vibrational frequency calculations to check the minimum nature of these species.

Cartesian Coordinates of the optimized structures

The presented data were obtained and visualized with Avogadro 1.2.0.

***E*-35**

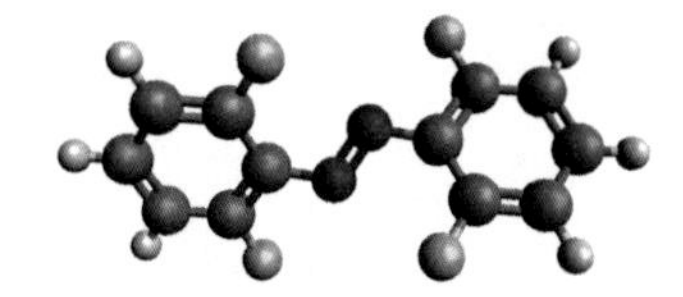

N	-0.40303	0.47486	0.07373
F	-2.19399	2.14180	1.11476
F	-1.50313	-1.92989	-1.12662
C	-4.52136	-0.32958	-0.03858
C	-4.02976	0.82755	0.54456
C	-2.66676	1.03248	0.55899
C	-1.75235	0.11016	0.03635
C	-2.29975	-1.04079	-0.54873
C	-3.66101	-1.26273	-0.59553
N	0.40306	-0.47496	0.07389
F	2.19418	-2.14182	1.11480
F	1.50295	1.92984	-1.12655
C	4.52133	0.32969	-0.03869
C	4.02984	-0.82747	0.54449
C	2.66685	-1.03248	0.55900
C	1.75236	-0.11023	0.03639
C	2.29966	1.04076	-0.54873
C	3.66090	1.26279	-0.59559
H	-5.58992	-0.50533	-0.06376
H	-4.67982	1.57253	0.98423
H	-4.02433	-2.16393	-1.07250
H	5.58988	0.50550	-0.06393
H	4.67997	-1.57241	0.98412
H	4.02414	2.16401	-1.07258

Zero-point correction =	0.158694 (Hartree/Particle)
Sum of electronic and zero-point Energies =	-968.881266
Sum of electronic and thermal Energies =	-968.867009
Sum of electronic and thermal Enthalpies =	-968.866065
Sum of electronic and thermal Free Energies =	-968.924727

***Z*-35**

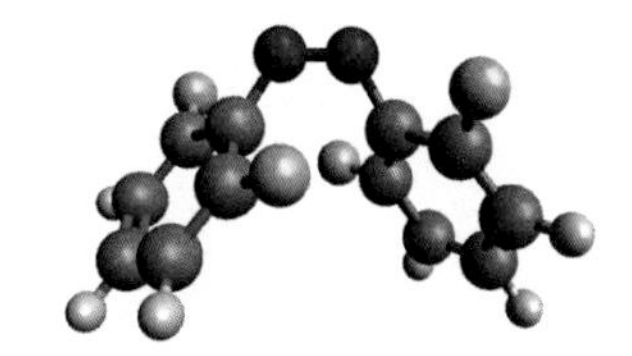

N	-0.61057	0.08454	-1.80342
F	-2.74104	-1.48848	-1.19685
F	-0.05577	2.13152	0.04833
C	-3.02429	0.78952	1.56990
C	-3.31005	-0.26214	0.71163
C	-2.47803	-0.48710	-0.36394
C	-1.34005	0.28005	-0.60120
C	-1.09887	1.33270	0.27817
C	-1.92103	1.59942	1.35351
N	0.61060	-0.08435	-1.80342
F	0.05567	-2.13141	0.04819
F	2.74115	1.48852	-1.19672
C	3.02426	-0.78970	1.56987
C	1.92096	-1.59952	1.35342
C	1.09882	-1.33269	0.27809
C	1.34007	-0.27999	-0.60121
C	2.47808	0.48710	-0.36388
C	3.31008	0.26201	0.71168
H	-3.67346	0.98309	2.41482
H	-4.16568	-0.90816	0.85937
H	-1.68616	2.43633	1.99838
H	3.67341	-0.98336	2.41479
H	1.68603	-2.43646	1.99823
H	4.16574	0.90798	0.85947

Zero-point correction =	0.158776 (Hartree/Particle)
Sum of electronic and zero-point Energies =	-968.871519
Sum of electronic and thermal Energies =	-968.857481
Sum of electronic and thermal Enthalpies =	-968.856537
Sum of electronic and thermal Free Energies =	-968.913562

***E*-60** (F4-alcohol)

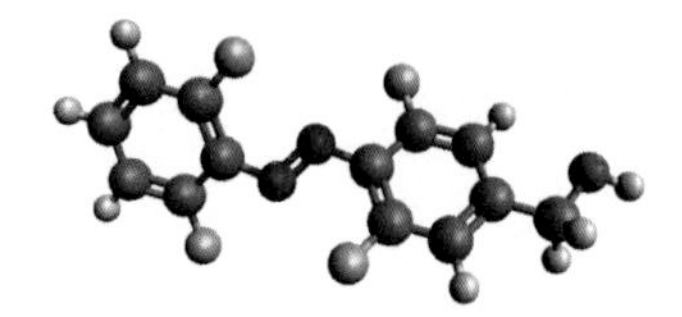

N	-1.13644	-0.48726	-0.03237
F	-2.93264	-2.18836	-1.01047
F	-2.22924	1.98375	1.03540
C	-5.24977	0.35818	-0.00856
C	-4.76183	-0.83086	-0.52712
C	-3.40060	-1.04772	-0.51572
C	-2.48258	-0.10938	-0.02936
C	-3.02669	1.07354	0.49153
C	-4.38642	1.30982	0.51090
N	-0.32387	0.45770	-0.07571
F	1.47524	2.20852	-0.92959
F	0.79129	-2.03104	0.98555
C	3.81258	-0.31066	0.08630
C	3.30841	0.87739	-0.41744
C	1.94245	1.06405	-0.44666
C	1.02306	0.10131	-0.01395
C	1.58030	-1.08430	0.49575
C	2.93940	-1.28909	0.55294
C	5.28890	-0.56929	0.13912
O	5.99539	0.56109	-0.30626
H	-6.31684	0.54437	-0.00465
H	-5.41342	-1.59160	-0.93659
H	-4.74604	2.23708	0.93825
H	3.96589	1.65419	-0.78163
H	3.30228	-2.22084	0.97188
H	5.56055	-0.82941	1.17264
H	5.50846	-1.44850	-0.48452
H	6.93295	0.35918	-0.30316

Zero-point correction = 0.191118 (Hartree/Particle)
Sum of electronic and zero-point Energies = -1083.302660
Sum of electronic and thermal Energies = -1083.285377

Sum of electronic and thermal Enthalpies =	-1083.284432
Sum of electronic and thermal Free Energies =	-1083.350858

***Z*-60** (F4-alcohol)

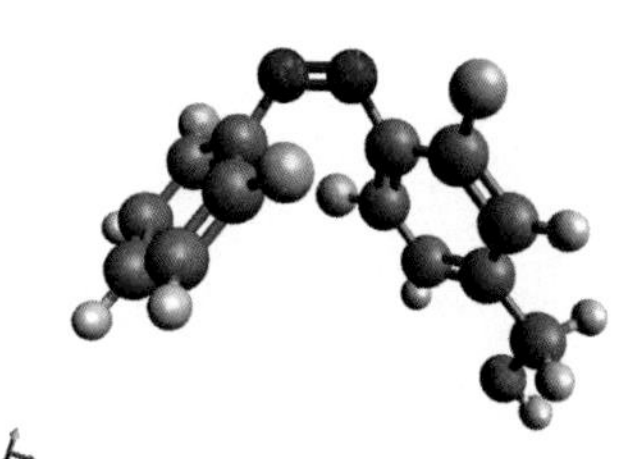

N	-1.67276	1.44171	-0.97286
F	-3.25425	-0.59762	-1.81372
F	-0.90100	1.33477	1.73579
C	-3.03876	-1.55537	1.64440
C	-3.40569	-1.57188	0.30687
C	-2.90937	-0.59598	-0.53024
C	-2.02543	0.38675	-0.09153
C	-1.69403	0.37253	1.26101
C	-2.18685	-0.57740	2.13199
N	-0.49845	1.76283	-1.17109
F	-0.14284	-0.99150	-1.70624
F	1.47236	2.91329	0.29750
C	2.92847	-0.37212	0.05914
C	1.93465	-1.03440	-0.64340
C	0.78887	-0.35139	-0.99752
C	0.58488	0.98859	-0.68397
C	1.62315	1.62812	-0.00806
C	2.76689	0.97265	0.38147
C	4.19336	-1.07259	0.46043
O	4.09406	-2.44710	0.18055
H	-3.42611	-2.31157	2.31592
H	-4.06894	-2.32531	-0.09777
H	-1.89854	-0.53298	3.17427
H	2.04397	-2.07462	-0.91744
H	3.52046	1.52133	0.93450
H	5.03112	-0.61536	-0.08634
H	4.36922	-0.89506	1.53118
H	4.94532	-2.85697	0.34542

Zero-point correction =	0.191383 (Hartree/Particle)
Sum of electronic and zero-point Energies =	-1083.292684
Sum of electronic and thermal Energies =	-1083.275695
Sum of electronic and thermal Enthalpies =	-1083.274750
Sum of electronic and thermal Free Energies =	-1083.338994

***E*-61** (F4-bisalcohol)

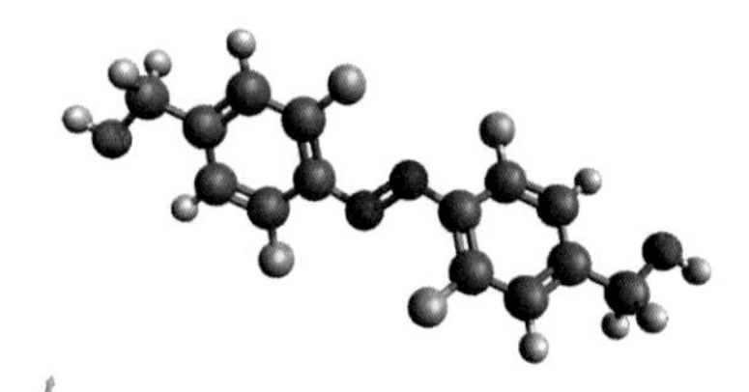

N	-0.40661	-0.47333	-0.03045
F	-2.20689	-2.23810	-0.86091
F	-1.52107	2.04849	0.94583
C	-4.54190	0.32225	0.05529
C	-4.03822	-0.88209	-0.40887
C	-2.67309	-1.07706	-0.41583
C	-1.75274	-0.10833	-0.00004
C	-2.30939	1.09346	0.46984
C	-3.66817	1.30754	0.50535
N	0.40660	0.47329	-0.03045
F	2.20685	2.23808	-0.86091
F	1.52112	-2.04853	0.94584
C	4.54191	-0.32223	0.05528
C	4.03820	0.88210	-0.40888
C	2.67306	1.07705	-0.41584
C	1.75274	0.10830	-0.00004
C	2.30942	-1.09349	0.46984
C	3.66820	-1.30754	0.50534
C	6.01712	-0.59158	0.08031
O	6.72650	0.55401	-0.32078
C	-6.01711	0.59162	0.08034
O	-6.72652	-0.55393	-0.32081
H	-4.69550	-1.66537	-0.75934
H	-4.02997	2.25230	0.89520
H	4.69547	1.66540	-0.75935

H	4.03001	-2.25229	0.89518
H	6.29892	-0.90035	1.09753
H	6.22491	-1.44217	-0.58565
H	7.66148	0.34241	-0.34880
H	-6.29891	0.90033	1.09760
H	-6.22489	1.44226	-0.58556
H	-7.66148	-0.34229	-0.34888

Zero-point correction =	0.223545 (Hartree/Particle)
Sum of electronic and zero-point Energies =	-1197.723889
Sum of electronic and thermal Energies =	-1197.703562
Sum of electronic and thermal Enthalpies =	-1197.702618
Sum of electronic and thermal Free Energies =	-1197.776885

***Z*-61** (F4-bisalcohol)

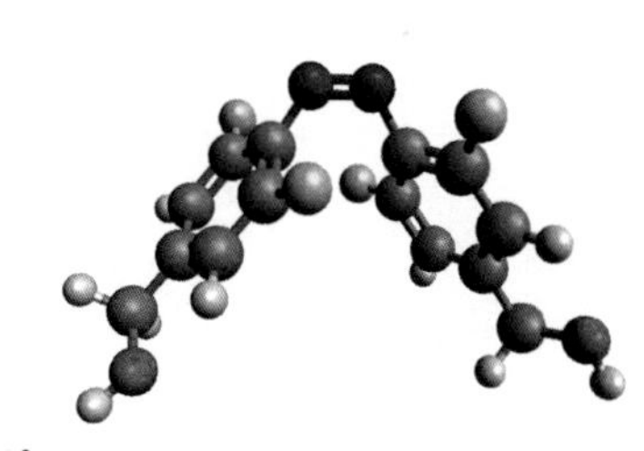

N	-0.55635	-2.37581	0.47117
F	-2.48458	-1.39235	2.10998
F	-0.27000	-1.01707	-1.99215
C	-3.04642	0.84408	-0.64875
C	-3.19325	0.22788	0.58525
C	-2.34087	-0.79955	0.92895
C	-1.30367	-1.22736	0.10616
C	-1.19893	-0.59456	-1.13142
C	-2.04739	0.41792	-1.51733
N	0.67780	-2.38095	0.46181
F	0.41414	-0.05963	2.05233
F	2.59400	-2.24634	-1.46043
C	3.18929	0.98378	0.01700
C	2.19689	1.00146	0.98389
C	1.34090	-0.07562	1.09289

C	1.43202	-1.19500	0.27171
C	2.46538	-1.18928	-0.66367
C	3.32060	-0.12352	-0.81631
C	4.15010	2.12737	-0.12808
O	3.71653	3.22417	0.63908
C	-3.94227	1.97506	-1.06198
O	-4.98836	2.12552	-0.13428
H	-3.96603	0.53851	1.27456
H	-1.91874	0.85755	-2.49975
H	2.08441	1.84464	1.65135
H	4.08033	-0.16957	-1.58796
H	5.14496	1.78705	0.19490
H	4.22715	2.39110	-1.19255
H	4.41245	3.88390	0.65220
H	-3.33625	2.89080	-1.12809
H	-4.32927	1.76801	-2.06999
H	-5.49077	2.91022	-0.36141

Zero-point correction = 0.223848 (Hartree/Particle)
Sum of electronic and zero-point Energies = -1197.713428
Sum of electronic and thermal Energies = -1197.693429
Sum of electronic and thermal Enthalpies = -1197.692485
Sum of electronic and thermal Free Energies = -1197.764435

***E*-63** (F4-aldehyde)

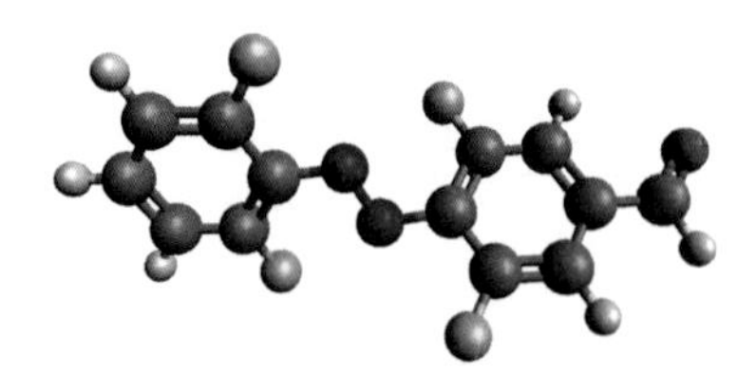

N	-0.99731	-0.39255	-0.15655
F	-2.66625	-2.16392	-1.20242
F	-2.26186	1.86974	1.17725
C	-5.15861	0.04616	0.11680
C	-4.58839	-1.04066	-0.52658

C	-3.21410	-1.12378	-0.58945
C	-2.36643	-0.14459	-0.05305
C	-2.99380	0.93304	0.59301
C	-4.36641	1.03150	0.68563
O	5.94955	-1.09042	0.38836
N	-0.25508	0.60111	-0.03646
F	1.41368	2.38492	-1.11120
F	0.98023	-1.74967	1.06987
C	5.36484	-0.14317	-0.06644
C	3.89541	0.01232	-0.05718
C	3.32957	1.15591	-0.60784
C	1.95982	1.29451	-0.58752
C	1.11308	0.31475	-0.06066
C	1.72915	-0.82080	0.49266
C	3.09492	-0.97634	0.50528
H	-6.23708	0.12569	0.17998
H	-5.18555	-1.82303	-0.97604
H	-4.79246	1.87925	1.20656
H	5.91489	0.69968	-0.53580
H	3.93989	1.93579	-1.04832
H	3.54403	-1.85656	0.94831

Zero-point correction =	0.167878 (Hartree/Particle)
Sum of electronic and zero-point Energies =	-1082.120131
Sum of electronic and thermal Energies =	-1082.103801
Sum of electronic and thermal Enthalpies =	-1082.102857
Sum of electronic and thermal Free Energies =	-1082.166360

***Z*-63** (F4-aldehyde)

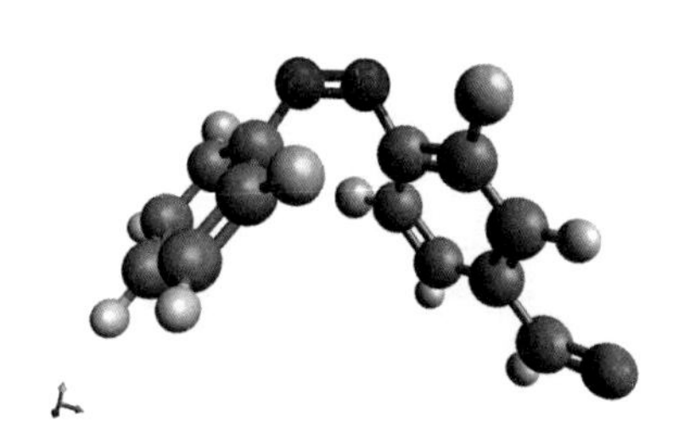

N	-1.55473	1.47024	-0.94634

F	-3.39142	-0.37187	-1.69772
F	-0.66175	1.27488	1.72344
C	-3.09140	-1.37498	1.74087
C	-3.52997	-1.34330	0.42511
C	-2.98065	-0.41826	-0.43587
C	-1.97325	0.46074	-0.04240
C	-1.57302	0.40274	1.29081
C	-2.11649	-0.49609	2.18502
O	5.01376	-1.21370	0.82515
N	-0.36367	1.67054	-1.19010
F	-0.27869	-1.06492	-1.81151
F	1.70926	2.61253	0.28002
C	4.05159	-1.60240	0.21764
C	2.88387	-0.76258	-0.11848
C	1.82334	-1.32351	-0.81882
C	0.73010	-0.53972	-1.11768
C	0.65928	0.79957	-0.74316
C	1.76027	1.33351	-0.06771
C	2.85631	0.57648	0.26287
H	-3.52054	-2.09084	2.43094
H	-4.29056	-2.01923	0.05659
H	-1.77234	-0.48872	3.21111
H	3.98249	-2.65237	-0.13752
H	1.83707	-2.36097	-1.13303
H	3.68629	1.00747	0.80857

Zero-point correction =	0.168066 (Hartree/Particle)
Sum of electronic and zero-point Energies =	-1082.110620
Sum of electronic and thermal Energies =	-1082.094537
Sum of electronic and thermal Enthalpies =	-1082.093593
Sum of electronic and thermal Free Energies =	-1082.155382

***E*-36** (F4-bisaldehyde)

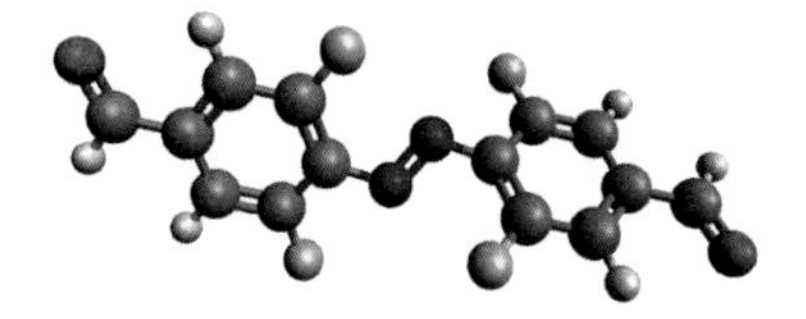

O	6.60295	-0.87770	0.57046
N	0.35521	0.51171	-0.16127
F	1.97939	2.29690	-1.27738
F	1.65294	-1.72114	1.13181
C	5.99762	0.01604	0.04197
C	4.52160	0.10587	-0.00534
C	3.92716	1.18849	-0.64144
C	2.55266	1.26580	-0.67313
C	1.73152	0.28025	-0.11408
C	2.37741	-0.79332	0.52570
C	3.74730	-0.88441	0.58989
O	-6.60295	0.87774	0.57046
N	-0.35522	-0.51174	-0.16125
F	-1.97940	-2.29691	-1.27737
F	-1.65292	1.72111	1.13183
C	-5.99762	-0.01601	0.04197
C	-4.52160	-0.10585	-0.00534
C	-3.92717	-1.18848	-0.64143
C	-2.55267	-1.26581	-0.67312
C	-1.73153	-0.28026	-0.11407
C	-2.37740	0.79331	0.52571
C	-3.74729	0.88441	0.58989
H	6.52801	0.85383	-0.45735
H	4.51798	1.96839	-1.10756
H	4.21972	-1.71563	1.09848
H	-6.52802	-0.85379	-0.45736
H	-4.51800	-1.96838	-1.10756
H	-4.21970	1.71564	1.09849

Zero-point correction = 0.176994 (Hartree/Particle)

Sum of electronic and zero-point Energies = -1195.358134

Sum of electronic and thermal Energies =	-1195.339714
Sum of electronic and thermal Enthalpies =	-1195.338770
Sum of electronic and thermal Free Energies =	-1195.407224

***Z*-36** (F4-bisaldehyde)

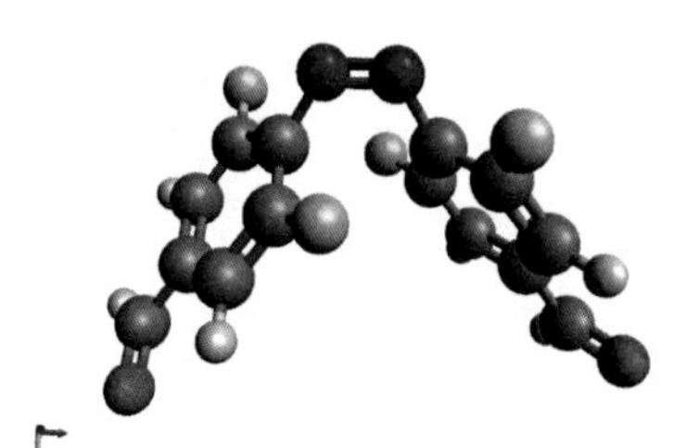

O	-4.02043	3.18303	-0.63003
N	-0.65979	-2.30540	-0.42812
F	-2.53704	-2.12398	1.52020
F	-0.40158	-0.05883	-2.08941
C	-4.06104	2.21044	0.07518
C	-3.14639	1.05540	-0.05022
C	-3.27934	-0.01824	0.82317
C	-2.42550	-1.09086	0.69586
C	-1.41196	-1.11748	-0.25914
C	-1.32229	-0.02620	-1.12830
C	-2.16899	1.05168	-1.03993
O	4.85831	2.42063	0.30892
N	0.57191	-2.30409	-0.43575
F	0.31860	-0.96179	2.02051
F	2.44232	-1.32525	-2.12831
C	3.98621	1.98994	1.01489
C	3.07200	0.89335	0.62986
C	2.09654	0.47547	1.52608
C	1.23886	-0.53717	1.15536
C	1.32522	-1.15658	-0.08894
C	2.34237	-0.73005	-0.94818
C	3.19974	0.28959	-0.61829
H	-4.81154	2.12043	0.88829
H	-4.03748	-0.02786	1.59786
H	-2.08098	1.88468	-1.72631
H	3.81016	2.40410	2.02987
H	1.99584	0.92752	2.50624
H	3.96284	0.62115	-1.31128

Zero-point correction =	0.177111 (Hartree/Particle)
Sum of electronic and zero-point Energies =	-1195.348938
Sum of electronic and thermal Energies =	-1195.330732
Sum of electronic and thermal Enthalpies =	-1195.329787
Sum of electronic and thermal Free Energies =	-1195.396804

***E*-62** (F3-alcohol)

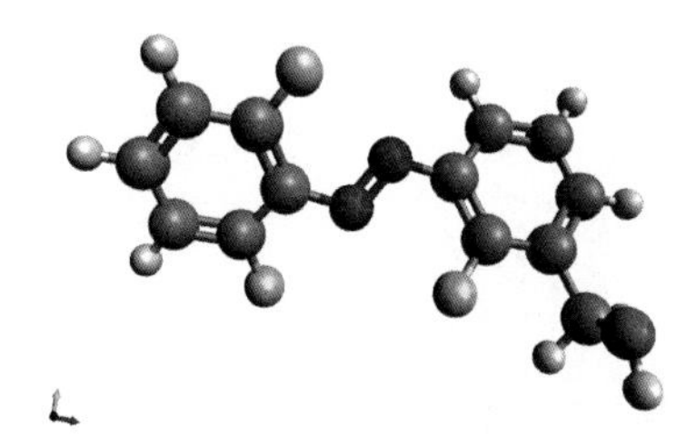

N	-0.78814	-0.18173	-0.19948
F	-2.31938	-2.13108	-1.17303
F	-2.21887	1.94743	1.15956
C	-4.96607	-0.16415	0.23680
C	-4.31572	-1.19548	-0.42216
C	-2.94353	-1.14136	-0.54362
C	-2.17606	-0.07728	-0.05647
C	-2.87989	0.93699	0.60790
C	-4.25101	0.90033	0.76224
N	-0.18031	0.90430	-0.24741
F	1.50832	-1.40896	0.37463
C	3.98690	0.99171	-0.59777
C	3.20495	2.10240	-0.87859
C	1.83477	2.01881	-0.73537
C	1.21677	0.83055	-0.34141
C	2.03575	-0.26871	-0.05188
C	3.41659	-0.20192	-0.17376
C	4.25666	-1.39040	0.17813
O	4.43798	-1.41149	1.58304
H	-6.04331	-0.19212	0.34675
H	-4.84905	-2.03926	-0.83996
H	-4.73506	1.70704	1.29770

H	5.06604	1.04468	-0.69524
H	3.66493	3.02933	-1.19788
H	1.18914	2.86694	-0.93031
H	3.76008	-2.30444	-0.16668
H	5.22100	-1.31014	-0.34062
H	4.75035	-2.28368	1.83543

Zero-point correction =	0.199739 (Hartree/Particle)
Sum of electronic and zero-point Energies =	-984.103535
Sum of electronic and thermal Energies =	-984.087247
Sum of electronic and thermal Enthalpies =	-984.086303
Sum of electronic and thermal Free Energies =	-984.150047

***Z*-62** (F3-alcohol)

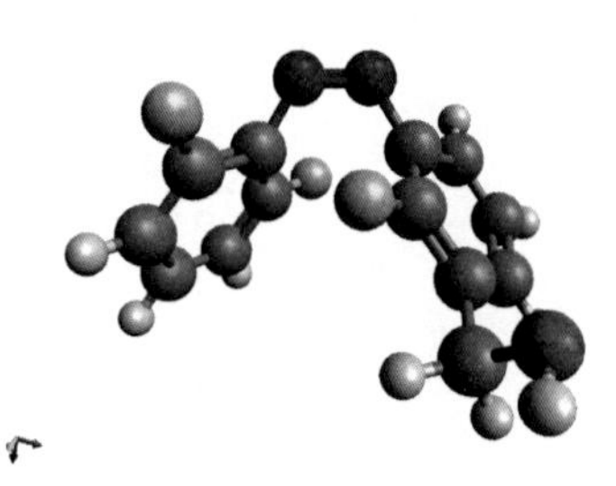

N	-1.29064	-0.81768	-1.72435
F	-2.35722	1.68523	-1.56640
F	-1.24901	-2.25998	0.68412
C	-2.91960	0.70466	1.84710
C	-2.92752	1.47315	0.69219
C	-2.35378	0.96075	-0.45186
C	-1.73594	-0.28617	-0.48318
C	-1.76712	-1.03036	0.69274
C	-2.34603	-0.55662	1.85234
N	-0.13581	-1.22048	-1.88007
F	0.58774	1.29475	-0.84325
C	2.97761	-0.75354	0.89226
C	2.64192	-2.01102	0.41014
C	1.60995	-2.14797	-0.49809
C	0.86986	-1.03802	-0.89247
C	1.24296	0.21276	-0.41400

C	2.29126	0.37962	0.47820
C	2.67003	1.75093	0.94391
O	3.48521	2.35101	-0.04609
H	-3.37305	1.09102	2.75151
H	-3.37153	2.45972	0.66140
H	-2.34296	-1.18373	2.73451
H	3.79862	-0.63862	1.59147
H	3.19581	-2.88387	0.73328
H	1.33779	-3.11531	-0.90291
H	1.76242	2.34154	1.11578
H	3.20668	1.66572	1.89791
H	3.53458	3.29357	0.13070

Zero-point correction =	0.199771 (Hartree/Particle)
Sum of electronic and zero-point Energies =	-984.093382
Sum of electronic and thermal Energies =	-984.077278
Sum of electronic and thermal Enthalpies =	-984.076334
Sum of electronic and thermal Free Energies =	-984.138587

***E*-64** (F3-aldehyde)

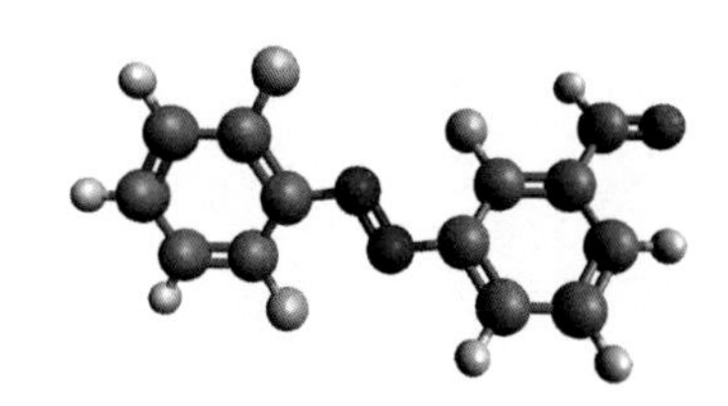

N	0.77302	-0.19810	0.12729
F	2.24008	-2.26648	0.90973
F	2.28131	2.07262	-0.89780
C	4.96300	-0.17683	-0.12290
C	4.27548	-1.27318	0.37263
C	2.89978	-1.21599	0.43783
C	2.16386	-0.08877	0.05043
C	2.90641	0.99157	-0.44996
C	4.28240	0.95490	-0.54412
N	0.15378	0.87942	0.20962

F	-1.47016	-1.38399	-0.69080
C	-4.03705	0.89316	0.40310
C	-3.28264	1.97568	0.80976
C	-1.90204	1.91625	0.71370
C	-1.24541	0.77949	0.24450
C	-2.03433	-0.30049	-0.17573
C	-3.42174	-0.25176	-0.10337
C	-4.24919	-1.39250	-0.55285
O	-5.45339	-1.39160	-0.50290
H	6.04395	-0.20542	-0.18653
H	4.78251	-2.16921	0.70563
H	4.79824	1.81490	-0.95158
H	-5.11986	0.89874	0.45296
H	-3.76260	2.86782	1.19213
H	-1.27889	2.75297	1.00775
H	-3.69614	-2.26485	-0.94376

Zero-point correction =	0.176277 (Hartree/Particle)
Sum of electronic and zero-point Energies =	-982.922560
Sum of electronic and thermal Energies =	-982.907063
Sum of electronic and thermal Enthalpies =	-982.906119
Sum of electronic and thermal Free Energies =	-982.967934

***Z*-64** (F3-aldehyde)

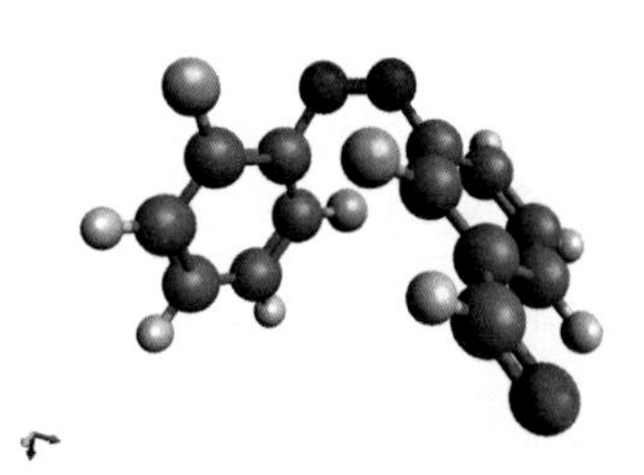

N	-1.36847	-1.27144	-1.39370
F	-2.59443	1.12345	-1.79775
F	-0.97983	-1.98862	1.29453
C	-2.73429	1.11085	1.79540
C	-2.91613	1.52664	0.48447
C	-2.42798	0.73976	-0.53636

C	-1.72700	-0.43961	-0.29848
C	-1.58304	-0.82764	1.03073
C	-2.07387	-0.07430	2.07713
N	-0.20883	-1.65353	-1.56392
F	0.44576	1.07357	-1.34277
C	3.13316	-0.32687	0.61986
C	2.83126	-1.67279	0.53798
C	1.71877	-2.08897	-0.17694
C	0.87028	-1.16588	-0.77608
C	1.20594	0.18202	-0.70417
C	2.32877	0.61913	-0.01308
C	2.67036	2.05583	0.06006
O	3.62127	2.47641	0.66869
H	-3.11827	1.71576	2.60739
H	-3.42978	2.44739	0.24016
H	-1.93551	-0.42912	3.09021
H	4.00161	0.03215	1.15959
H	3.46620	-2.40632	1.01913
H	1.47304	-3.14037	-0.26954
H	1.99269	2.73937	-0.48235

Zero-point correction =	0.176385 (Hartree/Particle)
Sum of electronic and zero-point Energies =	-982.912119
Sum of electronic and thermal Energies =	-982.896855
Sum of electronic and thermal Enthalpies =	-982.895911
Sum of electronic and thermal Free Energies =	-982.955907

Molecular Orbitals

The orbital energies were obtained from Gaussian09, and the corresponding 3D representation of the electron probability distributions were visualized by Avogadro 1.2.0.

Table 10: HOMO-1 (H-1), HOMO (H) and LUMO (L) Kohn-Sham orbital energies (in eV) of the *E*-isomers and *Z*-isomers of the compounds **36**, **60-64** (compared with TFAB **35**) calculated on the PBE0-D3/def2-TZVP level of theory. In all cases, the HOMO is the n-orbital. The HOMO-LUMO gaps (ΔHL) were calculated independently for *Z*- and *E*-isomers.[91]

compound		$\varepsilon_{H-1}(Z)$	$\varepsilon_{H}(Z)$	$\varepsilon_{L}(Z)$	$\Delta_{HL}(Z)$	$\varepsilon_{H-1}(E)$	$\varepsilon_{H}(E)$	$\varepsilon_{L}(E)$	$\Delta_{HL}(E)$
TFAB	**35**	-7.507	-6.780	-2.333	4.447	-7.426	-6.602	-2.525	4.077
F4-alcohol	**60**	-7.373	-6.603	-2.208	4.395	-7.145	-6.449	-2.402	4.047
F4-bisalcohol	**61**	-7.260	-6.458	-2.105	4.353	-6.941	-6.299	-2.284	4.015
F4-aldehyde	**63**	-7.771	-7.080	-2.886	4.194	-7.730	-6.898	-3.060	3.838
F4-bisaldehyde	**36**	-8.082	-7.359	-3.277	4.082	-7.994	-7.163	-3.493	3.670
F3-alcohol	**62**	-7.380	-6.626	-2.165	4.461	-7.136	-6.493	-2.378	4.115
F3-aldehyde	**64**	-7.669	-6.937	-2.571	4.366	-7.436	-6.799	-2.722	4.077

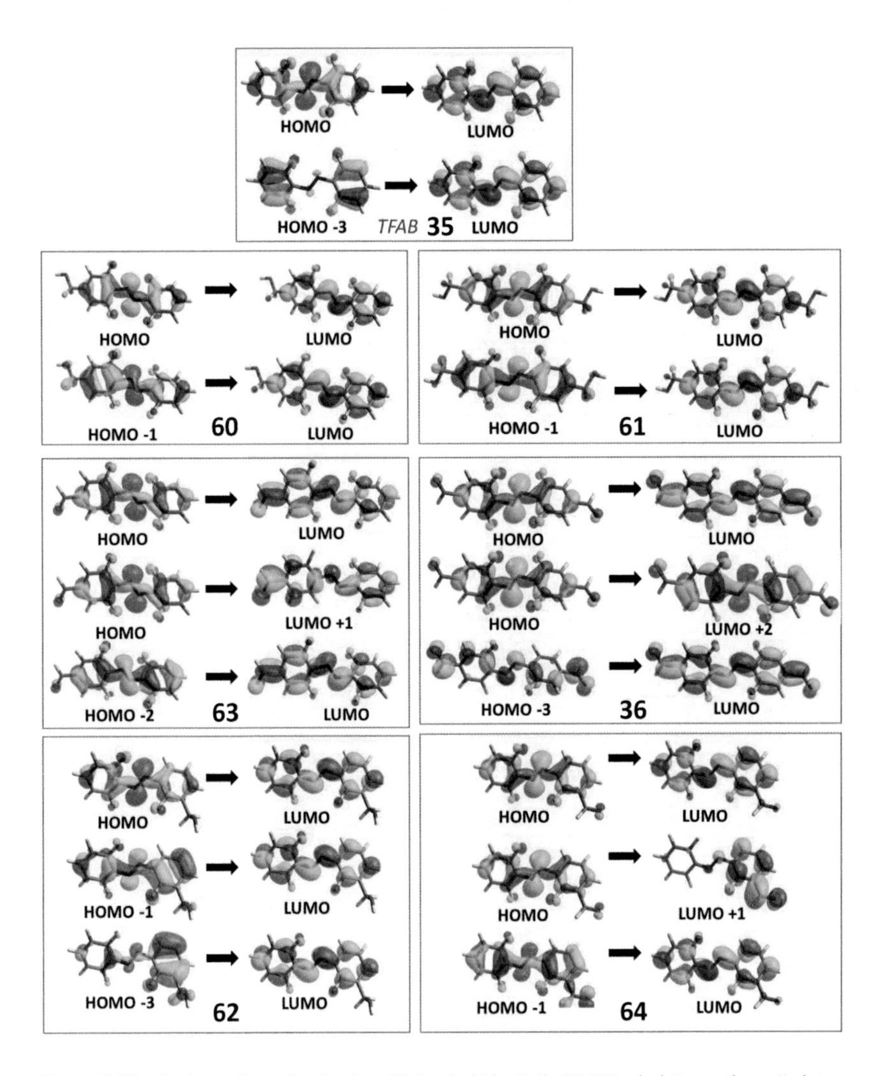

Figure 65: Visualized transitions of molecular orbitals, of which – in the TD-DFT calculations – the excited state 1 is composed for the corresponding *E*-isomer of molecules **35, 36**, **60-64**.

Table 11: Excitation features of the compounds **36**, **60-64**; TD-B3LYP/6-311G*/PCM (ACN) level of theory.

Compound	isomer	transition	λ [nm]	osc. strength	$\Delta\lambda_{n-\pi^*}$ [nm]
60	*Z*	n→π*	438	0.0539	48
	Z	π→π*	312	0.0177	
	E	n→π*	486	0.0656	
	E	π→π*	324	0.9371	
61	*Z*	n→π*	441	0.0637	45
	Z	π→π*	312	0.0193	
	E	n→π*	486	0.0739	
	E	π→π*	333	1.0980	
63	*Z*	n→π*	447	0.0647	61
	Z	π→π*	338	0.0001	
	E	n→π*	508	0.0873	
	E	π→π*	347	0.0003	
36	*Z*	n→π*	456	0.0780	75
	Z	π→π*	343	0.0001	
	E	n→π*	531	0.1143	
	E	π→π*	357	0.0003	
62	*Z*	n→π*	435	0.0374	50
	Z	π→π*	308	0.0216	
	E	n→π*	485	0.0446	
	E	π→π*	320	0.6390	
64	*Z*	n→π*	436	0.0373	53
	Z	π→π*	332	0.0016	
	E	n→π*	489	0.0414	
	E	π→π*	335	0.0054	

UV/Vis isomerization experiments

Stock solutions were prepared at 500 µM for each substance. To determine the final concentration at which an absorbance maximum of about 1.0 A.U. was reached, each compound was measured at a concentration of 50 µM and the thereupon calculated final concentrations are 38 µM (**36**) and 33 µM (**54**). The cuvettes with the samples were irradiated with light of different wavelengths (365 nm, 400 nm, 410 nm, 430 nm, 470 nm, 530 nm, 623 nm, 660 nm) directly before the measurement.

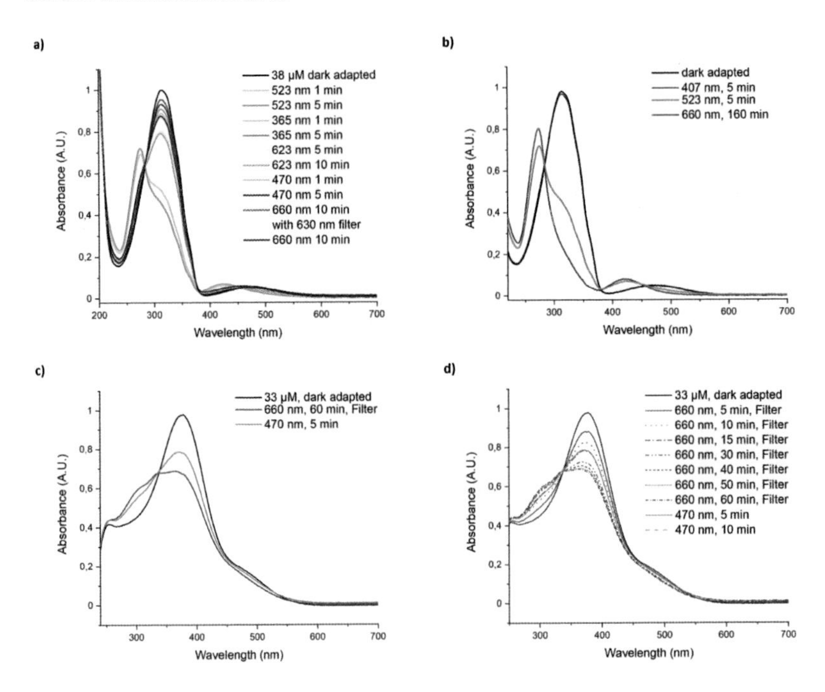

Figure 66: a) The absorption curves display a distinct isosbestic point at 284 nm of compound **36**. b) Photostationary state of dialdehyde **36** after irradiation at 407 nm, 523 nm and 660 nm. c,d) Absorption curves of **54** before and after irradiation, isosbestic point at 338 nm.

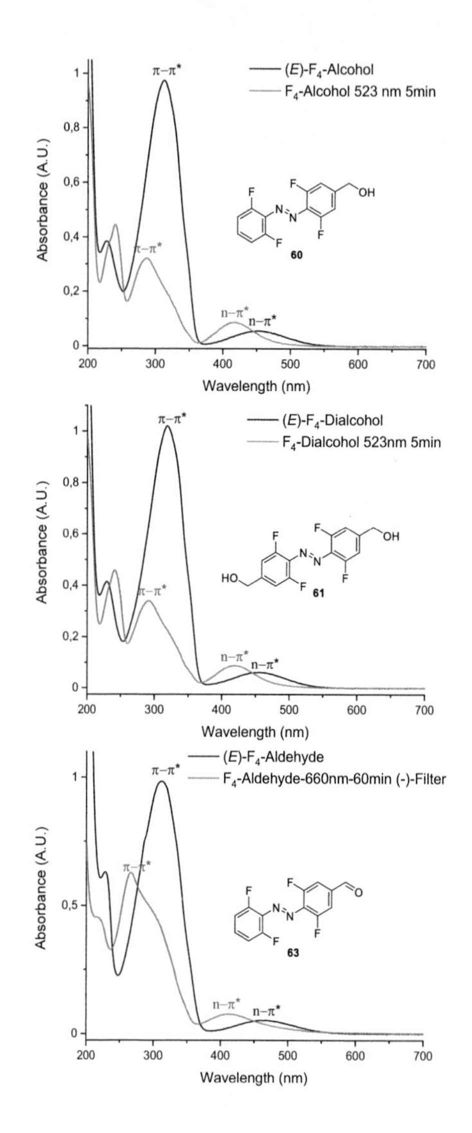
π–π*
(E)-F4-Alcohol
F4-Alcohol 523 nm 5min
Absorbance (A.U.)
Wavelength (nm)
60
n–π*
(E)-F4-Dialcohol
F4-Dialcohol 523nm 5min
61
(E)-F4-Aldehyde
F4-Aldehyde-660nm-60min (-)-Filter
63

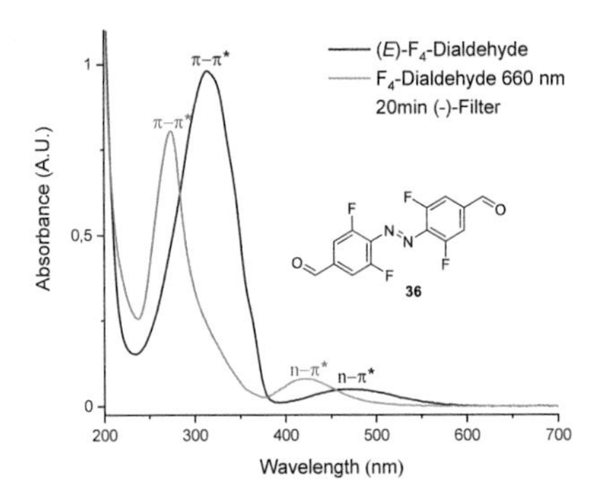

Figure 67: UV-Vis spectra measured for the compounds **60, 61, 63, 36** – comparison of the *E*-isomer ("dark state") and the irradiated mixture (with given wavelength and time) containing the majority of *Z*-isomer.

Calculation of the attenuation coefficients

A 1.5 mM solution of compound **36, 54, 60-64** was prepared by dissolving the dry substance in d_6-DMSO. The solution was transferred to a 1.5 mL quartz cuvette and the absorption (400 nm-700 nm) was measured. Subsequently the cuvette was irradiated at 523 nm (15 min) and the absorption was measured again. An aliquot (0.55 mL) of the irradiated solution was directly measured by NMR and the *E*/*Z* ratio was determined. The molar extinction coefficient **ε** (M^{-1} cm^{-1}) was calculated based on the Lambert-Beer Law: **A** is the absorbance; **c** is the concentration (M) and **l** the optical path-length (cm).

$$A = \varepsilon * c * l$$

The absorbance spectrum of the *Z*-isomer was calculated by the following formula:

$$A_{Z\ isomer} = \frac{A_{PSS,523nm} - \left(A_{E\ isomer} * \left(\% \frac{E\ isomer}{100}\right)\right)}{\left(\% \frac{Z\ isomer}{100}\right)}$$

Compound **60**

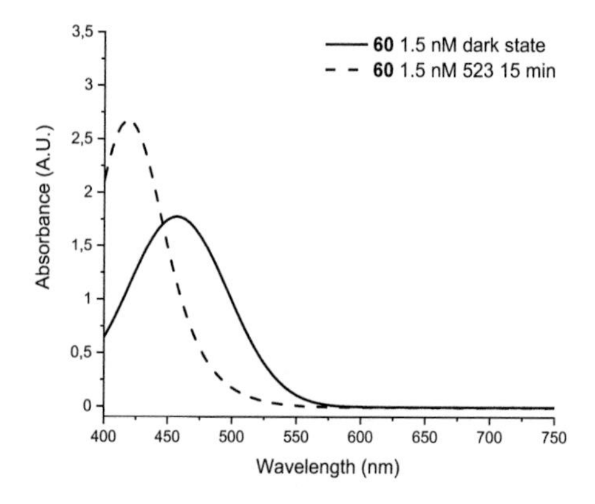

Figure 68: Absorbance spectrum of **60**. Raw data before calculation. y-axis: absorption (A.U.), x-axis wavelength (nm).

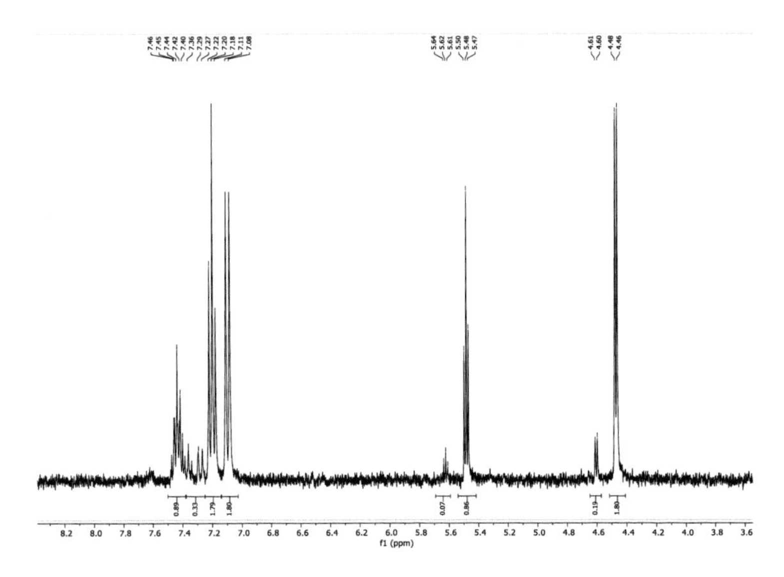

Figure 69: ^{1}H-NMR-spectrum (400 MHz, 16 scans, DMSO-d_6) of **60** (1.5 mM) after irradiation at 523 nm.

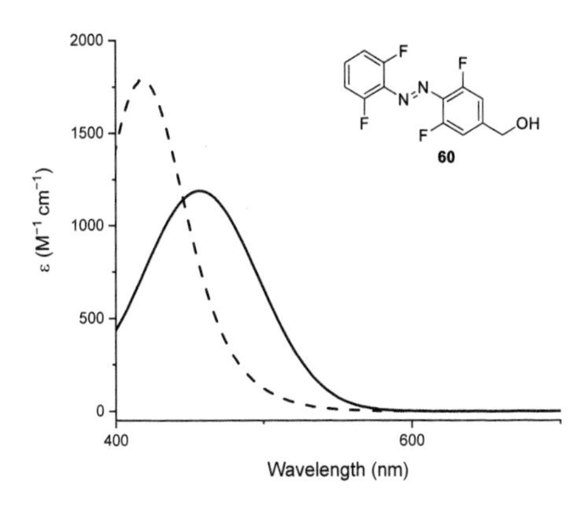

Figure 70: Molar extinction coefficient of **60**-*E* (solid) and **60**-*Z* (dashed).

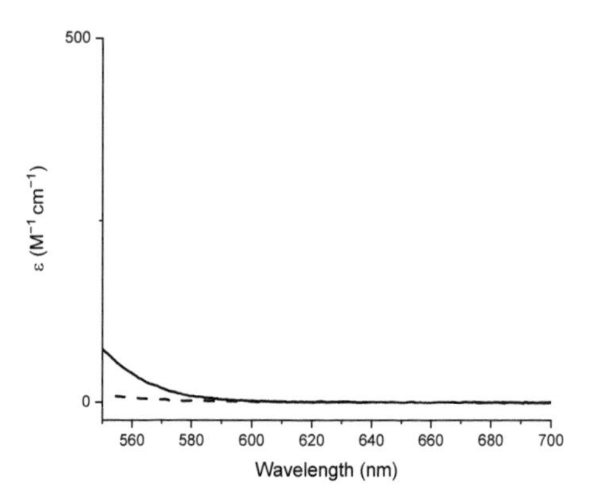

Figure 71: Molar extinction coefficient of **60**-*E* (solid) and **60**-*Z* (dashed), section between 550 nm – 700 nm.

Compound **61**

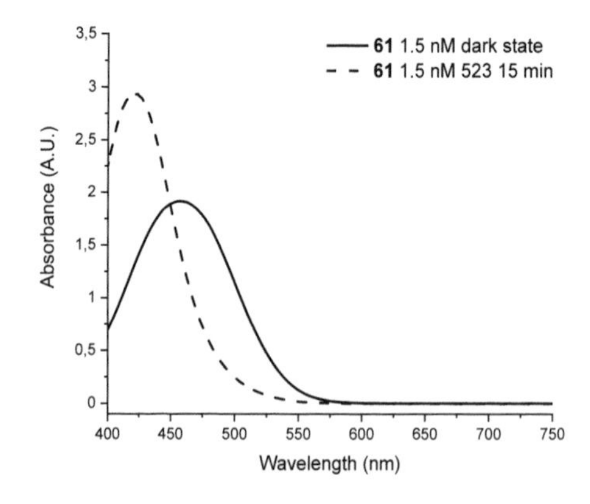

Figure 72: Absorbance spectrum of **61**. Raw data before calculation. y-axis: absorption (A.U.), x-axis wavelength (nm).

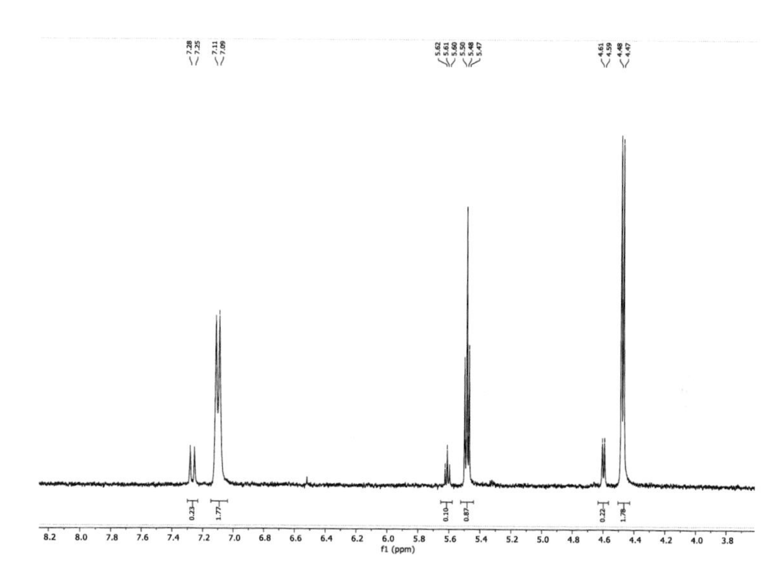

Figure 73: ^{1}H-NMR-spectrum (400 MHz, 32 scans, DMSO-d_6) of **61** (1.5 mM) after irradiation at 523 nm.

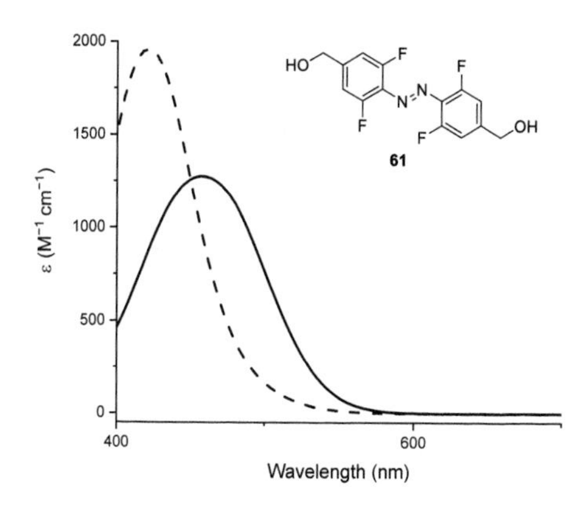

Figure 74: Molar extinction coefficient of **61**-*E* (solid) and **61**-*Z* (dashed).

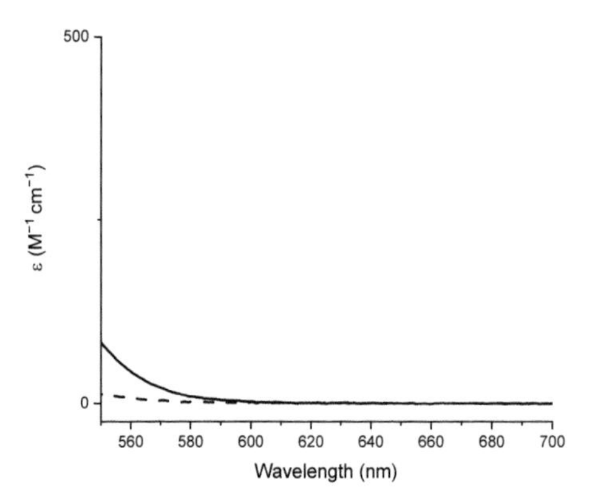

Figure 75: Molar extinction coefficient of **61**-*E* (solid) and **61**-*Z* (dashed), section between 550 nm – 700 nm.

Compound **63**

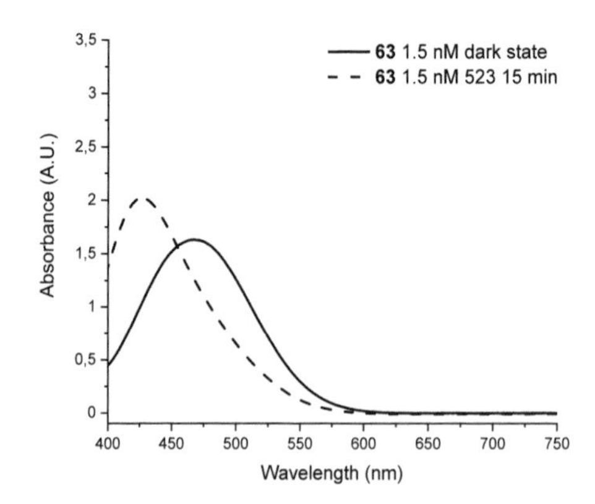

Figure 76: Absorbance spectrum of **63**. Raw data before calculation. y-axis: absorption (A.U.), x-axis wavelength (nm).

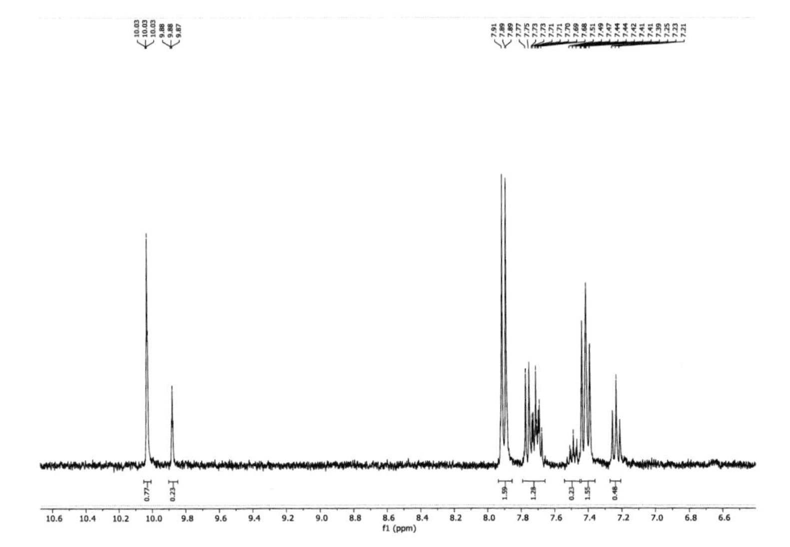

Figure 77: ^{1}H-NMR-spectrum (400 MHz, 32 scans, DMSO-d_6) of **63** (1.5 mM) after irradiation at 523 nm.

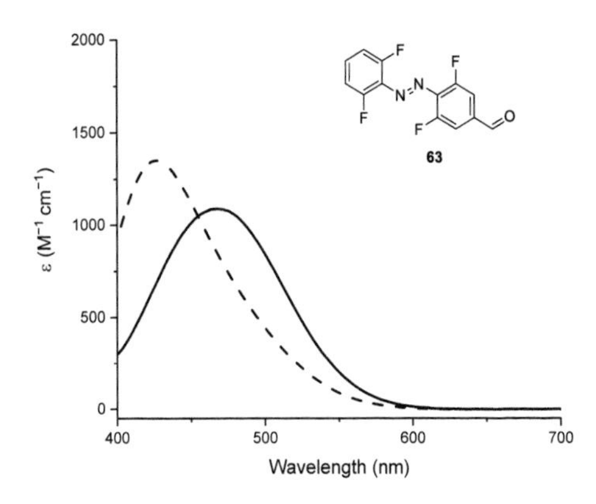

Figure 78: Molar extinction coefficient of **63**-*E* (solid) and **63**-*Z* (dashed).

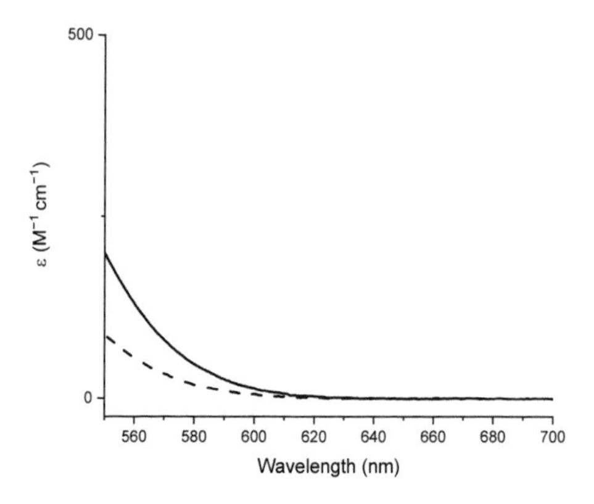

Figure 79: Molar extinction coefficient of **63**-*E* (solid) and **63**-*Z* (dashed), section between 550 nm – 700 nm.

Compound **36**

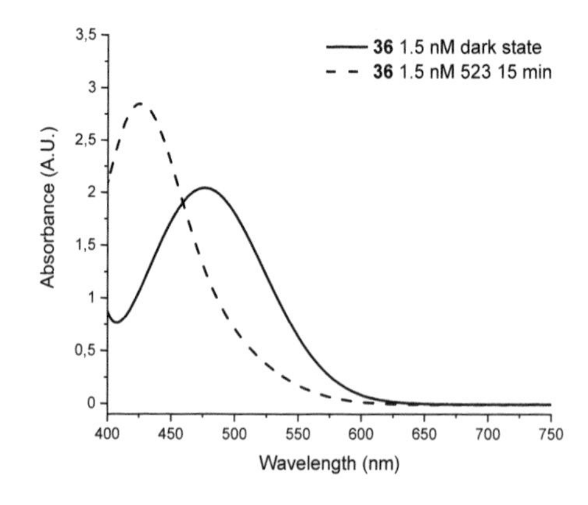

Figure 80: Absorbance spectrum of **36**. Raw data before calculation. y-axis: absorption (A.U.), x-axis wavelength (nm).

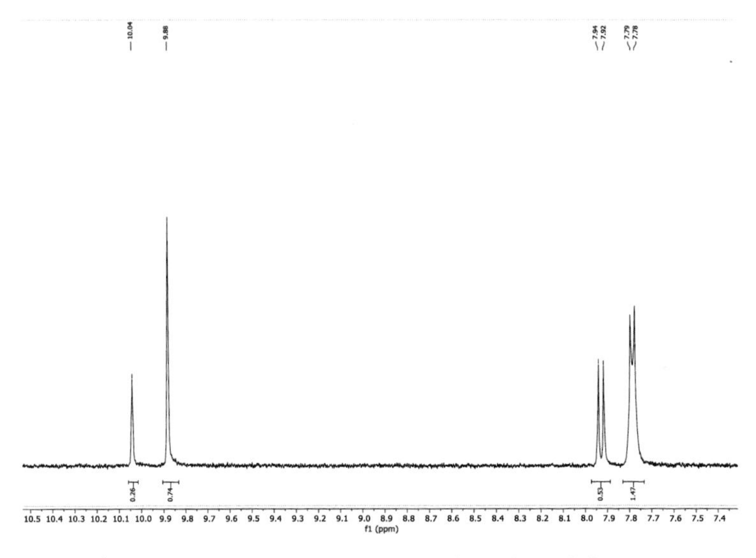

Figure 81: ^{1}H-NMR-spectrum (400 MHz, 32 scans, DMSO-d_6) of **36** (1.5 mM) after irradiation at 523 nm.

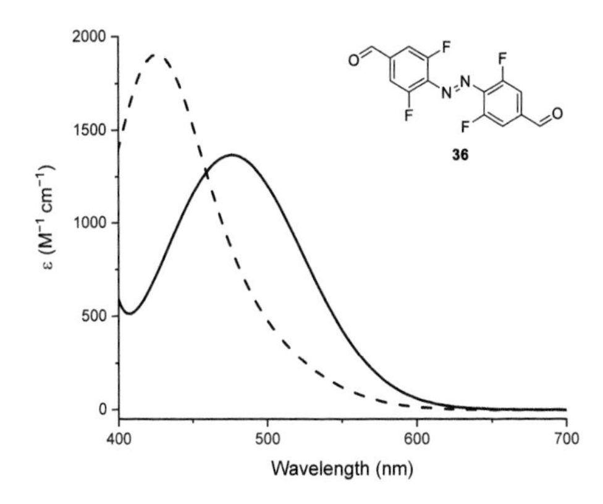

Figure 82: Molar extinction coefficient of **36**-*E* (solid) and **36**-*Z* (dashed).

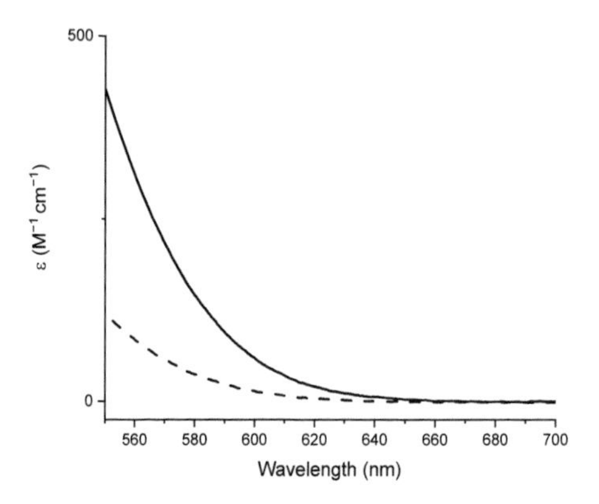

Figure 83: Molar extinction coefficient of **36**-*E* (solid) and **36**-*Z* (dashed) section between 550 nm – 700 nm.

Compound **62**

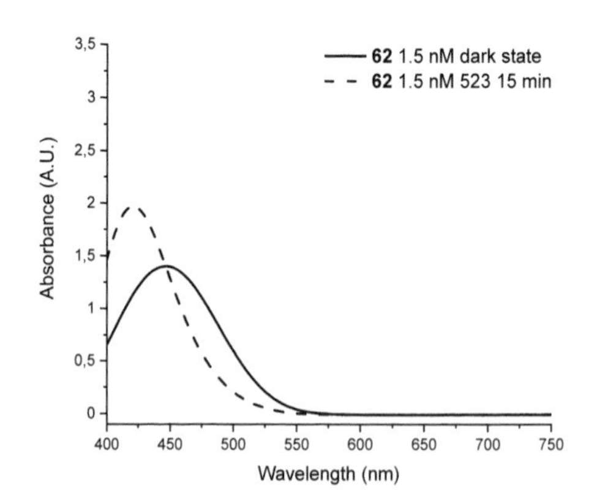

Figure 84: Absorbance spectrum of **62**. Raw data before calculation. y-axis: absorption (A.U.), x-axis wavelength (nm).

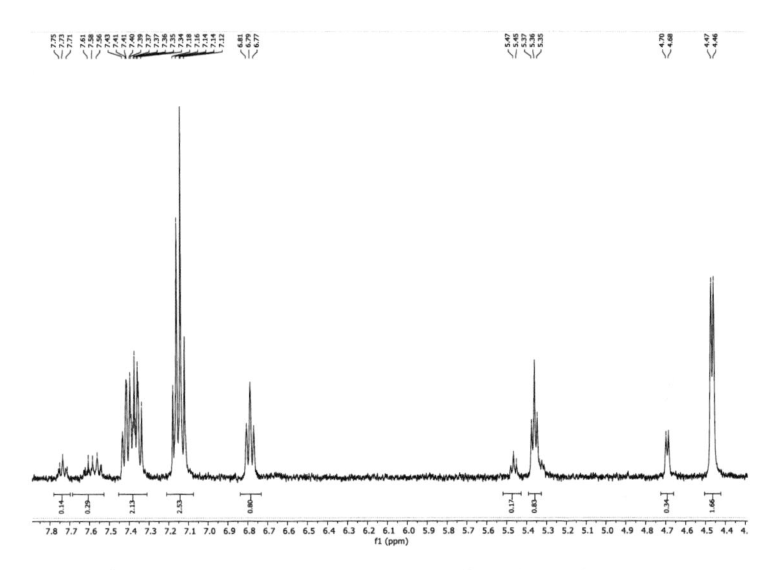

Figure 85: ^{1}H-NMR-spectrum (400 MHz, 32 scans, DMSO-d_6) of **62** (1.5 mM) after irradiation at 523 nm.

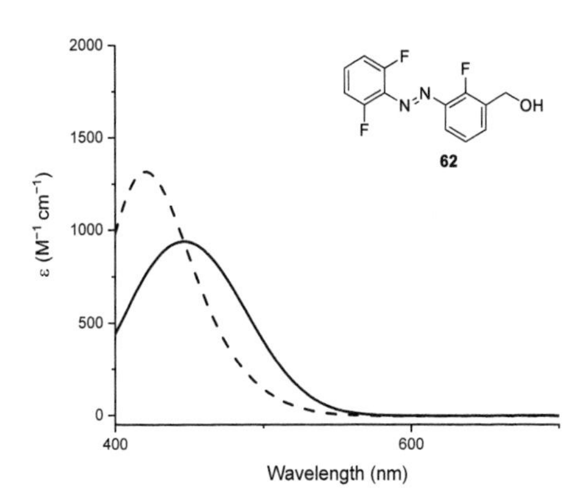

Figure 86: Molar extinction coefficient of **62**-*E* (solid) and **62**-*Z* (dashed).

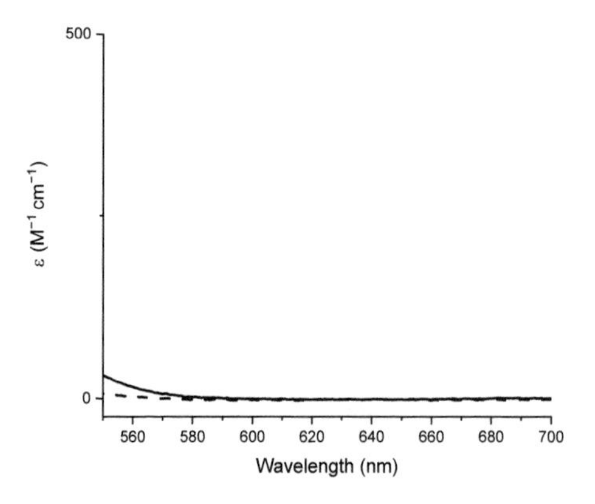

Figure 87: Molar extinction coefficient of **62**-*E* (solid) and **62**-*Z* (dashed), section between 550 nm – 700 nm.

Compound **64**

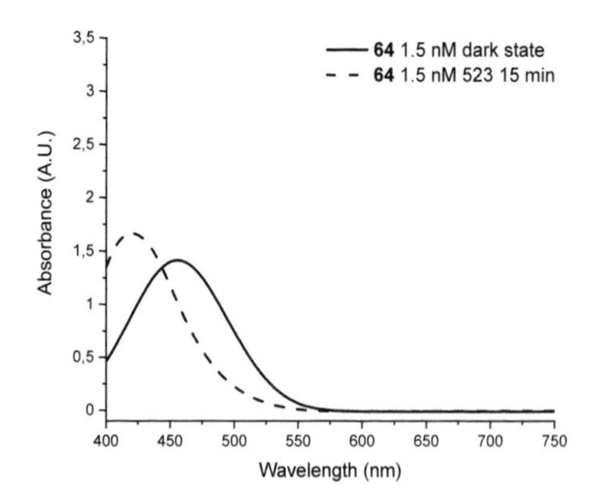

Figure 88: Absorbance spectrum of **64**. Raw data before calculation. y-axis: absorption (A.U.), x-axis wavelength (nm).

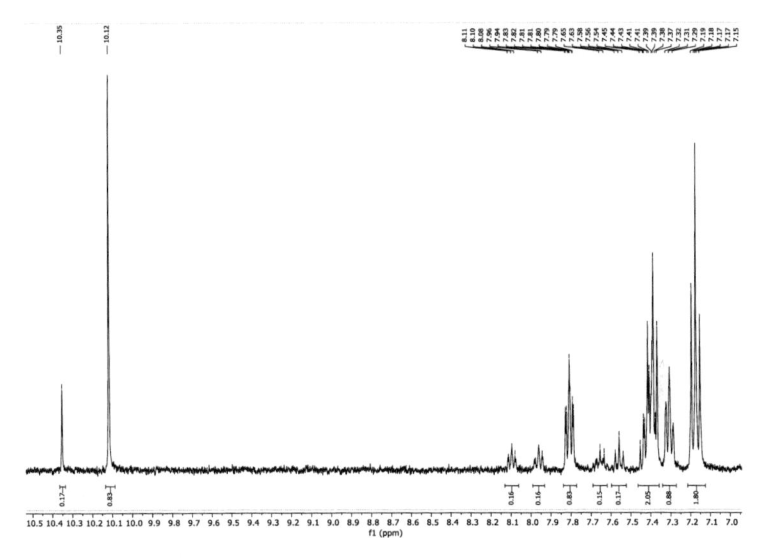

Figure 89: ^{1}H-NMR-spectrum (400 MHz, 32 scans, DMSO-d_6) of **64** (1.5 mM) after irradiation at 523 nm.

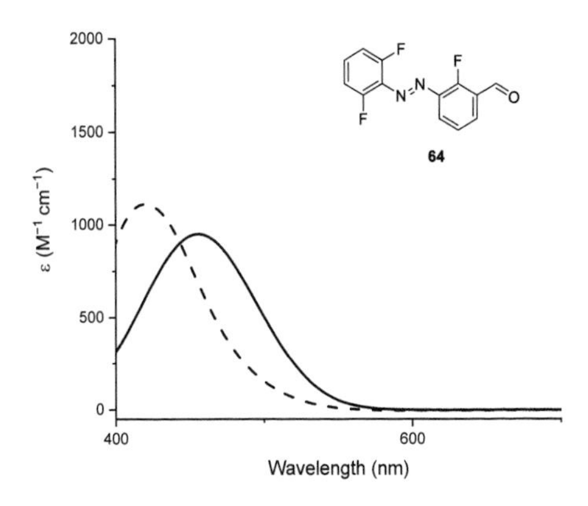

Figure 90: Molar extinction coefficient of **64**-*E* (solid) and **64**-*Z* (dashed).

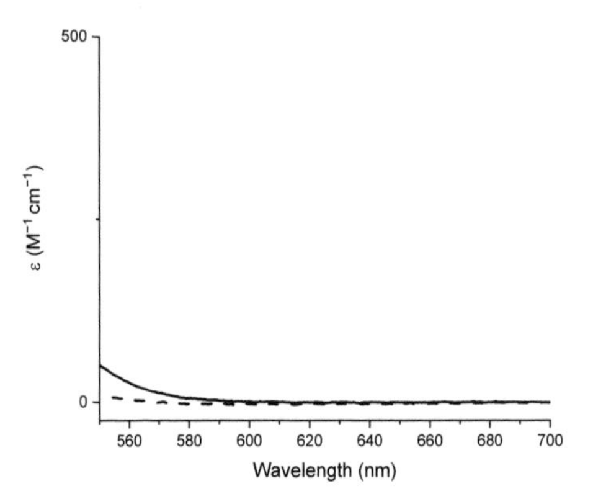

Figure 91: Molar extinction coefficient of **64**-*E* (solid) and **64**-*Z* (dashed), section between 550 nm – 700 nm.

Compound **54**

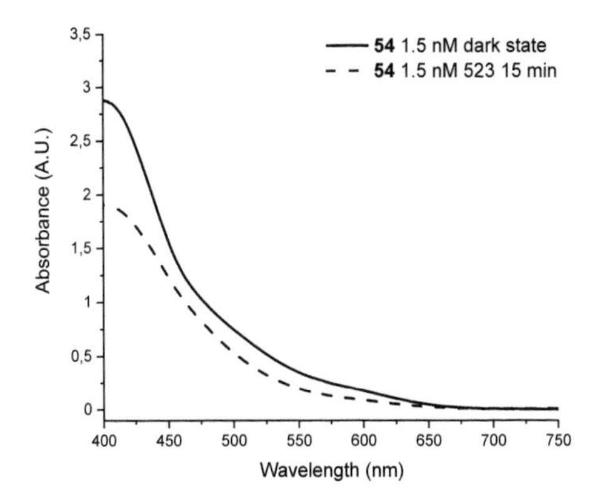

Figure 92: Absorbance spectrum of **54**. Raw data before calculation. y-axis: absorption (A.U.), x-axis wavelength (nm).

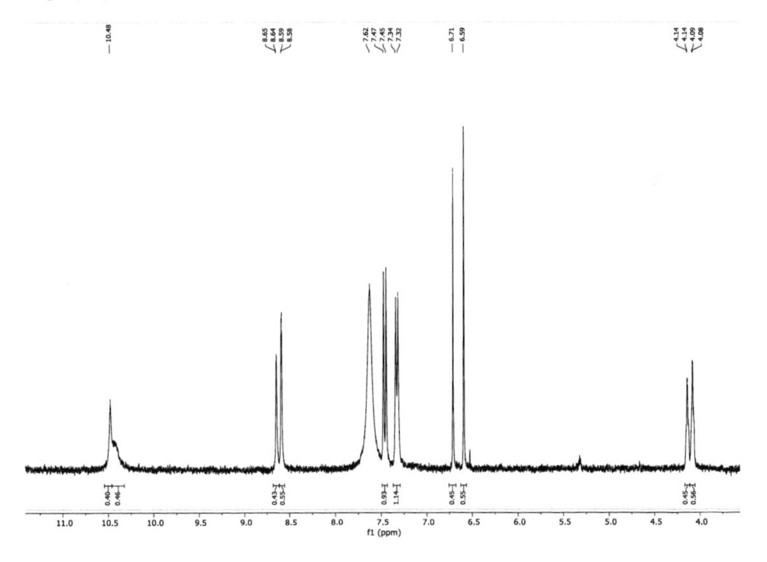

Figure 93: ^{1}H-NMR-spectrum (400 MHz, 32 scans, DMSO-d_6) of **54** (1.5 mM) after irradiation at 523 nm.

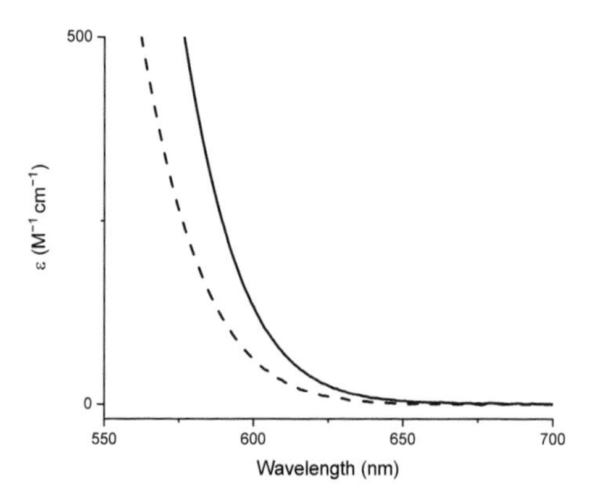

Figure 94: Molar extinction coefficient of **54**-*E* (solid) and **54**-*Z* (dashed).

Due to the high absorbance of compound **54** between 400 nm - 500 nm, a dilution to a final concentration of 100 µM was prepared to measure the range between 400 nm - 700 nm.

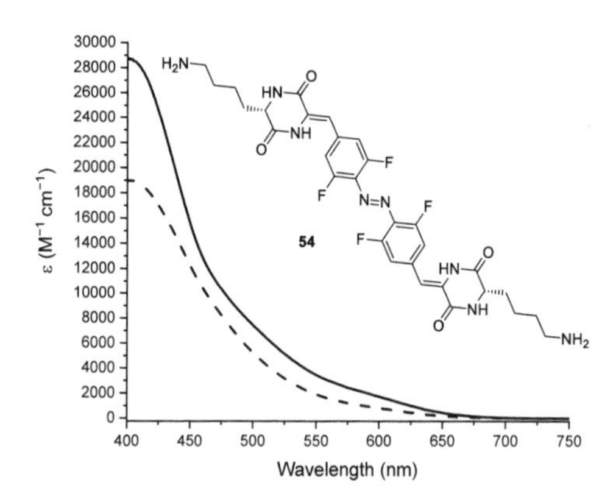

Figure 95: Molar extinction coefficient of **54**-*E* (solid) and **54**-*Z* (dashed).

Calculated and measured spectra

Comparison of the measured and calculated UV-Vis spectra of compounds **36**, **60-64** in the visible light range (>400 nm). Black curves correspond to the *E*-isomers, red curves to the *Z*-isomers.

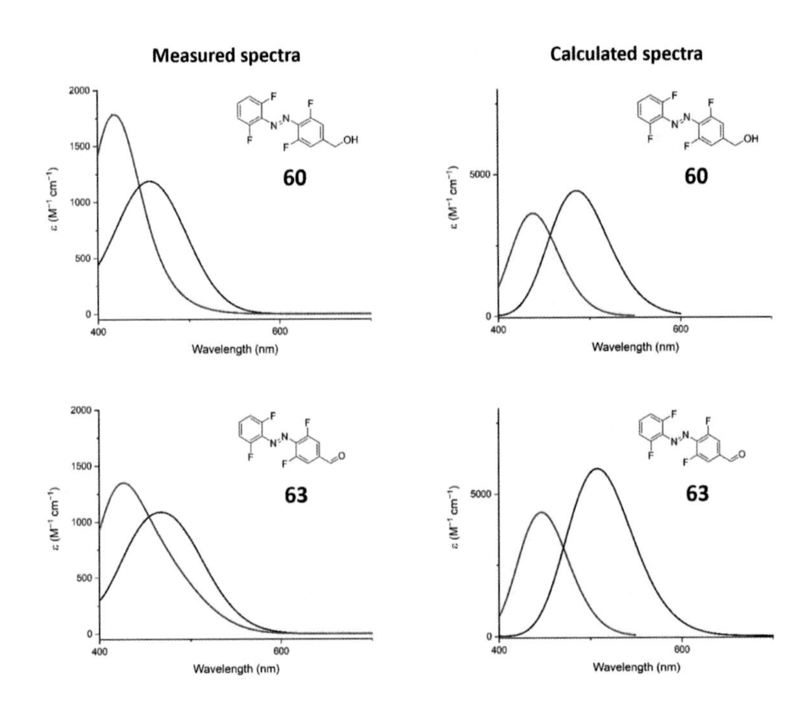

Figure 96: *Left:* Band separation of the *E*-isomers (black lines) and *Z*-isomers (red lines) of the compounds **60** and **63** in the visible light range (1.5 mM solutions in DMSO-d_6). In all cases, the y-axis depicts the molar attenuation coefficient ε (M^{-1} cm^{-1}). The spectra of pure *Z*-isomers (red lines) of the measured spectra have been calculated from spectra registered for samples irradiated for 15 min with 523 nm LED with concomitant determination of the *E/Z*-ratio by ^{1}H NMR. *Right:* calculated spectra.

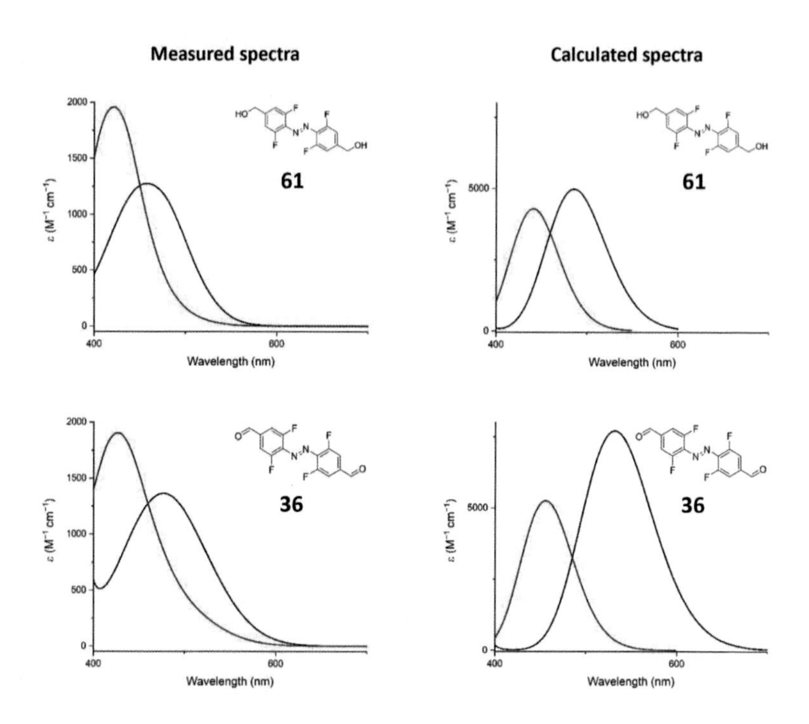

Figure 97: *Left:* Band separation of the *E*-isomers (black lines) and *Z*-isomers (red lines) of the compounds **36** and **61** in the visible light range (1.5 mM solutions in DMSO-d_6). In all cases, the y-axis depicts the molar attenuation coefficient ε (M^{-1} cm^{-1}). The spectra of pure *Z*-isomers (red lines) of the measured spectra have been calculated from spectra registered for samples irradiated for 15 min with 523 nm LED with concomitant determination of the *E/Z*-ratio by ^{1}H NMR. *Right:* calculated spectra.

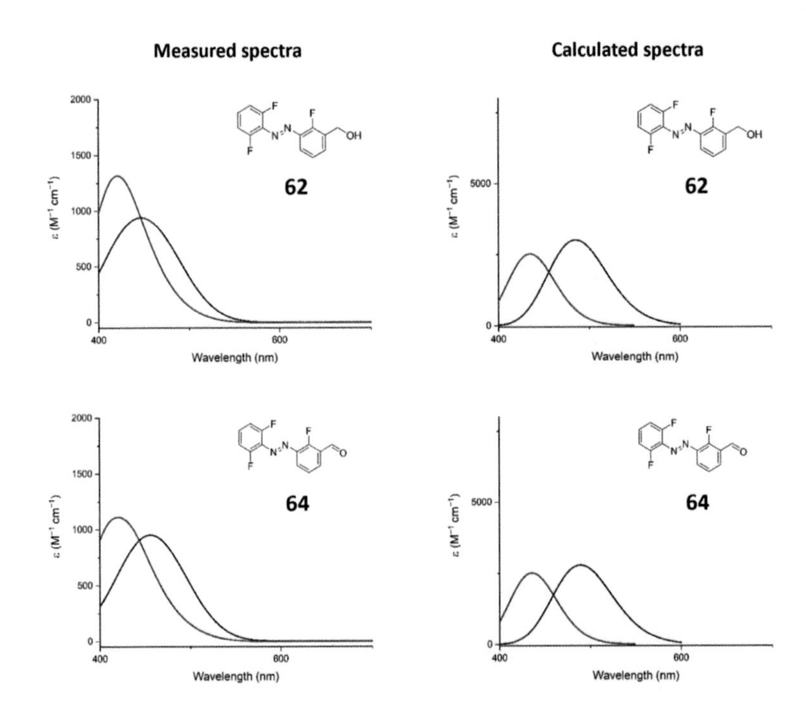

Figure 98: *Left:* Band separation of the *E*-isomers (black lines) and *Z*-isomers (red lines) of the compounds **62** and **64** in the visible light range (1.5 mM solutions in DMSO-d_6). In all cases, the y-axis depicts the molar attenuation coefficient ε (M^{-1} cm^{-1}). The spectra of pure *Z*-isomers (red lines) of the measured spectra have been calculated from spectra registered for samples irradiated for 15 min with 523 nm LED with concomitant determination of the *E/Z*-ratio by ^{1}H NMR. *Right:* calculated spectra.

Extinction coefficients

UV-Vis spectra of compound **36** (400 nm – 700 nm) were measured in MeCN at 20 °C, at concentrations of 100 μM, 167 μM, 250 μM, 333 μM, and 500 μM. The samples were prepared from a 500 μM stock solution in serial dilution.

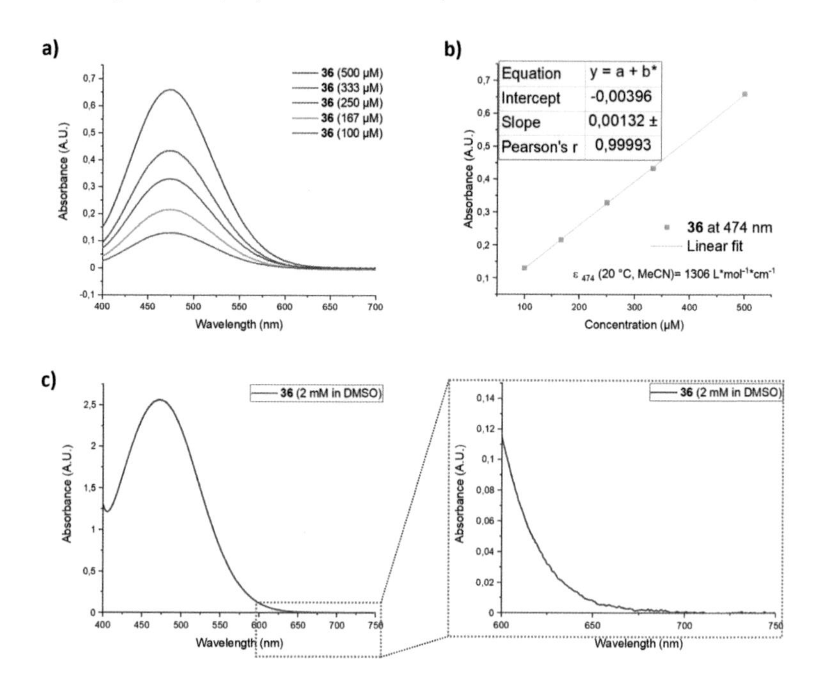

Figure 99: a) Absorption of dialdehyde **36** at various concentrations. b) The molar extinction coefficient of **36** is calculated from the slope of the shown linear fit. c) **36** was measured at a high concentration of 2 mM in DMSO to verify absorption >630 nm.

5.3.3 Biological stability

For prospective applications, the stability of compound **54** under physiological conditions had to be investigated. Reduction of azobenzene derivatives with thiol groups to arylhydrazines in biological systems is considered the most serious limitation for their *in vivo* application as photoswitches. Therefore, the stability of **54** in a reducing environment was evaluated: A solution with final concentrations of 0.1 mM of **54**, 10 mM reduced glutathione and 5 mM TCEP was prepared in 0.1 M pH 7.4 DPBS (no magnesium, no calcium) with 5% (v/v) DMSO. A 2 mM solution of compound **54** in DMSO and a 0.25 M solution of reduced glutathione in DPBS were prepared, the pH was adjusted to 7.4 with aqueous solution of NaOH. The mixture was kept at 25 °C and HPLC measurements were performed at 30 min intervals over a period of 12 h. For analysis, the peak area ratio of **54** and reduced glutathione was calculated, assuming that the area of the reduced glutathione is constant, due to its 100-fold higher concentration.

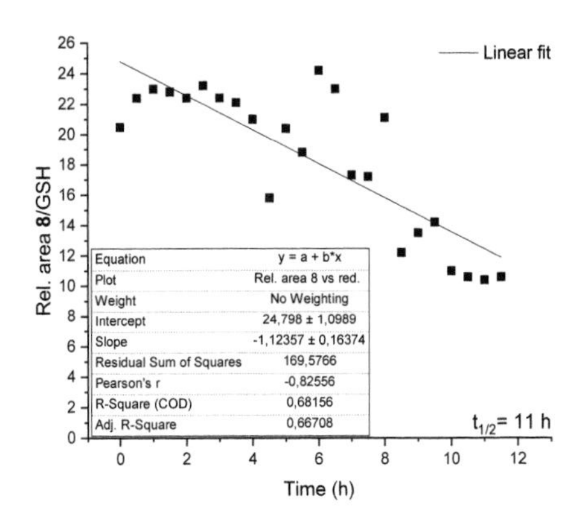

Figure 100: Biological stability of **54**. The calculated half-life $t_{1/2}$ is 11 h.

5.3.4 Viscosity experiments

To a 1.5 mL-vial (crimp top, 12×32 mm) was added the photochromic material **54** as a powder and 500 µL of the aqueous solution. This suspension was treated by ultrasound, followed by heating to 80 °C in a vial block. After equilibration at 80 °C for 5 min, the sample was heated to the boiling point by a heat gun. The hot solution became completely clear and upon cooling its viscosity increased.

Table 12: Results of the viscosity experiments in distilled water. *Melting point was determined in a single measurement.

Composition (x mg, 500 µL H_2O)	Approx. conc.	T_m °C	time at room temperature to achieve stability upon inversion
10	2 wt%	55*	> 5 min

Table 13: Results of the viscosity experiments in aqueous 200 mM NaCl solution. *Melting point was determined in a single measurement.

Composition (x mg, 500 µL 200 mM NaCl)	Approx. conc.	T_m °C	time at room temperature to achieve stability upon inversion
10	2 wt%	>99*	> 5 min
7.5	1.5 wt%	89 ± 1	> 5 min
5	1 wt%	93 ± 5	> 5 min

Table 14: Results of the viscosity experiments in Ringer's solution. *Melting point was determined in a single measurement.

Composition (x mg, 500 µL Ringer's solution)	Approx. conc.	T_m °C	time at room temperature to achieve stability upon inversion
10	2 wt%	83*	> 5 min
7.5	1.5 wt%	98*	> 5 min

5	1 wt%	84 ± 14	> 5 min
4	0.8 wt%	-	> 20 min
2.5	0.5 wt%	-	Not stable

5.3.5 Rheology and NMR relaxometry

Rheology

Rheological properties were investigated *via* oscillatory shear experiments on the strain-controlled rotational rheometer Ares G2 (TA Instruments, Eschborn, Germany) with a parallel-plate geometry (diameter: 30 mm). The sample of material **54**, with a concentration of 20 g/L (2 wt%) and a volume of 3 mL, was prepared according to the procedure described in 5.3.4 in a 10 mL-vial and poured onto the lower plate. The gap was set to 1 mm and the temperature controlled to 25 ± 0.1 °C by a Peltier element (Advanced Peltier System, TA Instruments). An oscillatory strain-sweep with a constant frequency of 1 Hz and a varying strain of γ = 0.01-1000 % was used to determine the linear viscoelastic regime (LVE) of the samples. In the linear viscoelastic regime, the measured shear moduli G' and G" are independent of the applied strain or stress.[206] A strain of 0.1 % was fixed for the subsequent frequency sweeps where the frequency was varied from 0.03 to 100 Hz.

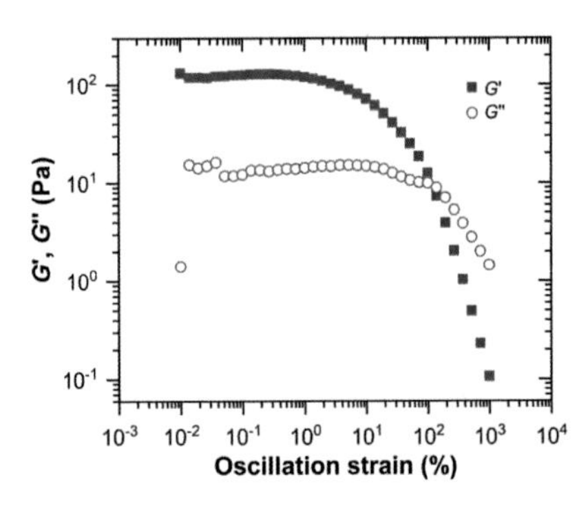

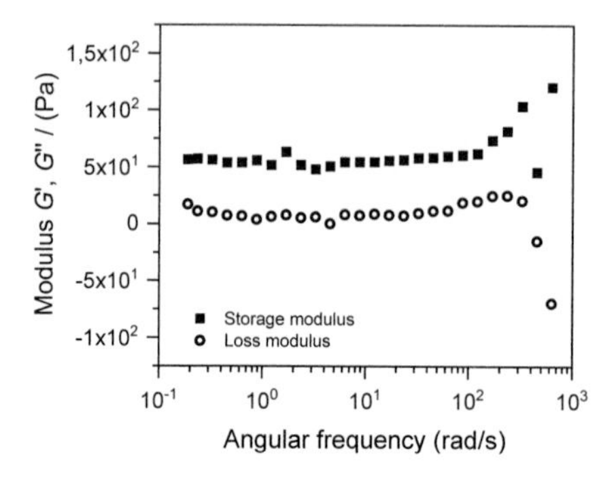

Figure 101: Rheological measurements of the viscous material composed of 2 wt% compound **54** in water. Strain sweep experiment (top); frequency sweep experiment (bottom).

^{1}H NMR T_2 relaxation

^{1}H NMR transverse relaxation (T_2) measurements were used to investigate the microstructure of the gel as it is directly linked to the molecular mobility (1–10 nm) of the components. The mobility can be quantitatively assessed by the decay of the transverse magnetization using time-domain NMR techniques. Herein, a magic sandwich echo (MSE) pulse sequence was used to refocus the initial transverse magnetization (100 ms) of rigid components in combination with a CPMG (Carr-Purcell-Meiboom-Gill) pulse sequence to refocus the magnetization of more mobile components up to 1 s.

The ^{1}H NMR T_2 relaxation measurements were performed on a bench-top 20 MHz minispec (Bruker, ND series). A flat NMR tube (diameter: 10 mm) was filled with 50 mg of **54** and 500 µL D_2O (sonicated for 1 h, then heated to boiling) to a maximum height of 10 mm, ensuring the highest homogeneity of the B1 magnetic field. The sample temperature was controlled to 39.5 °C by a BVT3000 unit (Bruker). To acquire a full relaxation curve over several orders of magnitude (10^{-3} – 10^3 ms), a combination of a MSE (magic sandwich echo) and two CPMG (Carr-Purcell-Meiboom-Gill) pulse sequences were used.[207] To avoid spin locking effects, a XY16 phase cycle was used for the CPMG.[208] The data points were acquired over 128 scans with a recycle delay of 800 ms. The MSE recorded the first 100 s, followed by two CPMGs with a pulse separation of 0.04 ms and 1 ms. The full relaxation curve is shown in Figure 102. The CPMG refocuses the magnetization of the more mobile components, e.g. solvent as HDO, and is evaluated with a biexponential fitting function (CPMG fit) to determine the contribution of mobile components.

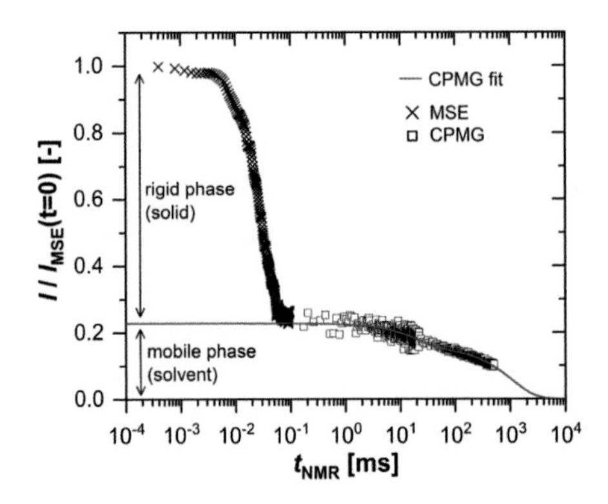

Figure 102: The distinct decay of the MSE clearly indicates a highly crystalline component in the sample.

As shown in Figure 102, a distinct decay of the FID by 80 % is observed in the first 100 ms, which indicates a highly polycrystalline nature of the microstructure and corroborates with the structures observed under the electron microscope (following section 5.3.6).

5.3.6 Microscopy images

Transmission Electron Microscopy

For the Transmission Electron Microscopy images, a 1 wt% viscous solution of sample **54** was prepared with Ringer's solution additionally containing a 1% (v/v) lead citrate solution (25 mg lead citrate in 0.1 M aq. NaOH) as described in the Gelation experiments 5.3.4. After cooling, the sample was equilibrated overnight at room temperature. Some samples were irradiated to confirm photomodulation of the material. Sample **A** was dark adapted, **B** was irradiated at 660 nm for 50 min and **C** was irradiated at 660 nm for 50 min, then regenerated by boiling as in the initial formation of the material. The resulting viscous solutions were diluted 1:10 with diH_2O and placed as small droplets on carbon-coated copper grids (400 mesh). The dilution of sample **B** was irradiated at 660 nm for additional 50 min to obtain sample **D**. The supernatant was removed carefully with a lint-free sheet and the grid was dried under atmospheric pressure. Examination was carried out by Heike Störmer on a Philips CM200 FEG transmission electron microscope, operated at 200 kV accelerating voltage.

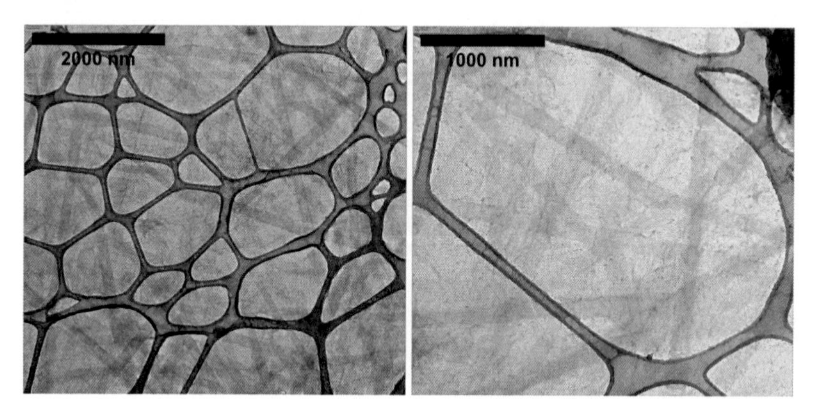

Figure 103: TEM sample A, dark adapted.

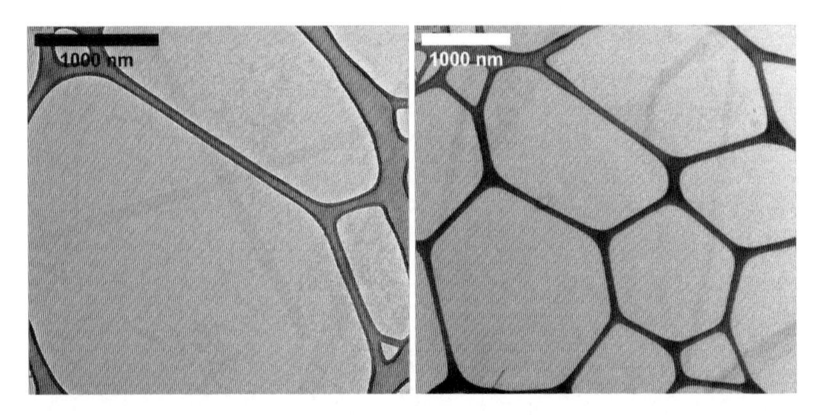

Figure 104: TEM sample B, irradiated at 660 nm for 50 min.

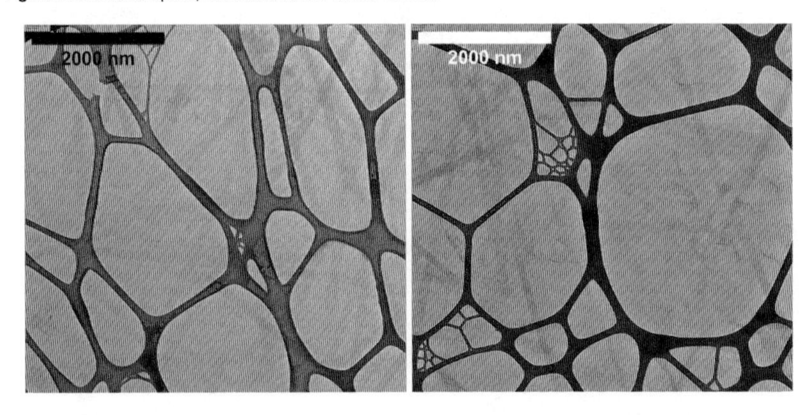

Figure 105: TEM sample C, irradiated at 660 nm for 50 min, then regenerated by boiling.

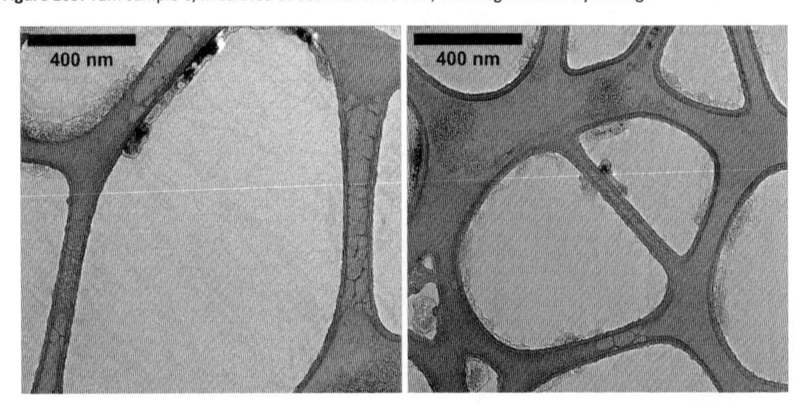

Figure 106: TEM sample D. The dilution of sample B was irradiated at 660 nm for additional 50 min to obtain sample D.

Scanning Electron Microscopy

For the Scanning Electron Microscopy images, a 1 wt% viscous solution sample of **54** was prepared with Ringer's solution and a 2 wt% viscous solution sample of **54** was prepared with diH_2O. The resulting material was freeze dried by lyophilization and then coated with a thin layer of platinum. Volker Zibat operated the microscope.

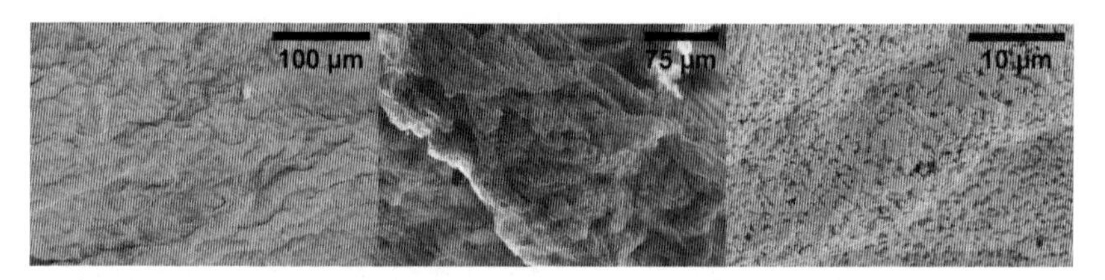

Figure 107: Freeze-dried samples of high viscosity materials prepared from 1 wt% **54** in Ringer's solution (no network visible).

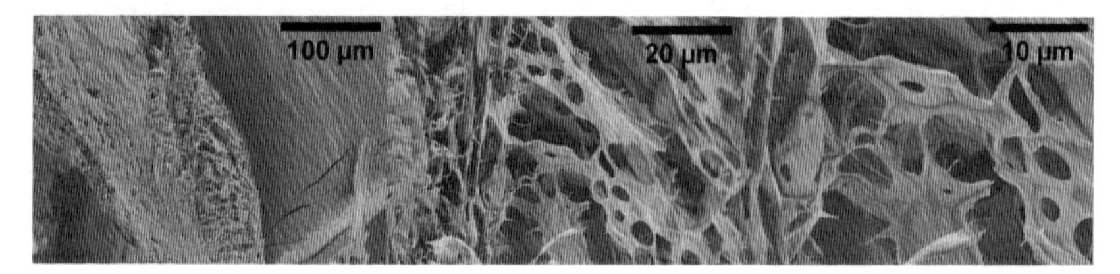

Figure 108: freeze-dried samples of high viscosity materials prepared from 2 wt% **54** in water (sponge-like network revealed).

5.4 Symmetric chiral TFAB gelator

5.4.1 Synthesis

Methyl (*S*)-3-(4-amino-3,5-difluorophenyl)-2-((*tert*-butoxycarbonyl)amino)propanoate (97)

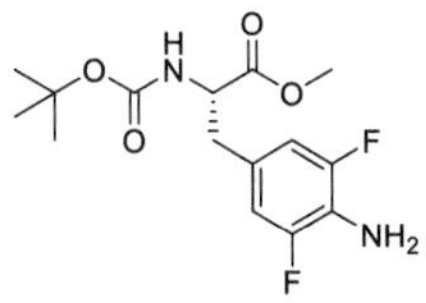

Dry *N,N*-dimethylformamide (18.2 mL) was added to zinc dust (1.79 g, 27.3 mmol, 3.00 eq) under an argon atmosphere followed by molecular iodine (347 mg, 1.37 mmol, 0.150 eq) and the mixture was stirred until the solution turned clear again. Methyl (2*R*)-3-iodo-2-[(2-methylpropan-2-yl)oxycarbonylamino]propanoate (3.00 g, 9.11 mmol, 1.00 eq) was added, followed by molecular iodine (347 mg, 1.37 mmol, 0.150 eq) and the solution was stirred for 15 min until it had cooled down to room temperature again. $Pd_2(dba)_3$ (209 mg, 228 µmol, 0.0250 eq), *S*-Phos (187 mg, 456 µmol, 0.0500 eq) and 4-bromo-2,6-fluoroaniline (2.66 g, 11.8 mmol, 1.30 eq) were added and the reaction mixture was stirred under an argon atmosphere for 3 d at room temperature. The solution was subsequently filtered over Celite®, concentrated under reduced pressure and the residue was purified by silica column chromatography (cH:EtOAc 5:1 with 1% Et_3N, R_f = 0.18) to yield 2.33 g of the product **97** (7.07 mmol, 78%) as light brown solid.

^{1}H NMR (400 MHz, CDCl$_3$): δ = 6.66 – 6.53 (m, 2H), 4.99 (d, *J* = 8.2 Hz, 1H), 4.50 (q, *J* = 6.5 Hz, 1H), 3.73 (s, 3H), 3.66 (s, 2H), 3.05 – 2.87 (m, 2H), 1.43 (s, 9H) ppm. **^{13}C NMR (101 MHz, CDCl$_3$)**: δ = 172.2, 155.2, 153.2 (d, *J* = 8.3 Hz), 150.8 (d, *J* = 8.4 Hz), 124.1 (d, *J* = 236.9 Hz), 112.0 (dd, *J* = 14.9, 7.2 Hz), 80.2, 54.5, 52.5, 37.6, 28.4 ppm. **^{19}F NMR (376 MHz, CDCl$_3$):** δ = -132.23 ppm. **TLC:** R_F = 0.18 developed in cH:EtOAc 5:1 with 1% Et_3N. **HRMS (FAB):** *m/z* calcd. for $C_{15}H_{20}N_2O_4F_2$ [M$^+$] = 330.1386 Da, found 330.1384 Da (Δ = - 0.37 ppm). **IR (ATR, $\tilde{\nu}$)** = 3455 (w), 3360 (s), 3002 (vw), 2990 (w), 2978 (w), 2953 (vw), 2932 (vw), 1731 (vs), 1687 (vs), 1676 (vs), 1650 (w), 1612 (w), 1591 (m), 1519 (vs), 1449 (s), 1438 (s), 1394 (w), 1368 (m), 1349 (m), 1332 (m), 1316 (m), 1281 (vs), 1248 (vs), 1221 (s), 1156 (vs), 1061 (m), 1051 (s), 1020 (vs), 990 (s), 962 (vs), 894 (w), 861 (w), 841 (vs), 813 (w), 786 (w), 776 (w), 761 (m), 737 (m), 721 (w), 623 (vs), 594 (vs), 579 (vs), 509 (m), 484 (m), 473 (m), 467 (m), 436 (w), 429 (w), 411 (s), 385 (m), 375 (m) cm^{-1}.

Dimethyl 3,3'-(((*E*)-diazene-1,2-diyl)bis(3,5-difluoro-4,1-phenylene))(2*S*,2'*S*)-bis(2-((*tert*-butoxycarbonyl)amino)propanoate) (98)

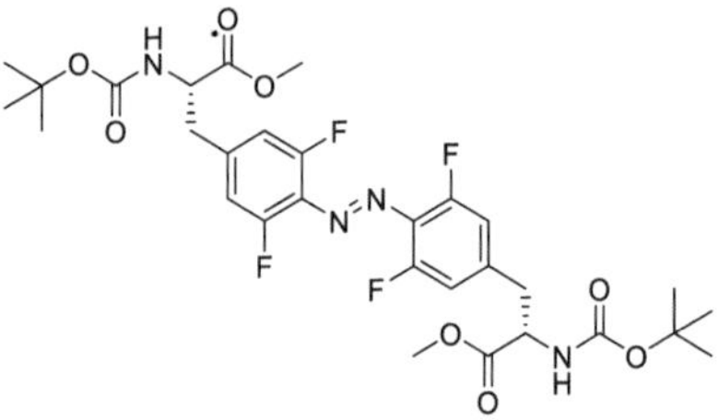

Methyl (*S*)-3-(4-amino-3,5-difluorophenyl)-2-((*tert*-butoxycarbonyl)amino)propanoate (**97**) (1.00 g, 3.02 mmol, 1.00 eq) was dissolved in 50 mL of dry DCM and 904 µL DBU (922 mg, 6.05 mmol, 2.00 eq) was added. The solution was stirred for 5 min at room temperature and then cooled down to −78 °C. NCS (808 mg, 6.05 mmol, 2.00 eq) was added in small portions and the solution was stirred for 10 min. The mixture was quenched with 100 mL aqueous $NaHCO_3$ solution. The organic layer was separated, washed with water (50 mL) and 1 M aqueous HCl (50 mL), and then dried over Na_2SO_4. The drying agent was filtered off and the solvent was removed under reduced pressure. Silica gel column chromatography was performed (cH:EtOAc 7:3) to yield 456 mg of a bright orange solid (708 µmol, 47%).

^{1}H NMR (400 MHz, CDCl3): δ = 6.85 (d, *J* = 9.8 Hz, 4H), 5.09 (d, *J* = 7.9 Hz, 2H), 4.61 (q, *J* = 6.5 Hz, 2H), 3.76 (s, 6H), 3.20 (dd, *J* = 13.8, 5.7 Hz, 2H), 3.07 (dd, *J* = 13.7, 6.0 Hz, 2H), 1.44 (s, 18H) ppm. **^{13}C NMR (101 MHz, $CDCl_3$)**: δ = 171.6,

156.9 (d, J = 4.8 Hz), 155.1, 154.3 (d, J = 4.6 Hz), 141.6, 130.7, 115.4 – 111.7 (m), 80.6, 54.1, 52.7, 38.4, 28.4. ppm. **^{19}F NMR (376 MHz, $CDCl_3$):** δ = -120.53, -120.79 ppm. **TLC:** R_F = 0.25 developed in cH:EtOAc 7:3. **HRMS (ESI):** *m/z* calcd. for $C_{30}H_{36}N_4O_8F_4$ $[M+Na]^+$= 679.6117 Da, found 679.2362 Da (Δ= - 0.26 ppm). **IR (ATR, ṽ)** = 3320 (w), 2982 (w), 2953 (w), 2939 (w), 2931 (w), 1748 (s), 1732 (s), 1681 (vs), 1626 (s), 1578 (m), 1517 (vs), 1438 (vs), 1394 (w), 1368 (m), 1350 (s), 1330 (m), 1292 (vs), 1272 (vs), 1254 (s), 1235 (m), 1225 (s), 1214 (s), 1160 (vs), 1061 (vs), 1050 (vs), 1040 (vs), 1013 (s), 1003 (s), 973 (w), 925 (w), 894 (w), 864 (w), 844 (vs), 813 (w), 790 (w), 781 (w), 758 (w), 742 (w), 730 (w), 704 (w), 684 (w), 676 (w), 664 (w), 637 (w), 615 (m), 598 (m), 558 (w), 547 (w), 511 (w), 494 (w), 470 (w), 448 (w), 432 (w), 418 (w) cm^{-1}. **UV-Vis (MeCN):** λ_{max} = 320 nm, 456 nm.

(2*S*,2'*S*)-3,3'-(((*E*)-diazene-1,2-diyl)bis(3,5-difluoro-4,1-phenylene))bis(2-((*tert*-butoxycarbonyl)amino)propanoic acid) (99)

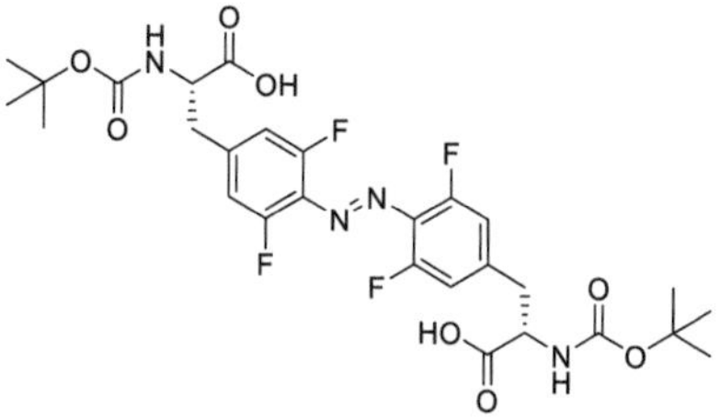

To a solution of compound **98** (400 mg, 609 µmol, 1.00 eq) in 1,4-dioxane (9 mL) was added an excess of lithium hydroxide (875 mg, 36.6 mmol, 60.0 eq) in water (9 mL) and the reaction mixture was stirred for 15 min at room temperature. The reaction was acidified by adding an aq. 2 M solution of HCl (ca. 20 mL). The crude product was extracted with EtOAc, washed with brine and dried over Na_2SO_4. The solvent was evaporated under reduced pressure and the crude product was purified by silica gel column chromatography (DCM:FA 99:1 to DCM:MeOH:FA 97:2:1) to yield 200 mg of the product **99** (318 µmol, 52%) as an orange solid.

^{1}H NMR (400 MHz, DMSO-*d*$_6$): δ = 12.79 (br s, 2H), 7.25 (d, *J* = 11.2 Hz, 4H), 7.19 (d, *J* = 8.6 Hz, 2H), 4.22 (ddd, *J* = 10.5, 8.5, 4.5 Hz, 2H), 3.16 (dd, *J* = 13.7, 4.5 Hz, 2H), 2.91 (dd, *J* = 13.7, 10.7 Hz, 2H), 1.31 (s, 18H). **^{13}C NMR (101 MHz, DMSO-*d*$_6$):** δ = 172.9, 155.5 (d, *J* = 4.2 Hz), 155.4, 153.0 (d, *J* = 4.5 Hz), 144.9, 129.1, 113.9 (d, *J* = 20.4 Hz), 78.2, 54.2, 36.3, 28.0 ppm. **^{19}F NMR (376 MHz, DMSO-*d*$_6$):** δ = -122.53 ppm. **TLC:** R_F = 0.13 developed DCM:MeOH:FA 97:2:1. **HRMS (FAB):** *m/z* calcd. for $C_{28}H_{33}N_4O_8F_4$ $[M+H]^+$ = 629.2229 Da, found 629.2228 Da (Δ= - 0.09 ppm). **IR (ATR, $\tilde{\nu}$)** = 3352 (w), 2990 (w), 2938 (w), 1707 (s), 1681 (vs), 1649 (w), 1626 (s), 1577 (m), 1519 (vs), 1436 (s), 1392 (w), 1368 (m), 1347 (w), 1319 (m), 1296 (m), 1268 (s), 1256 (s), 1230 (s), 1162 (vs), 1061 (s), 1050 (s), 1035 (s), 1017 (s), 990 (w), 952 (w), 931 (w), 902 (w), 867 (w), 846 (s), 779 (w), 752 (w), 735 (w), 708 (w), 632 (s), 619 (s), 611 (s), 588 (m), 562 (w), 510 (m), 486 (w), 460 (w), 443 (w), 425 (w), 404 (w), 392 (w), 377 (m) cm^{-1}.

Methyl *N*^{6}-(*tert*-butoxycarbonyl)-*N*^{2}-((*R*)-2-((*tert*-butoxycarbonyl)amino)-3-(4-((*E*)-(4-((*R*)-2-((*tert*-butoxycarbonyl)amino)-3-(((*R*)-6-((*tert*-butoxycarbonyl)amino)-1-methoxy-1-oxohexan-2-yl)amino)-3-oxopropyl)-2,6-difluorophenyl)diazenyl)-3,5-difluorophenyl)propanoyl)-L-lysinate (100)

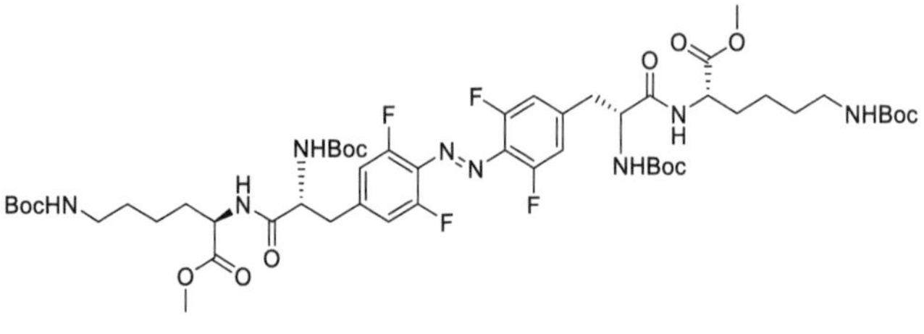

Starting material **99** (500 mg, 795 μmol, 1.00 eq), HBTU (603 mg, 1.59 mmol, 2.00 eq) and DIPEA (257 mg, 346 μL, 1.99 mmol, 2.50 eq) were dissolved in anhydrous *N,N*-dimethylformamide (2.75 mL) and stirred for 10 min at 20 °C under an argon atmosphere. H-Lys(Boc)-OMe·HCl (475 mg, 1.60 mmol, 2.01 eq) and DIPEA (257 mg, 346 μL, 1.99 mmol, 2.50 eq) were added and the resulting mixture was stirred for 3 h under an argon atmosphere until the starting material was consumed. The reaction was quenched with sat. NH_4Cl solution and extracted with EtOAc. The organic phase was washed with sat. NH_4Cl (3x) and brine (1x) and was dried over Na_2SO_4. The drying agent was filtered off and the solvent was removed under reduced pressure. The crude product was then purified by silica column chromatography (cH:EtOAc 1:1 to cH:EtOAc 1:4) to yield 640 mg of the product **100** (575 µmol, 72%) as a bright orange solid.

^{1}H NMR (400 MHz, DMSO-*d*$_6$): δ = 8.31 (d, *J* = 7.5 Hz, 2H), 7.26 (d, *J* = 11.3 Hz, 4H), 7.02 (d, *J* = 8.8 Hz, 2H), 6.76 (t, *J* = 5.7 Hz, 2H), 4.27 (dtd, *J* = 15.8, 9.2, 8.1, 4.8 Hz, 4H), 3.62 (s, 6H), 3.05 (dd, *J* = 13.5, 4.3 Hz, 2H), 2.90 (q, *J* = 6.4 Hz, 4H), 2.82 (dd, *J* = 13.6, 10.4 Hz, 2H), 1.83 – 1.55 (m, 4H), 1.36 (s, 12H), 1.30 (s, 9H), 1.29 (t, *J* = 14.8 Hz, 44H) ppm. **^{13}C NMR (101 MHz, DMSO-*d*$_6$)**: δ = 171.2, 155.5, 155.2, 153.0, 113.9 (d, *J* = 20.1 Hz), 78.2, 77.3, 54.8, 52.0, 51.8, 37.3, 30.6, 29.0, 28.2, 28.0, 22.6 ppm. **^{19}F NMR (376 MHz, DMSO-*d*$_6$):** δ = -122.59 ppm. **TLC:** R_F = 0.18 developed in cH:EtOAc 1:1. **HRMS (ESI):** *m/z* calcd. for $C_{52}H_{76}N_8O_{14}F_2$ $[M+Na]^+$= 1135.5309 Da, found 1135.5309 Da. **IR (ATR, ṽ)** = 3322 (w), 2975 (w), 2932 (w), 2860 (vw), 1740 (w), 1683 (vs), 1655 (vs), 1628 (m), 1577 (w), 1519 (vs), 1441 (s), 1390 (m), 1366 (s), 1323 (m), 1269 (s), 1249 (vs), 1228 (s), 1215 (s), 1163 (vs), 1044 (s), 1024 (m), 1013 (s), 849 (m), 779 (w), 751 (w), 738 (w), 637 (s), 613 (m), 598 (m), 569 (m), 534 (w), 513 (m), 496 (w), 486 (w), 470 (w), 459 (w), 449 (w), 441 (w), 433 (w), 421 (w), 409 (w), 394 (w), 381 (w) cm^{-1}.

Methyl ((*R*)-2-amino-3-(4-((*E*)-(4-((*R*)-2-amino-3-(((*R*)-6-amino-1-methoxy-1-oxohexan-2-yl)amino)-3-oxopropyl)-2,6-difluorophenyl)diazenyl)-3,5-difluorophenyl)propanoyl)-L-lysinate (101)

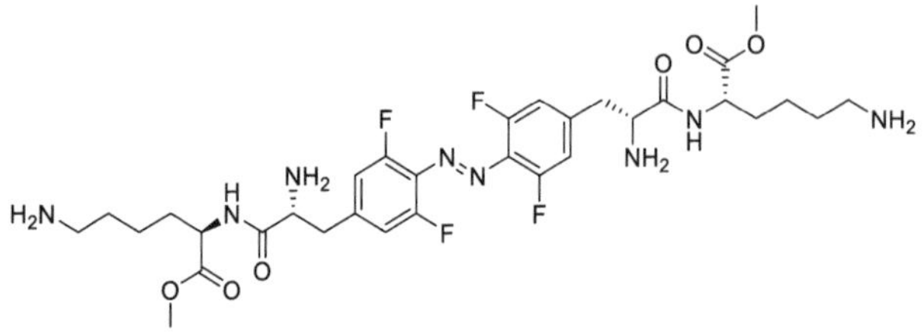

Compound **100** (500 mg, 449 µmol, 1.00 eq) was dissolved in dichloromethane (22.5 mL). 2,2,2-trifluoroacetic acid (22.5 mL) and triisopyropylsilane (1% v/v) were added at 20 °C. The mixture was stirred for 1 h. Subsequently, the solvent was co-evaporated with toluene to yield the product as the TFA salt of **101** as a dark red solid (320 mg, 449 µmol, quant.).

^{1}H NMR (400 MHz, DMSO-*d*$_6$): δ = 8.99 (d, *J* = 7.4 Hz, 2H), 8.32 (d, *J* = 5.2 Hz, 6H), 7.79 (s, 6H), 7.31 – 7.09 (m, 4H), 4.37 – 4.27 (m, 2H), 4.24 – 4.18 (m, 2H), 3.64 (s, 6H), 3.23 (dd, *J* = 14.1, 5.5 Hz, 2H), 3.10 (dd, *J* = 14.1, 7.8 Hz, 2H), 2.83 – 2.71 (m, 4H), 1.76 (ddd, *J* = 14.0, 9.2, 6.1 Hz, 2H), 1.65 (qd, *J* = 8.8, 4.7 Hz, 2H), 1.59 – 1.48 (m, 4H), 1.46 – 1.29 (m, 4H) ppm. **^{13}C NMR (101 MHz, DMSO-*d*$_6$):** δ = 171.8, 167.8, 158.4 (q, *J* = 35.7 Hz), 155.8, 114.5 (d, *J* = 2.7 Hz), 114.3 (d, *J* = 4.2 Hz), 52.7, 52.1, 52.0, 38.5, 36.4, 30.4, 26.5, 22.1 ppm. **^{19}F NMR (376 MHz, DMSO-*d*$_6$):** δ = -122.17 ppm. **HRMS (FAB):** *m/z* calcd. for $C_{32}H_{45}N_8O_6F_4$ $[M+H]^+$ = 713.3393 Da, found 713.3394 Da (Δ= 0.20 ppm). **IR (ATR, ṽ)** = 3072 (w), 2949 (w), 2874 (w), 1778 (w), 1730 (w), 1664 (vs), 1628 (s), 1578 (s), 1561 (m), 1527 (m), 1477 (w), 1441 (m), 1400 (w), 1361 (w), 1307 (w), 1177 (vs), 1135 (vs), 1044 (s), 1000 (m), 969 (m), 878 (w), 840 (s), 798 (s), 722 (s), 705 (s), 649 (m), 635 (m), 598 (m), 579 (m), 551 (m), 516 (m), 484 (m), 479 (m), 467 (m), 438 (m), 414 (m), 397 (m), 380 (m) cm^{-1}.

(3*S*,3'*S*,6*S*,6'*S*)-6,6'-((((*E*)-diazene-1,2-diyl)bis(3,5-difluoro-4,1-phenylene))-bis(methylene))bis(3-(4-aminobutyl)piperazine-2,5-dione) (55)

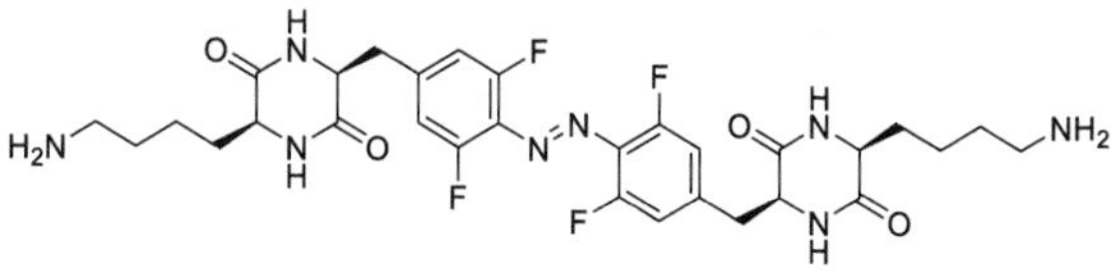

Compound **101** (836 mg, 715 µmol, 1.00 eq) was dissolved in butan-2-ol (65 mL). Subsequently, AcOH (336 mg, 320 µL, 5.59 mmol, 7.82 eq), 4-methylmorpholine (190 mg, 209 µL, 1.88 mmol, 2.63 eq) and DIPEA (285 mg, 384 µL, 2.20 mmol, 3.08 eq) were added. The mixture was heated to 120 °C for 2 h, then cooled down to room temperature. The solvent was removed under reduced pressure. The resulting residue was washed three times with cold MeCN to yield 435 mg of the product **55** (671 µmol, 94%) as an orange powdery solid.

^{1}H NMR (400 MHz, DMSO-*d*$_6$): δ = 8.22 (s, 4H), 7.66 (s, 4H), 7.17 (d, J = 11.1 Hz, 4H), 4.33 (d, J = 5.5 Hz, 2H), 3.82 (d, J = 5.6 Hz, 2H), 3.16 (dd, J = 13.7, 5.1 Hz, 2H), 3.06 (dd, J = 13.7, 5.2 Hz, 2H), 2.68 – 2.60 (m, 4H), 1.49 (td, J = 12.9, 9.8, 5.5 Hz, 2H), 1.38 (dt, J = 14.0, 7.1 Hz, 6H), 1.15 – 0.95 (m, 4H) ppm. **^{13}C NMR (101 MHz, DMSO-*d*$_6$)**: δ = 167.4, 166.5, 155.5, 153.0, 143.4, 129.4, 118.8, 114.6 (d, J = 20.8 Hz), 54.7, 53.5, 38.5, 37.8, 32.0, 26.6, 20.5 ppm. **^{19}F NMR (376 MHz, DMSO-*d*$_6$):** δ = -122.57 ppm. **HRMS (FAB):** *m/z* calcd. for $C_{30}H_{37}N_8O_4F_4$ $[M+H]^+$= 649.2868 Da, found 649.2867 Da (Δ= - 0.19 ppm). **IR (ATR, ṽ)** = 3186 (w), 3050 (w), 3041 (w), 2961 (w), 2953 (w), 2941 (w), 2931 (w), 2914 (w), 2894 (w), 2871 (w), 1666 (vs), 1626 (vs), 1575 (s), 1554 (m), 1545 (m), 1536 (m), 1528 (m), 1446 (vs), 1329 (s), 1255 (w), 1225 (w), 1198 (s), 1177 (s), 1132 (vs), 1043 (s), 936 (w), 914 (w), 834 (s), 799 (s), 771 (s), 764 (s), 747 (s), 721 (vs), 666 (m), 654 (s), 616 (m), 598 (m), 581 (m), 571 (m), 514 (s), 494 (s), 483 (s), 462 (s), 453 (s), 433 (s), 419 (s), 399 (s), 387 (s), 380 (m) cm^{-1}.

5.4.2 Gelation experiments

To a 1.5 mL-vial (crimp top, 12×32 mm) was added the photochromic material **55** as a powder and 500 µL of the appropriate aqueous solution as solvent. This suspension was sonicated for 2 min, followed by heating to 80 °C in a vial block. After equilibration at 80 °C for 5 min, the sample was heated to the boiling point by a heat gun. The results are described in Table 15.

Table 15: Hydrogelation attempts of **55** under various conditions. No gelation was observed under the tested conditions.

Mass of **55** in 500 µL of solvent	Approx. conc.	Solvent	Comment
10	2 wt%	H_2O	liquid
10	2 wt%	Ringer's solution	liquid
10	2 wt%	PBS	liquid, not completely dissolved
10	2 wt%	PBS, pH = 6	liquid, not completely dissolved
10	2 wt%	PBS, pH = 8	liquid
10	2 wt%	PBS, pH = 10	viscous, not completely dissolved
5	1 wt%	PBS, pH = 10	viscous, not completely dissolved
2.5	0.5 wt%	PBS, pH = 10	viscous

5.5 Symmetrical chiral gelator – model system for the addition of dopants

5.5.1 Synthesis

A. Leistner, D. G. Kistner, C. Fengler and Z. L. Pianowski, *RSC Adv.*, 2022, **12**, 4771.

DOI: 10.1039/D1RA09218A

Reversible photodissipation of composite photochromic azobenzene-alginate supramolecular hydrogels

Author contributions

The following syntheses towards compound **56**, PSS determination, UV-Vis spectroscopy, Gelation experiments (Table 16 and Table 18) and light induced gel-to-sol transition were performed by David Georg Kistner in his bachelor thesis under the supervision of Anna-Lena Leistner. The complete characterization and interpretation was done by Anna-Lena Leistner. Rheological and NMR relaxation experiments were performed by Christian Fengler.

L-Phe-(4-NO_2)-OH, (*S*)-4-Nitrophenylalanine (102)

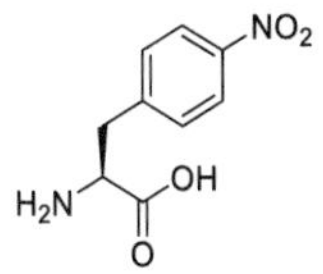

To stirring concentrated sulfuric acid (40 mL) at 50 °C was added L-Phenylalanine (**70**, 30.0 g, 182 mmol, 1.00 eq). After the starting material was dissolved, the mixture was cooled to 0 °C and concentrated nitric acid (29.2 g, 21.0 mL, 463 mmol, 2.55 eq) was added dropwise over the course of 2 h. The reaction mixture was then stirred for another 30 min at 0 °C and then allowed to warm to 20 °C and stirred for another 1 h. The reaction mixture was poured onto ice/water (200 mL), and 25% aq. ammonia solution was added to adjust the pH to 7-8 and precipitate the crude product. The precipitate was collected by filtration, washed three times with cooled water (20 mL) and dried. The crude product was recrystallized from water and dried under high vacuum to give 23.1 g of colorless crystals (110 mmol, 61%)

^{1}H NMR (400 MHz, D_2O with KOH): δ = 8.17 (d, J = 8.6 Hz, 1H), 7.44 (d, J = 8.6 Hz, 1H), 3.55 (t, J = 5.6 Hz, 1H), 3.07 (dd, J = 13.4, 5.9 Hz, 1H), 2.97 (dd, J = 13.4, 7.2 Hz, 1H) ppm. **^{13}C NMR (101 MHz, D_2O with KOH):** δ = 181.8, 146.8, 146.4, 130.2, 123.6, 57.3, 40.8 ppm. **TLC:** R_f = 0.45 (developed in 79% DCM, 20% MeOH, 1% Et_3N). **HRMS (EI+):** *m/z* calcd. for $C_9H_{10}N_2O_4$ [M] = 210.0635 Da, found 210.0637 Da (Δ = 0.7 ppm). **IR (ATR, $\tilde{\nu}$)** = 3291 (w), 3272 (w), 3259 (w), 3245 (w), 3234 (w), 3210 (w), 3200 (w), 3193 (w), 3109 (w), 3084 (w), 3065 (w), 3053 (w), 3043 (w), 2997 (w), 2982 (w), 2944 (w), 2901 (w), 2885 (w), 2877 (w), 2776 (w), 2752 (w), 2738 (w), 2725 (w), 2646 (w), 1697 (w), 1643 (w), 1611 (s), 1568 (s), 1534 (vs), 1514 (vs), 1494 (m), 1442 (m), 1417 (s), 1344 (vs), 1312 (s), 1293 (m), 1242 (w), 1207 (w), 1191 (w), 1176 (w), 1140 (w), 1105 (m), 1071 (m), 1013 (w), 946 (w), 877 (m), 863 (s), 844 (w), 815 (w), 768 (m), 744 (s), 717 (m), 697 (vs), 653 (s), 630 (m), 615 (m), 567 (w), 524 (vs), 492 (s), 459 (w), 443 (w), 416 (m), 395 (w), 378 (w) cm^{-1}.

L-Phe-(4-NO_2)-OMe · HCl, (*S*)-4-Nitrophenylalanine methyl ester hydrochloride (103)

A suspension of 10.0 g L-Phe-(4-NO_2)-OH **102** (47.6 mmol, 1.00 eq.) in 40 mL methanol was cooled to 0 °C on an ice-water bath. Then 25.9 mL thionyl chloride (42.5 g, 357 mmol, 7.50 eq) were slowly added, the ice-water bath was removed, and the reaction mixture was stirred overnight at room temperature.

The solvent was removed under reduced pressure and the residue was dissolved again in 200 mL of methanol. The solution was added to vigorously stirred diethyl ether (200 mL) to precipitate the product, which was filtered off and washed with small amounts of diethyl ether. The product was dried under high vacuum yielding 9.39 g of a colorless powder (36.0 mmol, 76%).[209]

^{1}H NMR (400 MHz, DMSO-*d*$_6$): δ = 8.81 (s, 3H), 8.23 – 8.15 (m, 2H), 7.62 – 7.53 (m, 2H), 4.37 (dd, *J* = 7.5, 6.0 Hz, 1H), 3.68 (s, 3H), 3.40 – 3.23 (m, 2H) ppm. **^{13}C NMR (101 MHz, DMSO-*d*$_6$):** δ = 169.0, 146.8, 143.1, 131.0, 123.6, 52.7, 52.7, 35.3 ppm. **TLC:** R_f = 0.36 (developed in 97% DCM, 3% MeOH). **HRMS (EI+):** *m/z* calcd. for $C_{10}H_{12}N_2O_4$ [M] = 224.0792 Da, found 224.0791 Da (Δ = – 0.4 ppm). **IR (ATR, ṽ)** = 2982 (w), 2953 (w), 2907 (w), 2874 (w), 2847 (w), 1741 (vs), 1601 (w), 1541 (w), 1517 (s), 1506 (vs), 1490 (vs), 1451 (m), 1346 (vs), 1327 (m), 1309 (w), 1238 (vs), 1186 (m), 1146 (s), 1108 (m), 1060 (m), 980 (w), 949 (w), 932 (w), 868 (m), 858 (s), 844 (m), 812 (w), 751 (s), 741 (w), 700 (s), 654 (w), 507 (w), 490 (w), 405 (w) cm^{-1}.

Boc-L-Phe-(4-NO$_2$)-OMe, Methyl (*S*)-2-((*tert*-butoxycarbonyl)amino)-3-(4-nitrophenyl)propanoate (104)

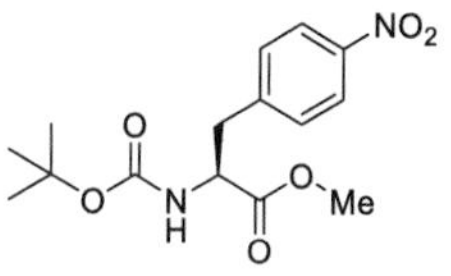

L-Phe-(4-NO$_2$)-OMe HCl **103** (8.50 g, 32.6 mmol, 1.00 eq) and 6.03 g of NaHCO$_3$ (71.7 mmol, 2.20 eq) were dissolved in 67 mL water. A solution of 7.83 g (Boc)$_2$O (35.9 mmol, 1.10 eq) in 67 mL 1,4-dioxane was added dropwise. The reaction was stirred for 21 h at room temperature and stopped when the reaction control *via* TLC showed full conversion. The solvent was removed under reduced pressure and the residue was then re-dissolved in 100 mL of water and 50 mL of EtOAc. The mixture was extracted three times with EtOAc (30 mL). The combined organic layers were washed with 5% aqueous KHSO$_4$ solution

(50 mL), 5% aqueous $NaHCO_3$ solution (20 mL), and brine (10 mL), then dried over Na_2SO_4. After filtration, the solvent was removed under reduced pressure. The product was dried under high vacuum yielding 9.89 g of a yellow powder (30.5 mmol, 94%).[210]

^{1}H NMR (400 MHz, $CDCl_3$): δ = 8.18 – 8.13 (m, 2H), 7.34 – 7.28 (m, 2H), 5.05 (d, J = 7.8 Hz, 1H), 4.63 (q, J = 6.8 Hz, 1H), 3.73 (s, 3H), 3.27 (dd, J = 13.8, 5.8 Hz, 1H), 3.12 (dd, J = 13.7, 6.4 Hz, 1H), 1.41 (s, 9H) ppm. **^{13}C NMR (101 MHz, $CDCl_3$):** δ = 171.6, 154.9, 147.5, 144.0, 130.3, 123.7, 80.4, 54.1, 52.5, 38.4, 28.3 ppm. **TLC:** R_f = 0.65 (developed in 50% cH, 50% EtOAc). **HRMS (FAB+):** *m/z* calcd. for $C_{15}H_{20}N_2O_6$ [M+H] = 325.1394 Da, found 325.1395 Da (Δ = 0.2 ppm). **IR (ATR, $\tilde{\nu}$)** = 3356 (w), 2983 (w), 1728 (s), 1687 (s), 1676 (s), 1605 (w), 1598 (w), 1517 (vs), 1460 (w), 1451 (w), 1438 (w), 1415 (vw), 1391 (w), 1368 (w), 1343 (vs), 1320 (m), 1298 (s), 1269 (vs), 1251 (s), 1231 (s), 1193 (m), 1156 (vs), 1102 (m), 1057 (m), 1051 (m), 1033 (m), 1013 (m), 994 (m), 970 (w), 931 (w), 887 (w), 857 (s), 840 (s), 819 (w), 795 (w), 775 (w), 752 (s), 731 (w), 700 (m), 652 (m), 625 (m), 608 (m), 551 (w), 524 (w), 514 (w), 492 (w), 466 (w), 436 (w), 416 (w), 398 (w) cm^{-1}.

Boc-L-Phe-(4-NH_2)-OMe, Methyl (*S*)-3-(4-aminophenyl)-2-((*tert*-butoxycarbonyl)amino)propanoate (105)

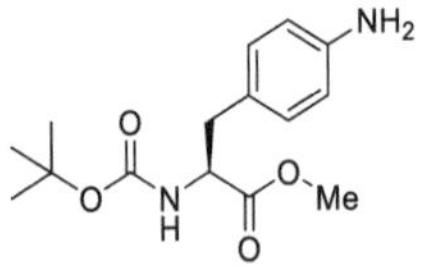

To a solution of 9.00 g Boc-L-Phe-(4-NO_2)-OMe **104** (27.7 mmol, 1.00 eq) in 50 mL MeOH was added 5% Pd/C (181 mg). The flask was set under vacuum and purged with argon three times. After evacuating the flask once more, hydrogen gas was added using a balloon. The reaction mixture was stirred vigorously. After 4 h and 24 h, additional 180 mg of 5% Pd/C were added. After 3 more hours, the reaction control indicated full conversion and the reaction mixture was filtered through Celite®. The solvent was removed under reduced pressure and the product was dried under high vacuum yielding 8.00 g of a highly viscous, orange oil (27.2 mmol, 98%).[211]

^{1}H NMR (400 MHz, CDCl$_3$): δ = 6.94 – 6.86 (m, 2H), 6.65 – 6.57 (m, 2H), 4.94 (d, J = 8.3 Hz, 1H), 4.50 (q, J = 6.6 Hz, 1H), 3.70 (s, 3H), 3.60 (br, 2H), 2.97 (q, J = 6.5, 4.7 Hz, 2H), 1.42 (s, 9H) ppm. **^{13}C NMR (101 MHz, CDCl$_3$):** δ = 172.7, 155.3, 145.5, 130.3, 125.8, 115.4, 80.0, 54.7, 52.3, 37.6, 28.5 ppm. **TLC:** R_f = 0.45 (developed in 50% cH, 50% EtOAc). **HRMS (FAB+):** *m/z* calcd. for $C_{15}H_{22}N_2O_4$ [M] = 294.1574 Da, found 294.1573 Da (Δ = – 0.4 ppm). **IR (ATR, $\tilde{\nu}$)** = 3437 (vw), 3424 (vw), 3411 (vw), 3366 (w), 3002 (vw), 2976 (w), 2952 (w), 2932 (w), 1737 (m), 1697 (vs), 1625 (m), 1517 (vs), 1500 (vs), 1438 (m), 1391 (m), 1366 (vs), 1278 (s), 1249 (s), 1215 (s), 1160 (vs), 1052 (s), 1016 (s), 992 (m), 922 (w), 856 (m), 824 (s), 802 (m), 778 (s), 759 (m), 728 (m), 654 (s), 637 (s), 626 (s), 608 (s), 565 (s), 534 (vs), 489 (s), 463 (s), 445 (s), 432 (s), 419 (s), 409 (s), 392 (s), 375 (s) cm^{-1}.

(*S*)-sym-(Boc)$_2$-PAP-OMe, Dimethyl 3,3'-(((*E*)-diazene-1,2-diyl)bis(4,1-phenylene))(2*S*,2'*S*)-bis(2-((*tert*-butoxycarbonyl)amino)propanoate) (106)

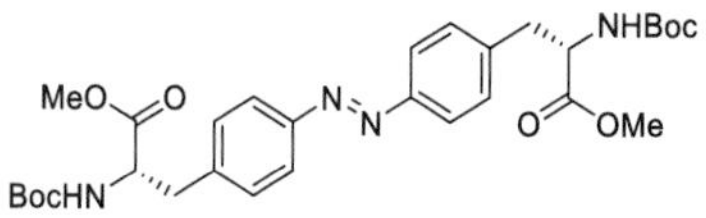

NHBoc
MeO O N N O OMe
BocHN

Boc-L-Phe-(4-NH$_2$)-OMe **105** (7.75 g, 26.4 mmol, 1.00 eq) was dissolved in 400 mL of dry DCM and 7.85 mL DBU (8.01 g, 52.6 mmol, 2.00 eq) was added. The solution was stirred for 5 min at room temperature and then cooled down to –78 °C. NCS (7.03 g, 52.6 mmol, 2.00 eq) was added in small portions and the solution was stirred for 10 min. The mixture was quenched with 300 mL of aqueous $NaHCO_3$ solution. The organic layer was separated, washed with water (100 mL) and 1 M aqueous HCl (100 mL), and then dried over Na_2SO_4. The drying agent was filtered off and the solvent was removed under reduced pressure. Silica gel column chromatography was performed (73% cH, 27% EtOAc, to 70% cH, 30% EtOAc) to yield 5.01 g of an orange solid (8.99 mmol, 68%).

^{1}H NMR (400 MHz, DMSO-*d*$_6$): δ = 7.80 (d, *J* = 8.2 Hz, 4H), 7.45 (d, *J* = 8.1 Hz, 4H), 7.38 (d, *J* = 8.2 Hz, 2H), 4.25 (ddd, *J* = 10.0, 8.1, 5.2 Hz, 2H), 3.63 (s, 6H), 3.10 (dd, *J* = 13.8, 5.1 Hz, 2H), 2.96 (dd, *J* = 13.8, 10.1 Hz, 2H), 1.32 (s, 18H) ppm. **^{13}C NMR (101 MHz, DMSO-*d*$_6$):** δ = 172.4, 155.4, 150.7, 141.5, 130.2, 122.4, 78.3, 54.9, 51.9, 36.3, 28.1 ppm. **TLC:** R_f = 0.1 (developed in 73% cH, 27% EtOAc). **HRMS (FAB+):** *m/z* calcd. for $C_{30}H_{40}N_4O_8$ [M] = 584.2841 Da, found 584.2840 Da (Δ = – 0.2 ppm). **IR (ATR, $\tilde{\nu}$)** = 3376 (w), 2982 (w), 2953 (w), 1759 (m), 1740 (s), 1691 (vs), 1680 (vs), 1604 (vw), 1514 (vs), 1458 (w), 1436 (m), 1422 (w), 1391 (w), 1368 (m), 1353 (w), 1329 (w), 1293 (s), 1248 (vs), 1218 (s), 1162 (vs), 1147 (vs), 1105 (m), 1057 (m), 1037 (m), 1024 (s), 1010 (s), 989 (m), 956 (w), 922 (w), 892 (w), 868 (w), 847 (m), 782 (w), 759 (w), 722 (w), 698 (w), 684 (w), 677 (w), 667 (w), 656 (w), 642 (w), 586 (m), 561 (vs), 527 (m), 492 (w), 479 (w), 466 (w), 439 (w), 404 (w), 399 (w) cm^{-1}. **UV-Vis (MeCN):** λ_{max} = 232 nm, 331 nm.

(*S*)-sym-(Boc)$_2$-PAP-OH, (2*S*,2'*S*)-3,3'-(((*E*)-diazene-1,2-diyl)bis(4,1-phenylene))bis(2-((*tert*-butoxycarbonyl)amino)propanoic acid) (107)

(*S*)-sym-(Boc)$_2$-PAP-OMe **106** (4.00 g, 6.84 mmol, 1.00 eq) was dissolved in 100 mL 1,4-dioxane and a solution of 9.83 g LiOH (409 mmol, 60 eq) in 100 mL water was added. After stirring for 15 min, 2 M aqueous HCl (225 mL) was added, and the mixture was extracted twice with EtOAc (250 mL). The combined organic layers were washed with brine (2×250 mL) and dried over anhydrous Na_2SO_4. After filtering off the drying agent, the solvent was removed under reduced pressure and the crude product was dried under vacuum. Silica gel column chromatography was performed starting with 1% FA in DCM, then gradually MeOH was added (1%, 1.5%, 2%, 3%) while keeping 1% FA in DCM. The product was obtained as an orange solid (3.37 g, 6.05 mmol, 88%).

^{1}H NMR (400 MHz, DMSO-*d*$_6$): δ = 7.80 (d, *J* = 8.1 Hz, 4H), 7.46 (d, *J* = 8.1 Hz, 4H), 7.17 (d, *J* = 8.4 Hz, 2H), 4.22 – 4.11 (m, 2H), 3.13 (dd, *J* = 13.8, 4.6 Hz, 2H), 2.93 (dd, *J* = 13.9, 10.4 Hz, 2H), 1.32 (s, 18H) ppm. **^{13}C NMR (101 MHz, DMSO-*d*$_6$):** δ = 173.4, 155.4, 150.7, 142.0, 130.2, 122.3, 78.1, 54.9, 36.3, 28.1, 27.8 ppm. **TLC:** R_f = 0.19 (developed in 1% FA, 1% MeOH, 98% DCM). **HRMS (FAB+):** *m/z* calcd. for $C_{28}H_{36}N_4O_8$ [M] = 556.2528 Da, found 556.2529 Da (Δ = 0.2 ppm **IR (ATR, ṽ)** = 3361 (w), 2983 (w), 2938 (w), 2931 (w), 1713 (s), 1686 (vs), 1604 (w), 1517 (vs), 1446 (w), 1421 (w), 1391 (m), 1367 (m), 1324 (m), 1305 (m), 1293 (m), 1249 (s), 1235 (s), 1154 (vs), 1105 (w), 1051 (m), 1026 (w), 1014 (m), 986 (w), 936 (m), 895 (w), 857 (s), 836 (s), 816 (w), 778 (m), 748 (m), 667 (w), 643 (m), 623 (m), 603 (s), 569 (vs), 526 (m), 504 (w), 496 (w), 473 (w), 466 (w), 435 (m), 422 (w), 411 (w), 404 (w), 390 (m), 384 (m) cm^{-1}. **UV-Vis (MeCN):** λ_{max} = 234 nm, 331 nm.

Sym-(Boc)Lys$_2$-(Boc)PAP-OMe, Dimethyl 2,2'-(((2*S*,2'*S*)-3,3'-(((*E*)-diazene-1,2-diyl)bis(4,1-phenylene))bis(2-((*tert*-butoxycarbonyl)amino)propanoyl))-bis(azanediyl))(2*S*,2'*S*)-bis(6-((*tert*-butoxycarbonyl)amino)hexanoate) (108)

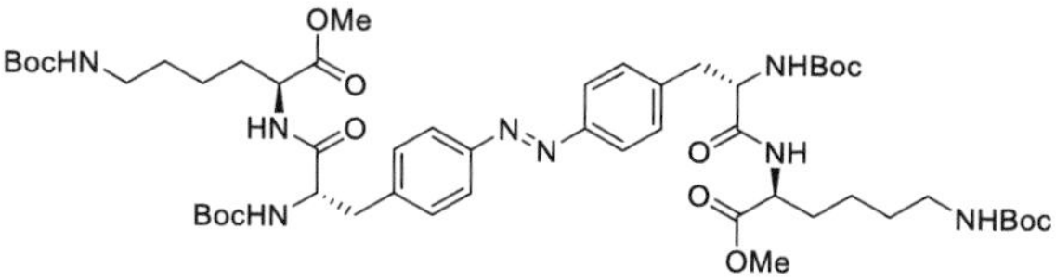

To a solution of 3.00 g (*S*)-sym-(Boc)$_2$-PAP-OH **107** (5.39 mmol, 1.00 eq) in anhydrous DMF (19 mL) were added 4.09 g HBTU (10.8 mmol, 2.00 eq) and 2.29 mL DIPEA (1.74 g, 13.5 mmol, 2.50 eq). The mixture was stirred for 10 min at room temperature under an argon atmosphere before 3.22 g methyl (2*S*)-2-amino-6-(*tert*-butoxycarbonylamino)hexanoate hydrochloride (H-L-Lys(Boc)-OMe·HCl, 10.8 mmol, 2.01 eq) and 2.29 mL DIPEA (1.74 g, 13.5 mmol, 2.50 eq)

were added. After stirring for 2.5 h under an argon atmosphere, the reaction was quenched by addition of 400 mL aqueous NH_4Cl solution which caused precipitation of an orange solid. The reaction mixture was extracted with 500 mL of EtOAc and the organic layer was washed with saturated aqueous NH_4Cl solution (2×400 mL) and with brine (400 mL). The organic layer was dried over Na_2SO_4, filtered and the solvent was removed under reduced pressure. The crude product was purified by silica gel column chromatography (R_f = 0.23 in 3% MeOH, 97% DCM, starting with pure DCM, then gradual addition of 0.5%, 1%, 2%, and 3% MeOH). The product **108** was obtained as an orange solid (2.97 g, 2.85 mmol, 53%).

^{1}H NMR (400 MHz, DMSO-*d$_6$*): δ = 8.31 (d, *J* = 7.5 Hz, 2H), 7.80 (d, *J* = 8.0 Hz, 4H), 7.49 (d, *J* = 8.1 Hz, 4H), 6.97 (d, *J* = 8.7 Hz, 2H), 6.77 (d, *J* = 6.0 Hz, 2H), 4.27 (qd, *J* = 9.3, 8.5, 4.7 Hz, 4H), 3.62 (s, 6H), 3.11 – 2.77 (m, 8H), 1.67 (dt, *J* = 34.1, 7.6 Hz, 4H), 1.36 (s, 18H), 1.29 (s, 18H), 1.47 – 1.10 (m, 8H) ppm. **^{13}C NMR (101 MHz, DMSO-*d$_6$*):** δ = 172.5, 171.7, 155.6, 155.2, 150.6, 142.0, 130.3, 122.2, 78.1, 77.3, 55.3, 51.9, 51.8, 38.2, 37.3, 30.7, 29.1, 28.3, 28.1, 22.6 ppm. **TLC:** R_f = 0.23 (developed in 3% MeOH, 97% DCM). **HRMS (EI+):** *m/z* calcd. for $C_{52}H_{80}N_8O_{14}$ [M+H] = 1041.5867 Da, found 1041.5851 Da (Δ = – 1.5 ppm). **IR (ATR, $\tilde{\nu}$)** = 3333 (w), 3327 (w), 2976 (w), 2931 (w), 2864 (vw), 1742 (m), 1683 (vs), 1655 (vs), 1514 (vs), 1455 (m), 1443 (m), 1391 (m), 1366 (s), 1330 (w), 1306 (m), 1289 (m), 1268 (s), 1248 (vs), 1211 (s), 1163 (vs), 1132 (s), 1106 (m), 1045 (m), 1014 (s), 864 (w), 844 (m), 809 (w), 778 (w), 758 (w), 734 (w), 708 (w), 642 (s), 635 (m), 623 (m), 619 (m), 575 (s), 558 (m), 527 (m), 507 (w), 493 (w), 486 (w), 475 (w), 463 (w), 455 (w), 436 (m), 397 (w), 380 (w) cm^{-1}. **UV-Vis (MeCN):** λ_{max} = 234 nm, 331 nm.

Sym-PAP-Lys-OMe·TFA$_n$, Dimethyl 2,2'-(((2*S*,2'*S*)-3,3'-(((*E*)-diazene-1,2-diyl)bis(4,1-phenylene))bis(2-aminopropanoyl))bis(azanediyl))(2*S*,2'*S*)-bis(6-aminohexanoate) bis(trifluoroacetate) (109)

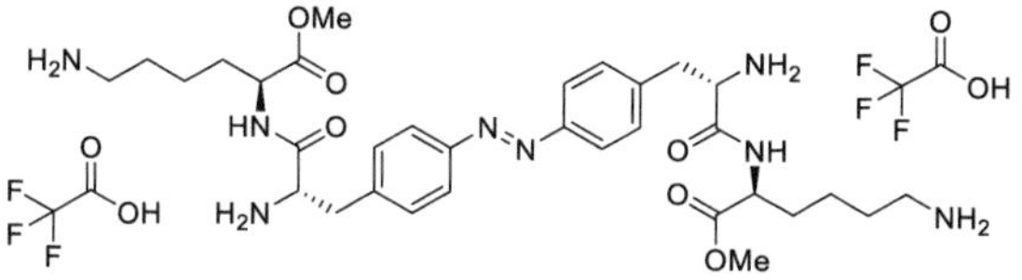

To a solution of 2.41 g sym-(Boc)Lys$_2$-(Boc)PAP-OMe **108** (2.31 mmol, 1.00 eq) in 110 mL DCM was added 110 mL TFA and 1.10 mL TIPS. The mixture was stirred at room temperature for 1 h before 100 mL of toluene were added. Approximately 80% of the solvent was removed under reduced pressure. Another 100 mL of toluene were added, and the solvent was evaporated under reduced pressure. After drying under high vacuum, the product was obtained as an orange solid (3.35 g, 2.31 mmol, 100%).

^{1}H NMR (400 MHz, DMSO-d_6): δ = 8.98 (d, J = 7.5 Hz, 2H), 8.46 – 8.20 (m, 6H), 7.86 (s, 6H), 7.84 (d, J = 8.3 Hz, 4H), 7.49 (d, J = 8.2 Hz, 4H), 4.31 (td, J = 8.3, 5.5 Hz, 2H), 4.16 (d, J = 7.6 Hz, 2H), 3.62 (s, 6H), 3.15 (ddd, J = 54.8, 14.1, 6.7 Hz, 4H), 2.77 (dp, J = 11.6, 5.8 Hz, 4H), 1.82 – 1.21 (m, 12H) ppm. **^{13}C NMR (101 MHz, DMSO-d_6):** δ = 171.7, 168.1, 158.4 (q, J = 33.7 Hz), 151.2, 138.6, 130.7, 122.7, 116.5 (q, J = 295.8 Hz), 53.1, 52.1, 51.9, 38.5, 36.7, 30.5, 26.6, 22.1 ppm. **HRMS (FAB+):** m/z calcd. for $C_{32}H_{48}N_8O_6$ [M+H] = 641.3770 Da, found 641.3771 Da (Δ = 0.3 ppm). **IR (ATR, $\tilde{\nu}$)** = 3162 (w), 3132 (w), 3099 (w), 3078 (w), 3067 (w), 3058 (w), 3050 (w), 3040 (w), 3024 (w), 3013 (w), 2989 (w), 2955 (w), 2945 (w), 2927 (w), 2910 (w), 2894 (w), 2885 (w), 2871 (w), 2816 (w), 1731 (w), 1664 (vs), 1560 (m), 1550 (m), 1545 (m), 1527 (m), 1500 (m), 1475 (w), 1438 (m), 1431 (m), 1424 (m), 1357 (w), 1307 (w), 1295 (w), 1252 (w), 1177 (vs), 1126 (vs), 1013 (m), 997 (m), 984 (m), 907 (w), 888 (w), 836 (s), 798 (vs), 742 (w), 721 (vs), 705 (s), 660 (m), 643 (w), 598 (m), 581 (m), 562 (m), 552 (m), 517 (m), 493 (w), 480 (m), 472 (m), 465 (m), 458 (m), 433 (m), 414 (m), 399 (m), 388 (m) cm^{-1}. **UV-Vis (H_2O):** λ_{max} = 335 nm, 426 nm. **UV-Vis (MeCN):** λ_{max} = 327 nm.

PAP-DKP-Lys$_2$, (3*S*,3'*S*,6*S*,6'*S*)-6,6'-((((*E*)-diazene-1,2-diyl)bis(4,1-phenylene))-bis(methylene))bis(3-(4-aminobutyl)piperazine-2,5-dione) bis(trifluoroacetate) (56)

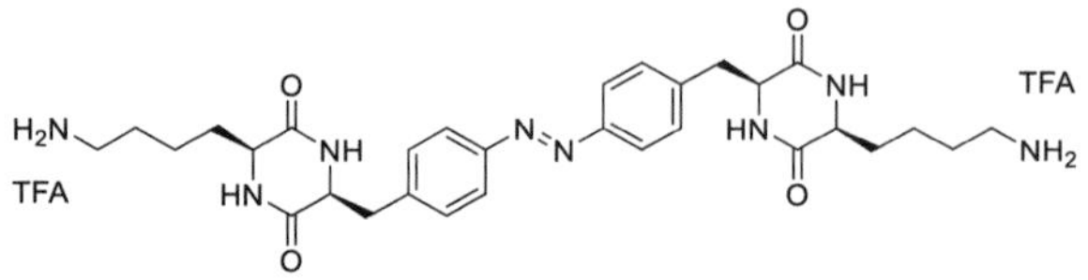

To a solution of the crude sym-PAP-Lys-OMe·TFA$_n$ **109** (3.14 g, 2.16 mmol, 1.00 eq) in 192 mL 2-butanol were added 966 µL glacial AcOH (1.01 g, 16.9 mmol, 7.82 eq), 631 µL *N*-methylmorpholine (575 mg, 5.68 mmol, 2.63 eq), and 1.13 mL DIPEA (860 mg, 6.65 mmol, 3.08 eq). The mixture was heated to reflux (120 °C) and stirred for 2 h during which an orange solid precipitated. The mixture was cooled down to room temperature and then concentrated by removing approximately half of the solvent under reduced pressure. After cooling down to room temperature, the solid was filtered off and washed with small amounts of ice-cold 2-butanol. The residue was dried under high vacuum to yield 1.33 g of an orange powder (1.65 mmol, 77%). For analytical purposes, purification was done by preparative HPLC with the following settings: 15 mL/min, 11 min gradient 20-30% MeCN in bidest. H_2O with 0.1% TFA, detection at 330 nm, retention time of 9.5 min. After lyophilization, the pure PAP-(DKP-Lys)$_2$ **(56)** was obtained as the TFA salt.

^{1}H NMR (400 MHz, DMSO-*d*$_6$): δ = 8.22 (d, *J* = 2.1 Hz, 2H), 8.12 (d, *J* = 2.2 Hz, 2H), 7.81 – 7.76 (m, 4H), 7.71 (s, 6H), 7.42 – 7.36 (m, 4H), 4.27 (td, *J* = 4.7, 2.3 Hz, 2H), 3.69 (t, *J* = 5.9 Hz, 2H), 3.21 (dd, *J* = 13.6, 4.4 Hz, 2H), 3.01 (dd, *J* = 13.6, 5.2 Hz, 2H), 2.65 – 2.53 (m, 4H), 1.36 – 1.19 (m, 6H), 1.06 – 0.86 (m, 6H) ppm. **^{13}C NMR (101 MHz, DMSO-*d*$_6$):** δ = 167.2, 166.5, 158.2 (q, *J* = 30.6 Hz), 151.0, 140.3, 131.3, 122.2, 55.2, 53.4, 38.5, 38.0, 32.4, 26.4, 20.6 ppm. **HRMS (FAB+):** *m/z* calcd. for $C_{30}H_{40}N_8O_4$ [M+H] = 577.3245 Da, found 577.3244 Da (Δ = – 0.3 ppm). **IR (ATR, ṽ)** = 3187 (w), 3180 (w), 3167 (w), 3148 (w), 3140 (w), 3089 (w), 3075 (w), 3048 (w), 3004 (w), 2961 (w), 2929 (w), 2894 (w), 1664 (vs), 1561 (w), 1543 (w), 1534 (w), 1523 (w), 1499 (w), 1459 (m),

1432 (m), 1334 (m), 1303 (w), 1200 (s), 1180 (s), 1130 (vs), 1016 (w), 915 (w), 834 (s), 799 (s), 772 (m), 721 (s), 694 (w), 639 (w), 630 (w), 612 (w), 601 (w), 575 (w), 550 (w), 527 (w), 518 (w), 473 (m), 459 (w), 438 (s), 432 (s), 418 (m), 387 (w), 380 (w) cm^{-1}. **UV-Vis (H_2O):** λ_{max} = 335 nm, 426 nm.

2,2',4,4',6,6'-hexafluoroazobenzene; F_6-AB (69)

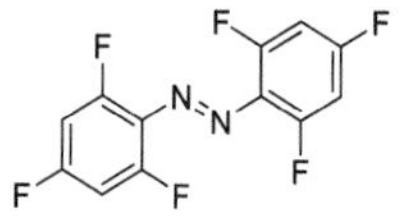

Synthesis according to Knie *et al.*[61a]

To a solution of 500 mg 2,4,6-trifluoroaniline (3.40 mmol, 1.00 eq) in 34 mL DCM was added a freshly ground mixture of $KMnO_4$ (1.70 g, 10.8 mmol, 3.16 eq.) and $FeSO_4 \cdot 7H_2O$ (1.70 g, 6.11 mmol, 1.80 eq). The mixture was heated to reflux (55 °C) and stirred overnight. It then was filtered through Celite®, washed with 200 mL DCM, and the filtrate was dried over anhydrous Na_2SO_4. After filtering off the drying agent, the solvent was removed under reduced pressure yielding orange crystals (R_f = 0.35 in 90% cH, 10% DCM). Silica gel column chromatography was performed to purify the product (100% cH to 82% cH, 18% DCM). The product was obtained as orange needle-like crystals yielding 110 mg (0.379 mmol, 22%).

^{1}H NMR (400 MHz, $CDCl_3$): δ = 6.89 – 6.78 (m, 4H) ppm. **^{13}C NMR (101 MHz, $CDCl_3$):** δ = 163.30 (dt, J = 255.5, 15.1 Hz), 156.65 (ddd, J = 263.4, 15.1, 6.6 Hz), 129.1 – 128.75 (m), 102.05 – 101.31 (m) ppm. **^{19}F NMR (376 MHz, $CDCl_3$):** δ = −102.65 (t, J = 8.2 Hz), −116.41 (d, J = 8.3 Hz) ppm. **HRMS (EI+):** *m/z* calcd. for $C_{12}H_4N_2F_6$ [M] = 290.0273 Da, found 290.0271 Da (Δ = – 0.19 ppm). **IR (ATR, ṽ)** = 3109 (w), 3089 (w), 2924 (w), 1638 (m), 1609 (s), 1591 (vs), 1516 (w), 1451 (vs), 1339 (vs), 1300 (m), 1221 (m), 1176 (s), 1126 (vs), 1064 (m), 1048 (vs), 1007 (s), 999 (vs), 847 (vs), 782 (s), 749 (m), 696 (w), 687 (w), 680 (w), 657 (m),

623 (s), 588 (w), 574 (w), 513 (s), 477 (w), 439 (vs), 397 (m), 375 (w) cm^{-1}. **UV-Vis (MeCN):** λ_{max} = 227 nm, 309 nm.

2,2',3,3',4,4',5,5',6,6'-decafluoroazobenzene; F_{10}-AB (22)

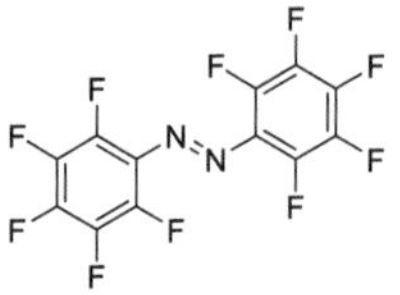

To a solution of 500 mg 2,3,4,5,6-pentafluoroaniline (2.73 mmol, 1.00 eq) in 34 mL DCM was added a freshly ground mixture of $KMnO_4$ (1.37 g, 8.65 mmol, 3.17 eq) and $FeSO_4 \cdot 7H_2O$ (1.37 g, 4.92 mmol, 1.80 eq). The mixture was heated to reflux (55 °C) and stirred overnight. It then was filtered through Celite®, washed with 200 mL DCM, and the filtrate was dried over anhydrous Na_2SO_4. After filtering off the drying agent, the solvent was removed under reduced pressure yielding orange crystals (R_f = 0.3 in 100% cH). Silica gel column chromatography was performed for purification (100% cH to 90% cH, 10% DCM. The product was obtained as orange needle-like crystals yielding 164 mg (0.453 mmol, 33%).

^{13}C NMR (101 MHz, $CDCl_3$): δ = 144.3, 142.8, 141.7, 140.2, 139.5, 136.9, 128.4, 29.9 ppm. **^{19}F NMR (376 MHz, $CDCl_3$):** δ =−148.1 – 148.3 (m), −148.3 – 148.5 (m), −160.0 – 162.9 (m) ppm. **HRMS (EI+):** *m/z* calcd. for $C_{12}N_2F_{10}$ [M] = 361.9896 Da, found 361.9897 Da. **IR (ATR, ṽ)** = 2951 (w), 2922 (w), 2853 (w), 1642 (w), 1503 (vs), 1405 (s), 1380 (m), 1319 (m), 1307 (m), 1269 (m), 1145 (m), 1111 (m), 1055 (w), 1033 (w), 1018 (m), 999 (vs), 976 (vs), 941 (vs), 921 (s), 854 (m), 798 (m), 779 (m), 747 (m), 738 (m), 703 (m), 670 (w), 654 (w), 645 (w), 609 (m), 560 (s), 527 (w), 511 (w), 499 (w), 489 (w), 480 (w), 458 (w), 441 (s), 418 (w), 398 (m), 384 (m) cm^{-1}. **UV-Vis (MeCN):** λ_{max} = 309 nm.

5.5.2 Photophysical properties

Photostationary states determined by ^{1}H NMR measurements

Photostationary states were determined by ^{1}H NMR measurements for compound **56** (16.7 mg/mL) equilibrated up to 30 min under the indicated light wavelength (λ_{max} of the respective LED light diode). For analysis, the NMR signals were assigned to the *E* and *Z* isomer, respectively. To determine the signals of the *Z* isomer, the dark spectrum was subtracted from the spectrum after 30 min irradiation at 365 nm (10 W LED).

(*E*)-PAP-DKP-Lys$_2$ (56): ^{1}H NMR (400 MHz, DMSO-*d*$_6$): δ = 8.22 (d, *J* = 2.1 Hz, 2H), 8.12 (d, *J* = 2.2 Hz, 2H), 7.81 – 7.76 (m, 4H), 7.71 (br s, 6H), 7.42 – 7.36 (m, 4H), 4.27 (tt, *J* = 4.9, 1.7 Hz, 2H), 3.69 (d, *J* = 5.9 Hz, 2H), 3.11 (ddd, *J* = 80.0, 13.6, 4.8 Hz, 4H), 2.58 (q, *J* = 6.7 Hz, 4H), 1.35 – 1.20 (m, 6H), 0.96 (tdt, *J* = 17.8, 7.9, 3.8 Hz, 6H) ppm.

(*Z*)-PAP-DKP-Lys$_2$ (56): ^{1}H NMR (400 MHz, DMSO-*d*$_6$): δ = 8.17 (d, *J* = 2.0 Hz, 2H), 8.11 (d, *J* = 2.1 Hz, 2H), 7.78 (br s, 6H), 7.12 – 7.05 (m, 4H), 6.76 – 6.71 (m, 4H), 4.18 (td, *J* = 4.6, 2.2 Hz, 2H), 3.68 (d, *J* = 5.9 Hz, 2H), 2.94 (ddd, *J* = 97.2, 13.6, 4.6 Hz, 4H), 2.69 (tq, *J* = 11.3, 5.7 Hz, 4H), 1.39 (p, *J* = 7.2 Hz, 4H), 1.34 – 1.21 (m, 2H), 1.10 – 0.90 (m, 6H) ppm.

Then three signals per isomer were identified that did not—or just barely—overlap with other signals. Those were integrated for each spectrum using the same intervals and divided by the number of protons. Each *E* isomer signal was assigned a *Z* isomer signal and the mole fraction of these pairs was calculated.

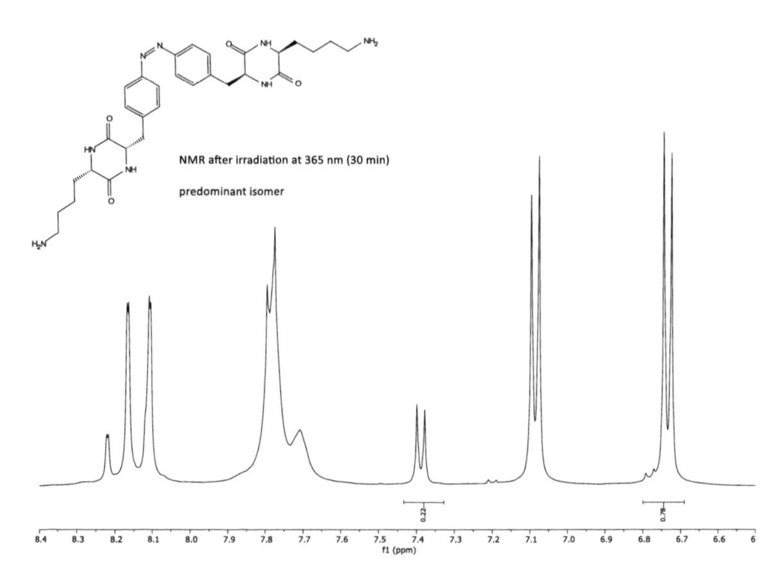

Figure 109: ^{1}H-NMR-spectrum (400 MHz, DMSO-d_6) of compound **56** after irradiation at 365 nm. A photostationary state of 76% *Z*-isomer is reached after 30 min irradiation time.

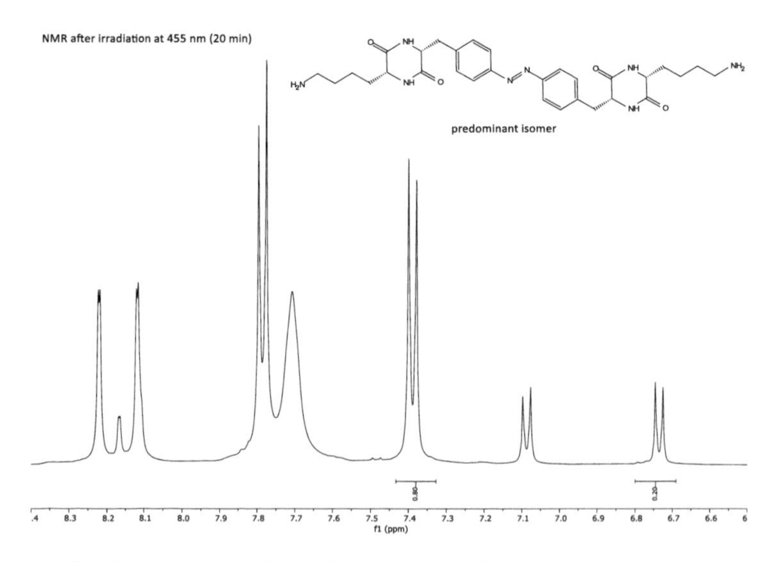

Figure 110: ^{1}H-NMR-spectrum (400 MHz, DMSO-d_6) of compound **56** after irradiation at 455 nm. A photostationary state of 20% *Z*-isomer is reached after 20 min irradiation time.

Thermal stability

The thermal stability was determined for compound **56**. Samples were prepared in H_2O AcOH and irradiated at 365 nm to yield a high PSS. The solution was kept at 20 °C (H_2O and AcOH) or at 60 °C (H_2O) and the isomer ratio was determined by HPLC in intervals. The obtained data was processed by calculating the ln(X_0/X_t), where X is the percentage of the respective *Z*-isomer and linear fitting (Equation (1)) of the obtained values. The calculated slope corresponds to the isomerization rate constant k which is used to calculate the half-life $t_{1/2}$.

(1) $$x_t = x_0 \cdot e^{-k \cdot t} \leftrightarrow ln\left(\frac{x_0}{x_t}\right) = k \cdot t$$

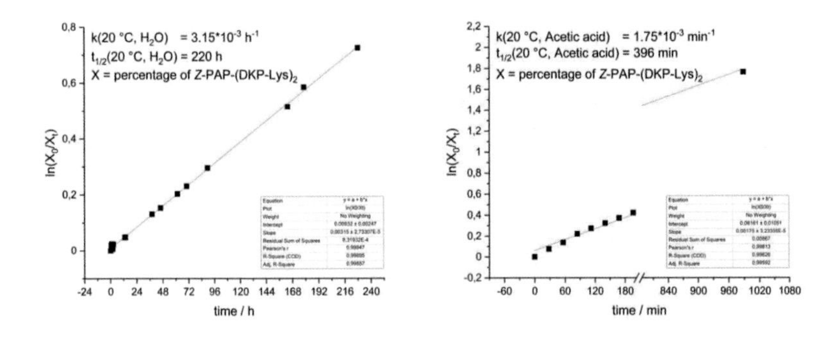

Figure 111: Linear fit of the decay of the *Z*-isomer of compound **56** at 20 °C in H_2O or AcOH respectively for first-order kinetics.

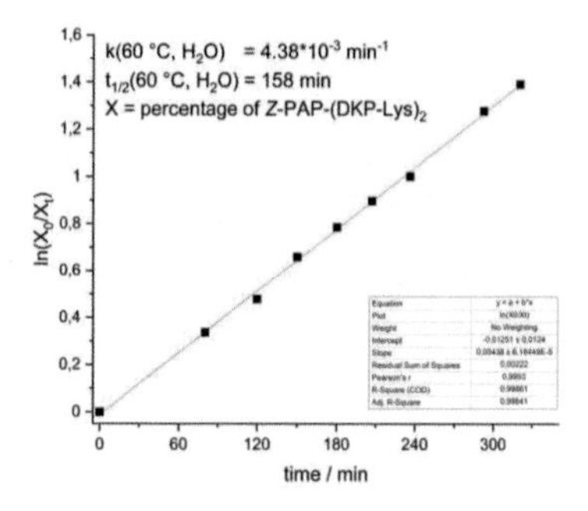

Figure 112: Linear fit of the decay of the *Z*-isomer of compound **56** at 60 °C in H_2O for first-order kinetics.

UV/Vis isomerization experiments

A 500 µM stock solution of compound **56** in MeCN was prepared and diluted to reach a final concentration of 50 µM. The cuvette with the sample was irradiated with light of varying wavelengths (365 nm, 455 nm) directly before the measurement.

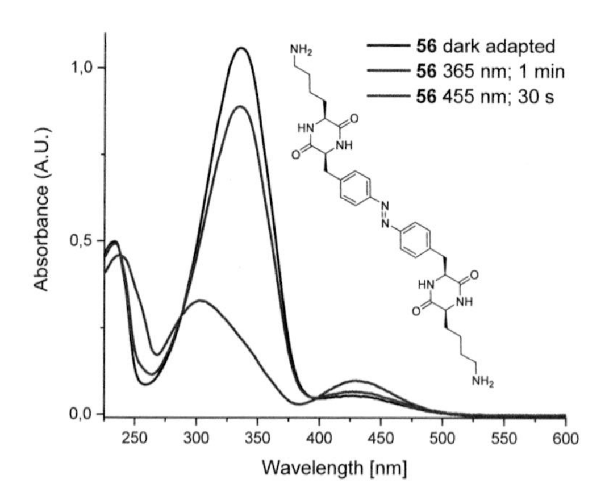

Figure 113: Absorption curves of compound **56** before and after irradiation. The shown curves after irradiation correspond to the photostationary states and distinct isosbestic points are displayed.

Isosbestic points are determined at 238, 287 and 397 nm. Only the isosbestic point at 287 nm was used for quantitative analysis *via* HPLC.

5.5.3 Gelation experiments

Method A

To a 1.5 mL-vial (crimp top, 12×32 mm) was added the photochromic material **56** (and the additives) as powder and 500 µL of the appropriate aqueous solution (Table 16-Table 19). This suspension was sonicated for 2 min, followed by heating to 80 °C in a vial block. After equilibration at 80 °C for 5 min, the sample was heated to the boiling point by a heat gun. The hot solution became completely clear and upon cooling the hydrogels formed. Successfully formed gels were prepared in triplicates.

The gel-to-sol transition temperature is characteristic for a particular gel composition and depends on the strength of the supramolecular self-assembly. Consequently, it can be used to estimate its relative stability in comparison to other gel samples. The value of this transition will be referred to as "melting temperature".

Melting temperatures were determined in triplicates. Gels were prepared as previously described. The vials were then put upside-down in a slowly stirred water bath (60 rpm) on a magnetic hotplate stirrer equipped with a thermometer. The water bath was heated (1.5 °C/min) until the gel started to melt and formed a sol.

Pure compound **56** formed stable gels in the range between 1.5 – 3.0 wt% in PBS buffer (pH 7.4). At lower concentration of 1.0 wt%, the gel was of low mechanical stability and collapsed upon shaking. The overall stability increased by addition of the additive sodium alginate. Here, stable gels were formed with the lowest concentration at 0.6 wt% gelator **56** and 0.6 wt% alginate. Though stable gels were also formed at lower gelator concentration of 0.4 wt% when higher amounts of alginate were added (0.8–1.2 wt%), these compositions are distinguished by their clear appearance.

Table 16: Hydrogelation experiments without additives by Method A.

Composition (x mg of **56**, 500 µL water)	Approx. conc.	Description	T_m °C
10	2.0 wt%	viscous orange liquid	-
2	0.4 wt%	viscous yellow liquid	-
1.5	0.3 wt%	viscous yellow liquid	-

Composition (x mg of **56**, 500 µL Ringer's solution)	Approx. conc.	Description	T_m °C
10	2.0 wt%	viscous orange liquid	-
2	0.4 wt%	viscous yellow liquid	-
1.5	0.3 wt%	viscous yellow liquid	-

Composition (x mg of **56** + 500 µL 200 mM aq. NaCl)	Approx. conc.	Description	T_m °C
10	2.0 wt%	viscous orange liquid	-
2	0.4 wt%	viscous yellow liquid	-
1.5	0.3 wt%	viscous yellow liquid	-

Composition (x mg of **56**, 500 µL PBS)	Approx. conc.	Description	T_m °C
15	3.0 wt%	opaque, orange gel	-
10	2.0 wt%	opaque, orange gel	78
7.5	**1.5 wt%**	**opaque, orange gel**	**77**
5	1.0 wt%	Unstable orange gel, sensitive to shaking	69
2	0.4 wt%	viscous yellow liquid	-
1.5	0.3 wt%	viscous yellow liquid	-

Table 17: Hydrogelation experiments with fluorinated azobenzenes as additives by Method A. *after 15 h equilibration time.

Composition (x mg + 500 µL PBS)		Approx. conc.		Description
Gelator **56**	F_6-AB **69**	Gelator **56**	F_6-AB **69**	
10	10	2.0 wt%	1 eq	inhomogeneous orange gel*
7.5	7.5	1.5%	1 eq	inhomogeneous orange gel*
5	5	1.0 wt%	1 eq	inhomogeneous yellow liquid
Gelator **56**	F_{10}-AB **22**	Gelator **56**	F_{10}-AB **22**	
10	10	2.0 wt%	1 eq	inhomogeneous orange gel*
7.5	7.5	1.5%	1 eq	inhomogeneous orange gel*
5	5	1.0 wt%	1 eq	viscous inhomogeneous yellow liquid
		Gelator **56**	Gelator **55**	
		0.96 wt%	1.05 wt%	viscous red liquid

Table 18: Hydrogelation experiments with sodium alginate as additive by Method A.

Composition (x mg + 500 µL H_2O)		Approx. conc.		Description	T_m °C
Gelator **56**	Alginate	Gelator **56**	Alginate		
5	5	1.0 wt%	1.0 wt%	opaque, orange gel	-

4	8	0.8%	1.6%	opaque, orange gel	-
4	4	0.8 wt%	0.8 wt%	opaque, orange gel	-
3	12	0.6 wt%	2.4 wt%	opaque, orange gel	-
3	9	0.6 wt%	1.8 wt%	almost clear, orange gel	79
3	6	0.6 wt%	1.2 wt%	almost clear, orange gel	76
3	**3**	**0.6 wt%**	**0.6 wt%**	**opaque, orange gel**	**83**
3	2	0.6 wt%	0.4 wt%	unstable orange gel, sensitive to shaking	-
3	1.5	0.6 wt%	0.3 wt%	unstable orange gel, sensitive to shaking	-
2	6	0.4 wt%	1.2 wt%	clear yellow gel	57
2	4	0.4 wt%	0.8 wt%	clear yellow gel	55
2	2	0.4 wt%	0.4 wt%	viscous yellow liquid	-
1.5	6	0.3 wt%	1.2 wt%	unstable yellow gel, sensitive to shaking	-
0	5	-	1.0 wt%	viscous colorless liquid	-
0	4	-	0.8 wt%	colorless liquid	-
0	3	-	0.6 wt%	colorless liquid	-

Table 19: Hydrogelation experiments at different pH by Method A. *

Composition (x mg + 500 µL Buffer)		Approx. Concentration		Description	pH*
Gelator **56**	Alginate	Gelator **56**	Alginate		
3	3	0.6 wt%	0.6 wt%	opaque, orange gel, not stable upon shaking	4
3	3	0.6 wt%	0.6 wt%	opaque, orange gel, not stable upon shaking	6
3	6	0.6 wt%	1.2 wt%	opaque, orange gel, not stable upon shaking	6
3	3	0.6 wt%	0.6 wt%	opaque, orange gel, not stable upon shaking, precipitate	7.4
3	6	0.6 wt%	1.2 wt%	opaque, orange gel, not stable upon shaking	7.4
3	3	0.6 wt%	0.6 wt%	opaque, orange gel	8
3	3	0.6 wt%	0.6 wt%	Slightly colored liquid, precipitate	10
0	3	-	0.6 wt%	colorless liquid	-

*Buffer preparation:

pH=4

Components	Molecular weight (g/mol)	Conc. (mg/L)	Conc. (mM)
Citric acid	192	856	4.46
Potassium Chloride (KCl)	75.0	200	2.67
Sodium Chloride (NaCl)	58.0	8000	138
Sodium Phosphate dibasic (Na_2HPO_4-$7H_2O$)	268	292	1.09

The final exact pH was adjusted to 3.96 by addition of diluted aqueous NaOH solution and measuring the pH on a pH meter (WTW pH 3310 with a SenTix® 41 electrode).

pH=6

Components	Molecular weight (g/mol)	Conc. (mg/L)	Conc. (mM)
Potassium Chloride (KCl)	75.0	200	2.67
Sodium Chloride (NaCl)	58.0	8000	138
Sodium Phosphate monobasic (NaH_2PO_4)	120	1051	8.76
Sodium Phosphate dibasic (Na_2HPO_4-$7H_2O$)	268	330	1.23

The final exact pH was adjusted to 5.96 by addition of diluted aqueous HCl solution and measuring the pH on a pH meter (WTW pH 3310 with a SenTix® 41 electrode).

pH=8

Components	Molecular weight (g/mol)	Conc. (mg/L)	Conc. (mM)
Potassium Chloride (KCl)	75.0	200	2.67
Sodium Chloride (NaCl)	58.0	8000	138
Sodium Phosphate monobasic (NaH_2PO_4)	120	468	3.90
Sodium Phosphate dibasic (Na_2HPO_4-$7H_2O$)	268	1635	6.10

The final exact pH was adjusted to 8.02 by addition of diluted aqueous HCl solution and measuring the pH on a pH meter (WTW pH 3310 with a SenTix® 41 electrode).

pH=10

Components	Molecular weight (g/mol)	Conc. (mg/L)	Conc. (mM)
Potassium Chloride (KCl)	75.0	200	2.67
Sodium Chloride (NaCl)	58.0	8000	138
Sodium Carbonate (Na_2CO_3)	106	6360	60.0
Sodium Bicarbonate ($NaHCO_3$)	84.0	3360	40.0

The final exact pH was adjusted to 10.05 by addition of diluted aqueous HCl solution and measuring the pH on a pH meter (WTW pH 3310 with a SenTix® 41 electrode).

Method B

Photoresponsive supramolecular smart materials find increasing interest in biomedical applications, for example as supramolecular hydrogels in drug delivery systems.[165] Here, limiting factors are biocompatible switching wavelengths on the one hand, but also biocompatible gel formation on the other hand. To address the latter, a novel gel formation protocol was developed. It is based on the mutual stabilization of gelator **56** and sodium alginate (comparable to the egg-box model[153]). The components were separately dissolved in water, and no gel formation was observed. The solution of gelator **56** was irradiated with light of 365 nm to build up the polar and bent *Z*-isomer, which hinders formation of the strong interactions described by the egg-box model. Mixing of both solutions produced a viscous mixture, in which the fraction of polar *Z*-isomer promoted homogenous mixing. Subsequent irradiation with blue light (455 nm) reduced the fraction of *Z*-isomer and resulted in the mutual stabilization of gelator **56** and alginate – a stable gel was formed.

To a 1.5 mL-vial (screw top) was added the photochromic material **56** as a powder and 250 µL of water. After complete dissolution, the solution was irradiated for 10 min at 365 nm. Subsequently, 250 µL of a 2× concentrated stock solution of sodium alginate in water was added and mixed by repetitive pipetting. Then, the mixture was irradiated at 455 nm for 10 min. Successfully formed gels were prepared in triplicates.

Melting temperatures were determined in triplicates. Gels were prepared as previously described. The vials were then put upside-down in a slowly stirred water bath (60 rpm) on a magnetic hotplate stirrer, equipped with a thermometer. The water bath was heated (1.5 °C/min) until the gel started to melt and formed a sol.

Table 20: Hydrogelation experiments by Method B.

Composition (x mg + 500 µL H_2O)		Approx. Concentration		Description	T_m °C
Gelator **56**	Alginate	Gelator **56**	Alginate		
3	3	0.6 wt%	0.6 wt%	opaque, orange gel	89
2	2	0.4 wt%	0.4 wt%	opaque, orange gel	85
1.5	**1.5**	**0.3 wt%**	**0.3 wt%**	**clear, yellow gel**	**80**

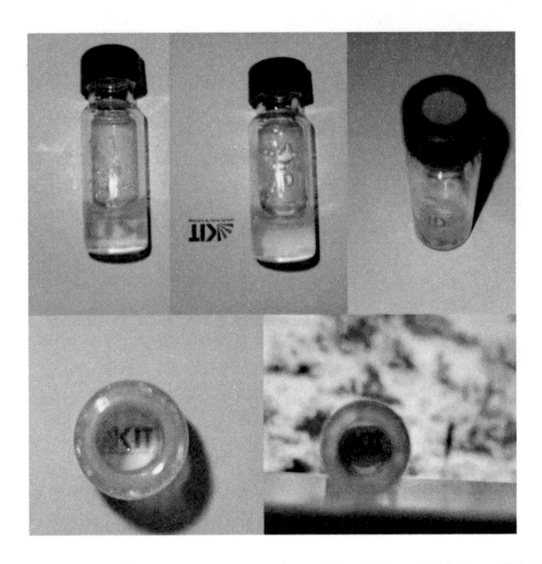

Figure 114: Pictures of a hydrogel formed by 0.3 wt% gelator **56** and 0.3 wt% alginate (Method B).

5.5.4 Light induced gel-to-sol transition

According to Method A described in the previous section, gels were prepared with 0.6 wt% PAP-DKP-Lys_2 **56** and 0.6 wt% alginate in water. After equilibrating overnight, the samples were irradiated with two LEDs at 365 nm (10 W) for

15 min. Subsequently, the gels were inverted, and one gel was irradiated at 455 nm for 15 min (A), while a second gel was kept in the dark (B, see Figure 115). Sample A solidified at the vial top, while sample B was a highly viscous liquid. After 30 min equilibration time, sample B was still liquid. Sample A stayed solid at the vial top.

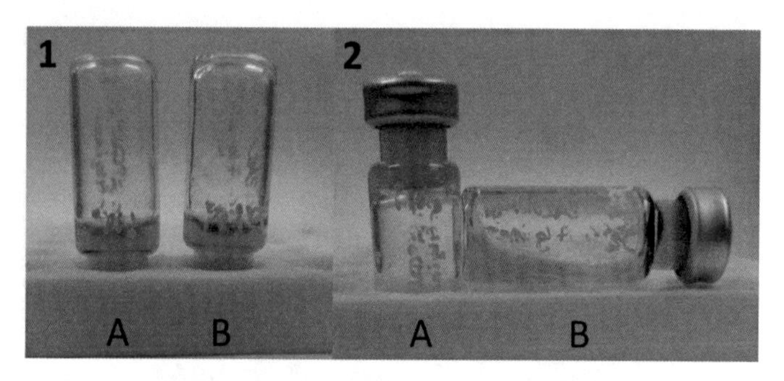

Figure 115: 1) Gels after irradiation at 365 nm. 2) Gel A was irradiated at 455 nm; Gel B was kept in the dark.

5.5.5 Light-induced rhodamine release

It was investigated how efficiently hydrogels, based on the gelator **56** and sodium alginate, release encapsulated guest molecules by means of diffusion (in darkness) or dissipation of the inner gel structure upon irradiation with UV light.

The chosen composition of 0.3 wt% of the gelator **56** and 0.3 wt% of alginate in 500 µL of water was prepared by Method B. As described in section 5.5.3, a stable gel is formed after irradiation of the mixed components at 455 nm. By this preparation method heat is avoided, which could have a damaging impact on some cargo substances. Preparation with cargo was done as follows:

The photochromic gelator **56** (1.5 mg, powder) and water (245 µL) were mixed in a 1.5 mL glass vial (screw top). This solution was irradiated at 365 nm (10 min), then a 100-fold concentrated stock solution (5 µL) of the chosen cargo rhodamine dissolved in EtOH was added. Subsequently, a 2-fold concentrated stock solution of alginate (250 µl) was added, thoroughly mixed and the final mixture was irradiated at 455 nm for 10 min to obtain the final gel. Before a release experiment, the hydrogels were kept overnight in darkness at room temperature. Concentration of the cargo rhodamine was adapted to the HPLC detection range. 250 µg total mass of rhodamine were incorporated into the gels.

Quantification of the passive diffusion – cargo "leaking" from hydrogels in darkness

500 µL of PBS buffer pH 7.4 was slowly added on top of a gel sample (on the wall of the vial) and immediately removed with a micropipette to wash away unbound or loosely bound guest molecules from the surface. Addition of fresh 500 µL of PBS buffer followed. The gel was incubated together with the buffer on the top in darkness. 500 µL of the liquid was collected after 5 min by gently turning the vial sideways and pipetting off the liquid from the side wall of the vial. Then, fresh 500 µL of PBS buffer was added on the side wall of the vial, incubated in darkness and removed after 5 min in the same way as described above. That process was repeated for a total duration of 40 min by collecting 9 subsequent volume aliquots. After that time, the gel remained visually unaffected.

Procedure of the light induced release

To measure the release process upon UV light irradiation, the procedure described above was repeated exactly, but after initial washing of the gel surface, the sample was placed in an irradiation chamber and illuminated with two 10 W LEDs (365 nm, from the distance of 5 cm).

Short breaks in irradiation were taken for the replacement of 500 µL aliquot with fresh 500 µL of PBS buffer every 5 min, but the overall irradiation time was 40 min. The irradiation time was sufficient to fully convert the gel samples into sol. All aliquots were weighted before the HPLC measurement to calculate the released amount of the substance. The concentration of the aliquots was calculated by a previously measured calibration curve (Figure 116).

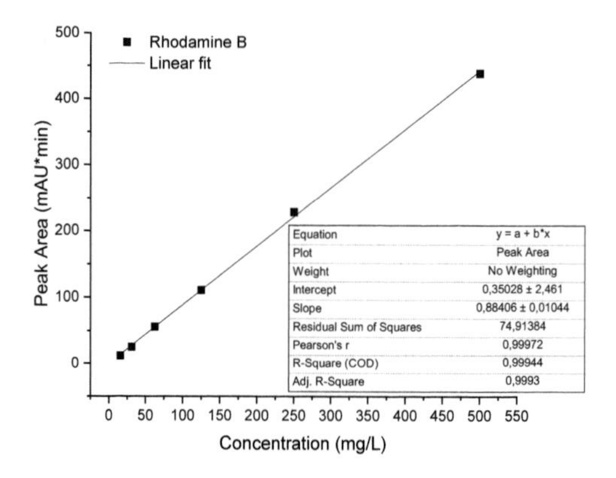

Figure 116: Calibration curve for the light induced release of rhodamine.

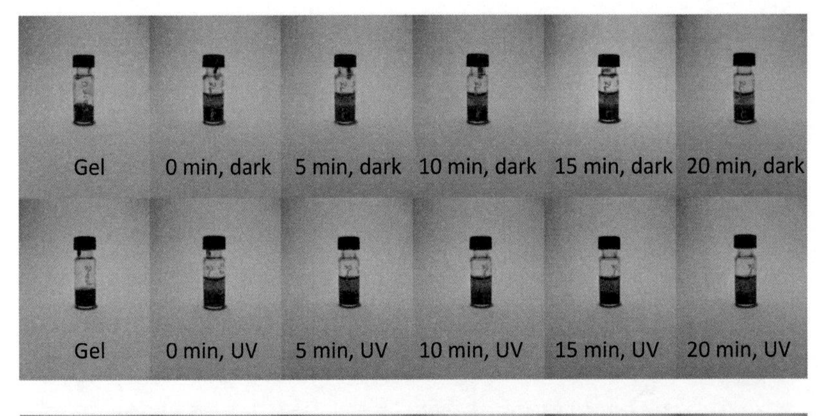

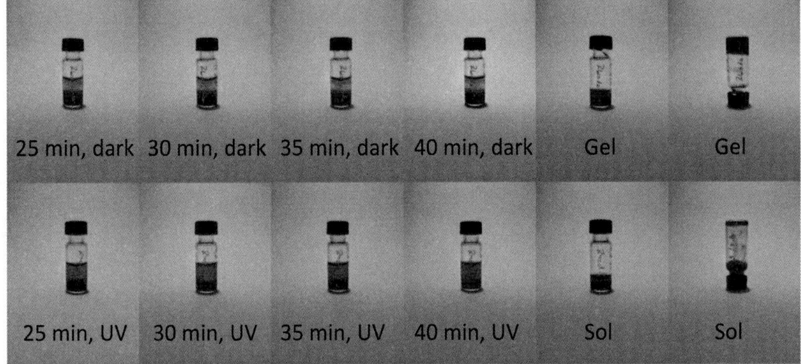

Figure 117: Visual development of the gels and supernatant during the release experiment.

As a result, there was only minor leaking in the dark equilibrated gel (51.4 µg in total during 40 min), while there was a distinct release (112.5 µg in total during 40 min) from the irradiated gel, which can be seen by the color development and the quantified amount of released rhodamine (Figure 117 and Figure 118).

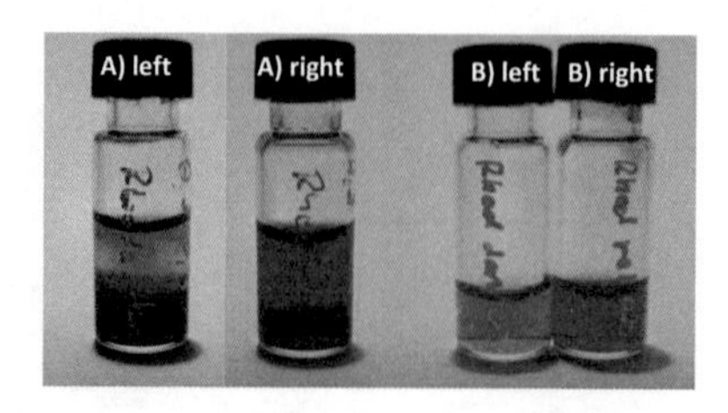

Figure 118: A) left: Dark equilibrated gel with supernatant after cycle 9; right: illuminated gel with supernatant after cycle 9. B) photo at ambient light; left: supernatant of cycle 9 in the dark; right: supernatant of the irradiated sample at cycle 9.

5.5.6 Diffusion of the gelator **56** after addition of Ca^{2+} ions

Sodium alginate forms gels by addition of divalent ions, for example Ca^{2+}. Therefore, the influence of Ca^{2+} ions on the composite hydrogelator **56** / alginate gels was assessed. For this purpose, two Gels at 0.6 wt% gelator **56** and 0.6 wt% alginate were prepared by Method A (see Figure 119).

Figure 119: Hydrogels at 0.6 wt% gelator **56** and 0.6 wt% alginate prepared by Method A.

On top of these gels was added 500 µL of a 10 wt% solution of $CaCl_2$ in water and equilibrated for 1 h.

Figure 120: 500 µL $CaCl_2$ solution (10 wt% in water) added on top of the gels for 1 h equilibration time.

Subsequently, the supernatant was removed and fresh $CaCl_2$ solution was added. Then, one vial was kept in the dark, while the second one was irradiated for 1 h (365 nm). Next, the supernatant was removed, and the procedure was repeated four times in total. The supernatant of the irradiated vials was colorful, while the dark vials were less colored (Figure 121).

Figure 121: Four irradiation / dark cycles. Left row: fresh $CaCl_2$ solution was added to the top. Right row: 1 h equilibration either at 365 nm or in the dark.

After four cycles, in both irradiated and dark equilibrated vial, a stable gel remained. The gel in the irradiated vial was less opaque and less colorful compared to the non-irradiated vial (Figure 122).

Figure 122: Remaining gel after the Ca^{2+} experiment. In each picture the right vial was irradiated. The gel in the irradiated vial is less opaque, the bottom plate of the vial is clearly visible (red arrow).

5.5.7 Cell viability assays of the hydrogelator **56**

HeLa cells were grown in DMEM (Dulbecco`s Modified Eagle Medium), which was modified with 10% FCS (fetal calf serum) and 1% penicillin/streptomycin solution (10,000 units/mL of penicillin and 10,000 µg/mL of streptomycin) in a humid incubator at 37 °C with 5% CO_2. Cells were detached from the surfaces with Trypsin-EDTA (0.25%) from Gibco®. Cells were washed with PBS (Phosphate-Buffered Saline) from Gibco®.

96-well plates with a flat bottom were prepared by filling all wells on the outer border with 200 µL PBS and the remaining wells with 100 µL of a cell suspension (30.000 cells/mL) in DMEM. The prepared plate was incubated overnight to ensure cell attachment to the well-bottom and cell growth.

For the dilution-series of each compound, a stock solution of DMEM, modified with 0.25% DMSO, was prepared, to ensure that all cells are treated with the same conditions. Consequently, the first sample of the dilution series was prepared by dissolving the substance in DMSO and adding a specific amount of this solution to a specific amount of non-modified DMEM so that a final concentration of 0.25% DMSO is reached.

To apply the substances to the 96-well plate, the DMEM was removed without disturbing the cells grown in the plate and adding subsequently 100 µL to each well. To ensure the same treatment to the control rows, the DMEM was removed from the wells and DMSO-modified DMEM (100 µL) was added to the corresponding wells. The 96-well plate was incubated for 48 h.

The positive control was treated with 5 µL of Triton™ X-100 detergent (10% solution (w/v)) per well for at least 5 min before adding 15 µL of MTT dye-solution (Cell Proliferation Kit I (MTT) from Roche) to all sample wells and incubating for 3 h in the dark. 100 µL of stop solution (Cell Proliferation Kit I (MTT) from Roche) was added after incubation to stop the reduction of MTT to formazan, thus preventing overreaction and enabling solubilization of formazan crystals. After 24 h of solubilization in the incubator, the plate was read out with a plate reader (BioTek® EPOCH^2, Gen5 Data Analysis) by measuring the absorption of each well at 595 nm.

The raw data was processed as followed by first subtracting the positive control (all cells are dead) from all measured values in one row to remove background absorption. Each concentration was measured sixfold per plate (Triplicates; 18 values in total), therefore the values for each concentration and the negative control (all cells alive) were averaged and the standard deviation was calculated. The cell viability was calculated as a percentage of the negative control and normalized by assuming the highest obtained viability as 100%.

Table 21: Cell viability of PAP-(DKP-Lys)$_2$ **56** against HeLa cell line in the dark state and after irradiation.

PAP-(DKP-Lys)$_2$ **365 nm irradiated**				PAP-(DKP-Lys)$_2$ **dark adapted**			
Conc.	Conc. [M]	Cell viability [%]	Stdev	Conc.	Conc. [M]	Cell viability [%]	Stdev
1 mM	1.00E-03	84	11	**1 mM**	**1.00E-03**	**51**	**15**
100 µM	1.00E-04	100	11	100 µM	1.00E-04	100	10
10 µM	1.00E-05	95	11	10 µM	1.00E-05	99	15
1 µM	1.00E-06	94	13	1 µM	1.00E-06	96	12
100 nM	1.00E-07	96	12	100 nM	1.00E-07	97	12
10 nM	1.00E-08	96	10	10 nM	1.00E-08	94	11
1 nM	1.00E-09	96	11	1 nM	1.00E-09	90	10
100 pM	1.00E-10	90	10	100 pM	1.00E-10	88	8

5.5.8 Microscopy images

Transmission Electron Microscopy (TEM):

For the Transmission Electron Microscopy images, two 1.5 wt% hydrogel samples of **56** in PBS buffer and two mixed samples (0.6 wt% **56** and 0.6 wt% alginate) in H_2O were prepared as described by method A (section 5.5.3). After cooling, the samples were equilibrated overnight at room temperature. One sample of each composition was irradiated at 365 nm (10 W LED) until liquefaction. The resulting samples were added as small droplets to carbon-coated copper grids (400 mesh). The supernatant was removed carefully with a lint-free sheet and the grid was dried under atmospheric pressure. Examination was carried out by Heike Störmer on a Philips CM200 FEG transmission electron microscope, operated at 200 kV accelerating voltage. All images were recorded defocused.

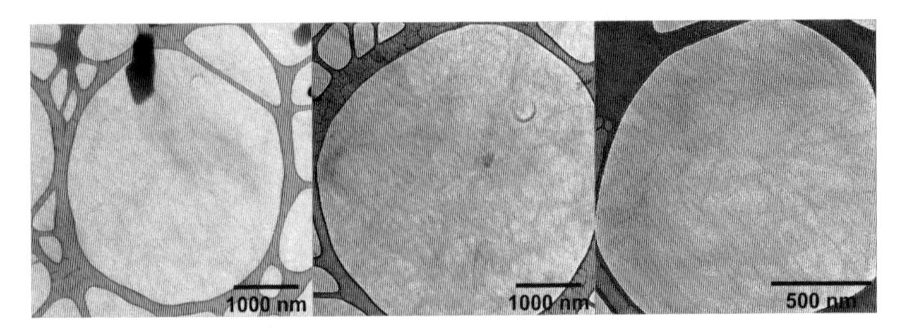

Figure 123: TEM images of hydrogelator **56** 1.5 wt% in PBS buffer; dark adapted.

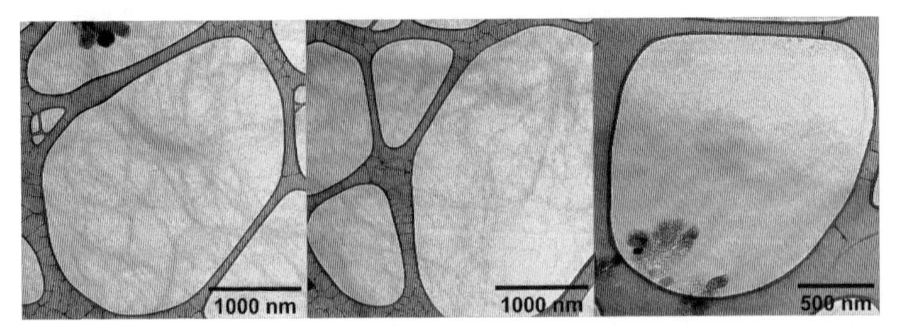

Figure 124: TEM images of hydrogelator **56** 1.5 wt% in PBS buffer; irradiated.

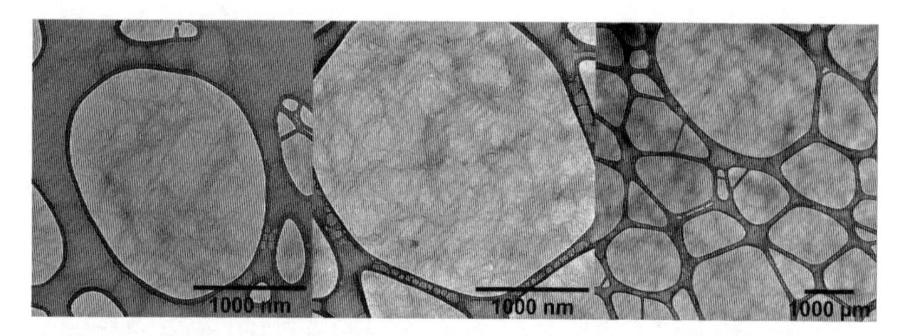

Figure 125: TEM images of 0.6 wt% hydrogelator **56** and 0.6 wt% alginate in H_2O; dark adapted.

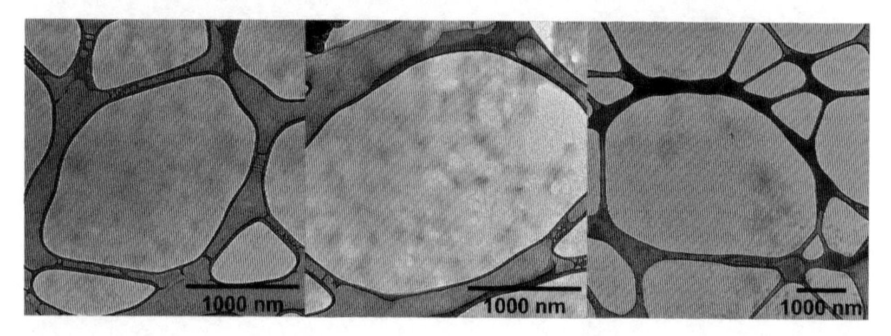

Figure 126: TEM images of 0.6 wt% hydrogelator **56** and 0.6 wt% alginate in H_2O; dark adapted.

Scanning Electron Microscopy (SEM):

For the Scanning Electron Microscopy images, a sample with 1.5 wt% hydrogelator **56** was prepared with PBS buffer and a sample with 0.6 wt% hydrogelator **56** and 0.6 wt% alginate was prepared with diH_2O. The resulting material was freeze dried by lyophilization and then coated with a thin layer of platinum. The microscope was operated by Volker Zibat.

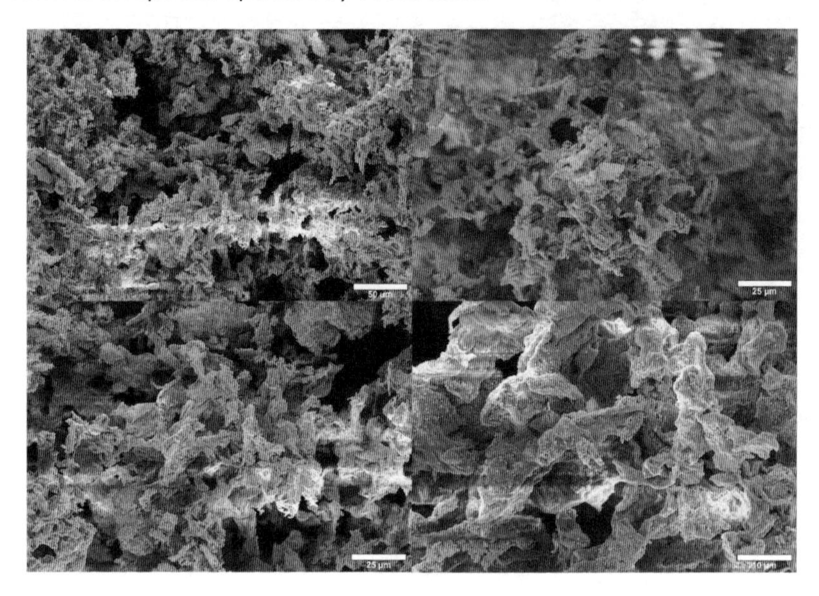

Figure 127: SEM images of 1.5 wt% hydrogelator **56** in PBS buffer.

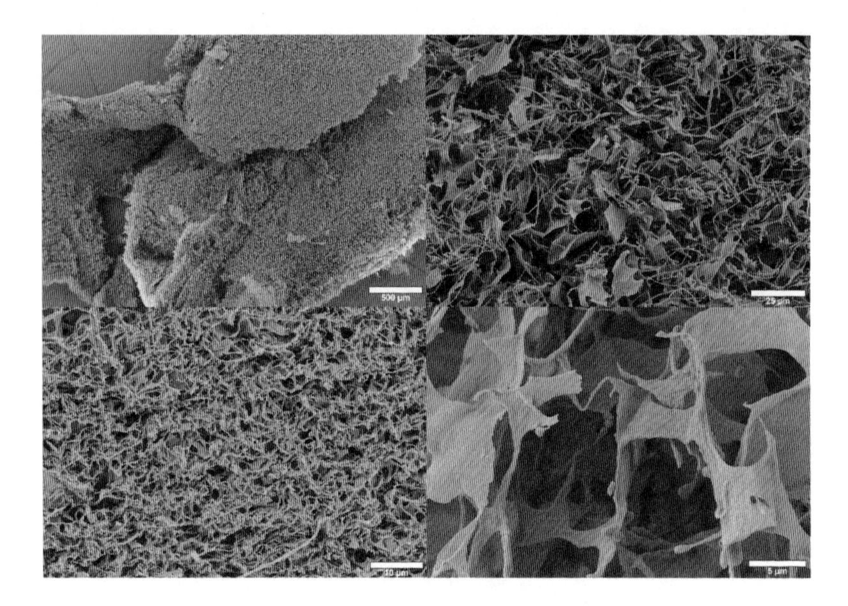

Figure 128: SEM images of 0.6 wt% hydrogelator **56** and 0.6 wt% alginate in H_2O.

5.5.9 Rheological experiments

Oscillatory shear experiments were conducted by Christian Fengler on a strain-controlled ARES-G2 (TA Instruments) rheometer with a parallel plate geometry (25 mm diameter and 1 mm gap). A volume of 0.5 mL of a sample consisting of 0.6 wt% hydrogelator **56** and 0.6 wt% alginate (prepared as described in method A; section 5.5.3) was poured onto the lower plate. The temperature was maintained at 20 °C by a Peltier element. A strain sweep was conducted by varying the strain amplitude in the range γ = 0.01 – 37 % at an angular frequency ω = 6.28 rad s^{-1} to determine the linear viscoelastic (LVE) regime. A frequency sweep was recorded in the range ω = 0.19 - 452 rad s^{-1} at γ = 0.5 % within the LVE regime.

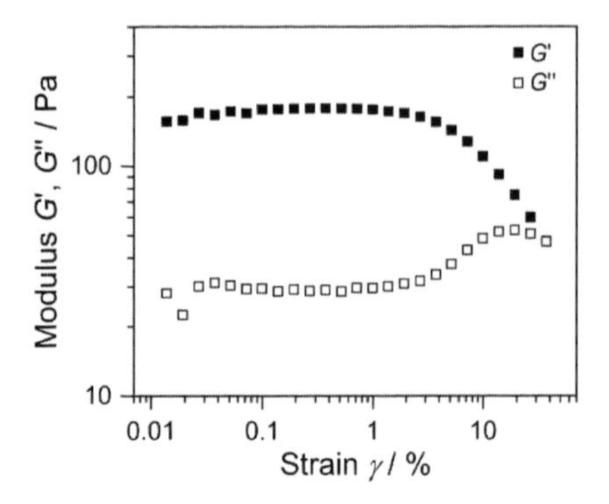

Figure 129: Strain sweep of a sample consisting of 0.6 wt% hydrogelator **56** and 0.6 wt% alginate at an angular frequency $\boldsymbol{\omega}$ = 6.28 rad s^{-1}.

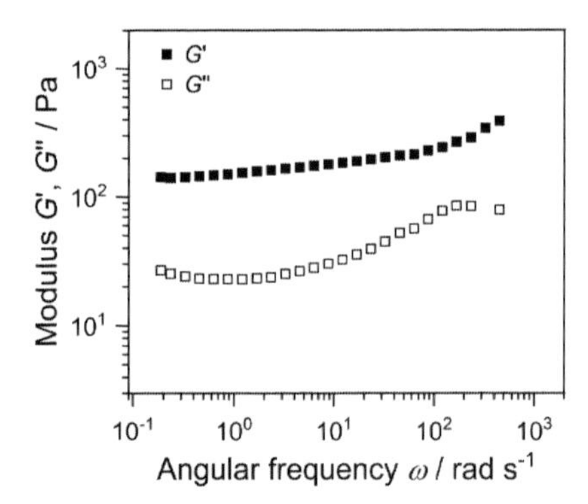

Figure 130: Elastic $\boldsymbol{G'}$ and loss modulus $\boldsymbol{G''}$ as a function of angular frequency of the compound **56**. Frequency sweep was performed at a strain amplitude of 0.5 % and 20 °C.

5.6 Transfer of the composite alginate system to PAP-DKP-Lys

5.6.1 Synthesis

The synthesis towards Phenylazophenyl (PAP)-DKP-Lys **52** followed the procedures developed and published by Pianowski *et al.* [121b] The ^{1}H NMR shifts of resynthesized **52** in Figure 131 are in accordance with the published data.

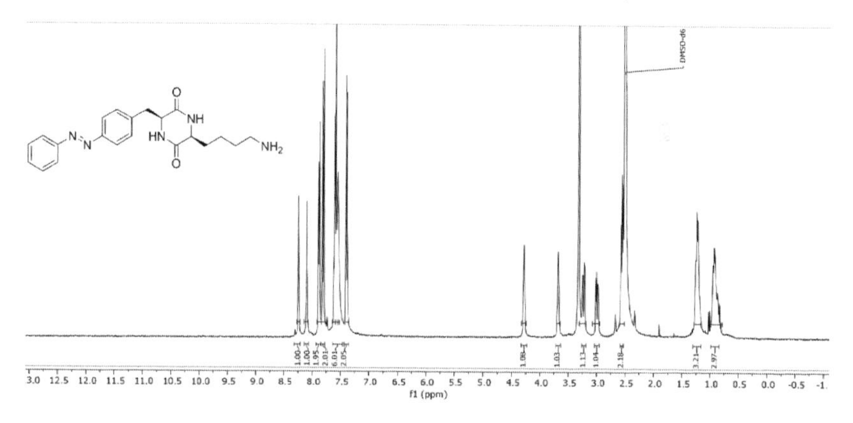

Figure 131: Structure and ^{1}H NMR spectrum of PAP-DKP-Lys **52**. ^{1}H NMR (400 MHz, DMSO) δ = 8.25 (d, J = 2.1 Hz, 1H), 8.10 (d, J = 2.1 Hz, 1H), 7.92 – 7.84 (m, 2H), 7.81 (d, J = 8.3 Hz, 2H), 7.65 – 7.54 (m, 3H), 7.55 (s, 3H), 7.40 (d, J = 8.3 Hz, 2H), 4.28 (t, J = 4.7 Hz, 1H), 3.67 (d, J = 6.1 Hz, 1H), 3.23 (dd, J = 13.5, 4.1 Hz, 1H), 2.99 (dd, J = 13.5, 5.1 Hz, 1H), 2.55 (t, J = 7.7 Hz, 2H), 1.29 – 1.17 (m, 3H), 0.99 – 0.78 (m, 3H) ppm.

5.6.2 Photophysical properties

Thermal stability

The thermal stability was determined for compound **52**. Samples at a concentration of 500 µM were prepared in AcOH as well as in H_2O and irradiated at 365 nm to yield a high PSS of the *Z*-isomer. The isomer ratio was determined by HPLC in intervals. The obtained data was processed by calculating the $\ln(X_0/X_t)$, where X is the percentage of the respective *Z*-isomer and linear fitting (Equation (1)) of the obtained values. The calculated slope corresponds to the isomerization rate constant k, which is used to calculate the half-life $t_{1/2}$.

$$x_t = x_0 \cdot e^{-k \cdot t} \leftrightarrow ln\left(\frac{x_0}{x_t}\right) = k \cdot t \quad (1)$$

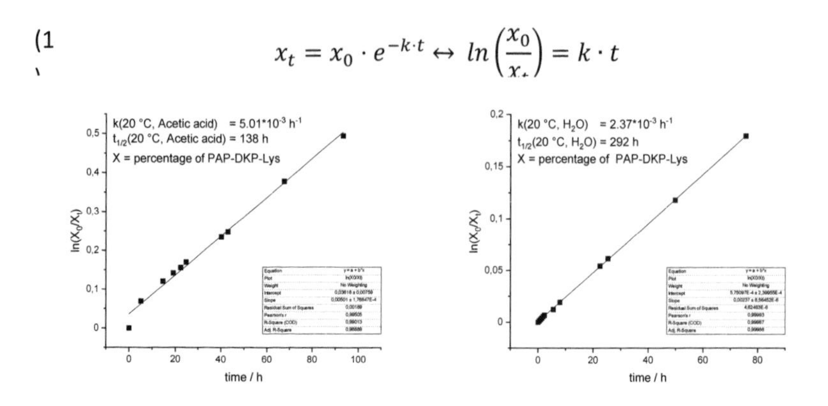

Figure 132: Linear fit of the decay of the *Z*-isomer of compound **52** at 20 °C in AcOH and H_2O for first-order kinetics.

5.6.3 Gelation experiments

To a 1.5 mL-vial (crimp top, 12×32 mm) was added the photochromic material **52** and the additive alginate as powder and 500 µL of the aqueous solution. This suspension was sonicated for 2 min followed by heating to 80 °C in a vial block. After equilibration at 80 °C for 5 min, the sample was heated to the boiling point by a heat gun. The hot solution became completely clear and upon cooling the hydrogels formed. Successfully formed gels were prepared in triplicates.

Table 22: Hydrogelation experiments with PAP-DKP-Lys **52** and sodium alginate.

Composition (x mg + 500 µL H_2O)		Approx. conc.		Description	T_m °C
Gelator **52**	Alginate	Gelator **52**	Alginate		
3	3	0.6 wt%	0.6 wt%	clear orange gel	>100
2	3	0.4 wt%	0.6 wt%	clear orange gel	-

1.5	3	0.3 wt%	0.6 wt%	clear orange gel	-
1	3	0.2 wt%	0.6 wt%	clear yellow gel	100
3	2	0.6 wt%	0.4 wt%	clear orange gel	-
3	1.5	0.6 wt%	0.3 wt%	clear orange gel	-
3	1	0.6 wt%	0.2 wt%	clear orange gel	>100
3	0.5	0.6 wt%	0.1 wt%	opaque orange gel	-
2	2	0.4 wt%	0.4 wt%	clear orange gel	-
1	1	0.2 wt%	0.2 wt%	clear yellow gel	96±2
0.5	0.5	0.1 wt%	0.1 wt%	weak, clear yellow gel	48±5

Pure compound **52** was investigated by Johannes Karcher. Compound **52** formed stable gels in the range between 2.0 – 3.0 wt% in deionized water and can be stabilized by addition of NaCl or htDNA.[121b] At a lower concentration of 2.0 wt%, the gel was of low mechanical stability and sensitive upon shaking. Therefore, the behavior of the gel upon addition of sodium alginate was investigated. >10–fold increased stability was observed when gelator **52** was combined with sodium alginate. The lowest critical gelation concentration was determined at 0.1 wt% gelator/ 0.1 wt% alginate and all formed gels were completely clear.

5.6.4 Microscopy images

Both microscopy methods operate under non-natural conditions (vacuum, dry samples), consequently the recorded images do not represent the state of a wet gel. Nevertheless, different gels and gel compositions can be compared, when prepared in the same manner.

Transmission Electron Microscopy

For the Transmission Electron Microscopy images, samples of 0.6 wt% **52** with 0.6 wt% alginate were prepared in 200 mM aqueous NaI as described in 5.6.3. After cooling, the samples were equilibrated overnight at room temperature. One sample was irradiated at 365 nm to isomerize the azobenzene photoswitch unit. The resulting gels were added as small droplets to carbon-coated copper grids (400 mesh). The supernatant was removed carefully with a lint-free sheet and the grid was dried under atmospheric pressure. Examination was carried out by Heike Störmer on a Philips CM200 FEG transmission electron microscope, operated at 200 kV accelerating voltage.

TEM images (Figure 133 and Figure 134) revealed a fibrous network structure in both gel samples. The fibers are multiple µm long and only few nm thick. Photoisomerization of the hydrogelator **52** did not result in liquefaction of the gel. This is supported by the TEM images where fibers were still visible in the irradiated sample.

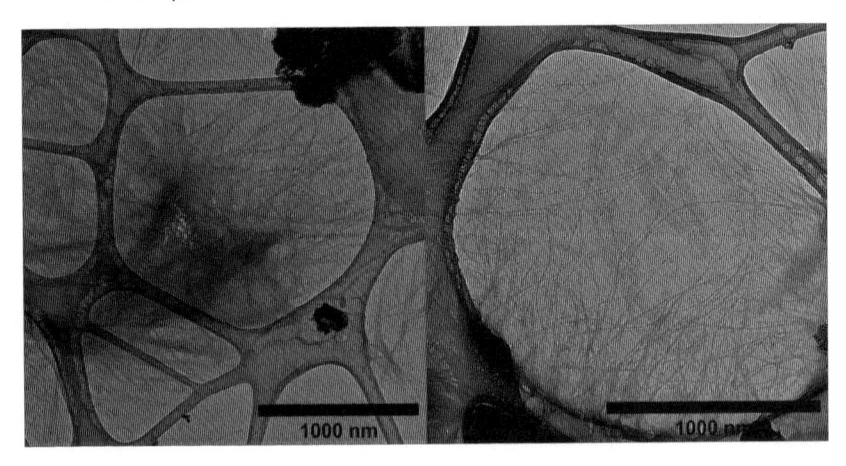

Figure 133: Transmission electron microscopy (TEM) of the hydrogel in darkness.

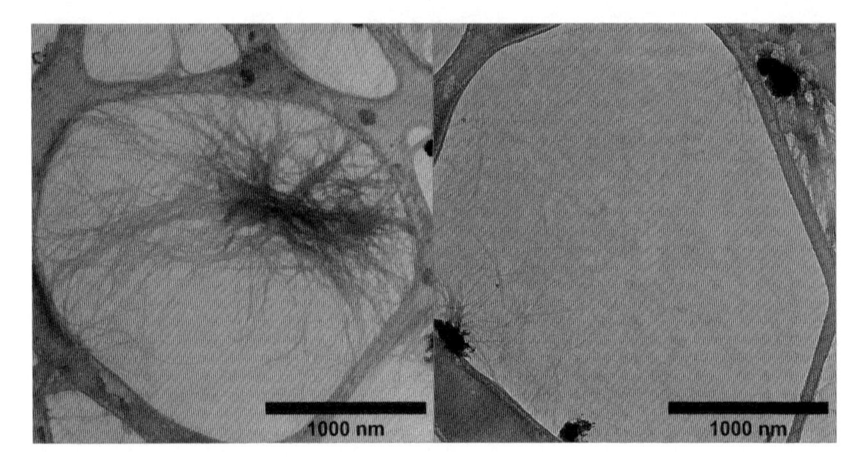

Figure 134: Transmission electron microscopy (TEM) of the hydrogel after irradiation at 365 nm.

Scanning Electron Microscopy

For the Scanning Electron Microscopy images, a sample of 0.6 wt% **52** with 0.6 wt% alginate was prepared with diH_2O. The resulting material was freeze dried by lyophilization and then coated with a thin layer of platinum. Volker Zibat operated the microscope.

The SEM images (Figure 135) match the findings of TEM investigation. Again, a well-developed network built from multi-µm long fiber-like structures is revealed.

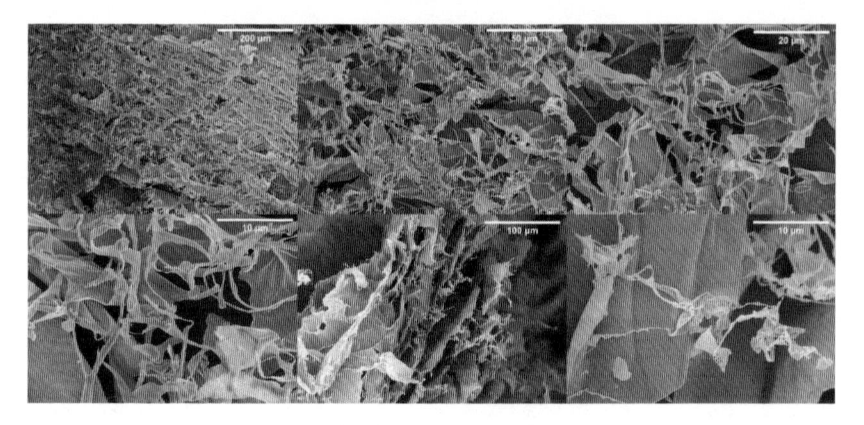

Figure 135: Scanning electron microscopy (SEM) of the freeze-dried hydrogel formed of **52**; various magnifications.

5.7 Transfer of the composite alginate system to F_2-PAP-DKP-Lys

The syntheses towards compound **57** were performed by Simon Ludwig in his bachelor thesis under the supervision of Anna-Lena Leistner and were based on the work of Johannes Karcher.[165] The following gelation experiments (5.7.2) were performed by Simon Ludwig in his bachelor thesis under the supervision of Anna-Lena Leistner.[212]

5.7.1 Synthesis

The ^{1}H NMR shifts of resynthesized **57** in Figure 136 were in accordance with the published data.

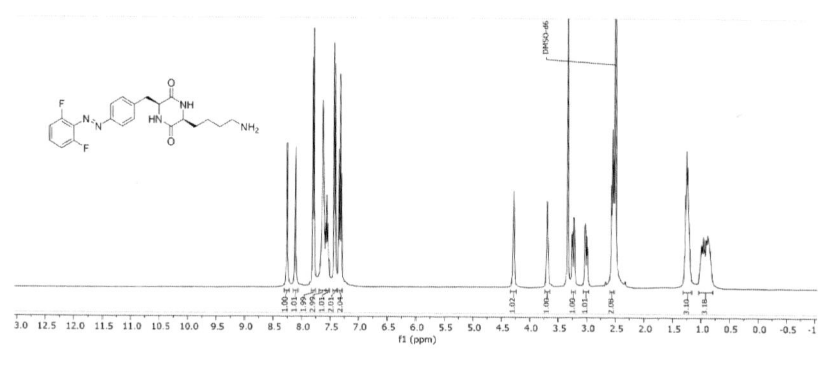

Figure 136: Structure and ^{1}H NMR spectrum of F_2-PAP-DKP-Lys **57**. ^{1}H NMR (400 MHz, DMSO) δ = 8.26 (d, J = 2.1 Hz, 1H), 8.11 (d, J = 2.1 Hz, 1H), 7.80 (d, J = 8.1 Hz, 2H), 7.63 (s, 3H), 7.56 (tt, J = 8.5, 6.3 Hz, 1H), 7.42 (d, J = 8.2 Hz, 2H), 7.32 (t, J = 8.9 Hz, 2H), 4.28 (t, J = 4.9 Hz, 1H), 3.69 (t, J = 5.9 Hz, 1H), 3.24 (dd, J = 13.4, 4.2 Hz, 1H), 3.01 (dd, J = 13.5, 5.1 Hz, 1H), 2.55 (t, J = 7.6 Hz, 2H), 1.25 (h, J = 7.5 Hz, 3H), 1.05 – 0.79 (m, 3H) ppm.

5.7.2 Gelation experiments

Method A

To a 1.5 mL-vial (crimp top, 12×32 mm) was added the photochromic material **57** and the additive alginate as powder and 500 µL of the according aqueous solution. This suspension was treated by ultrasonic waves for 2 min followed by heating to 80 °C in a vial block. After equilibration at 80 °C for 5 min, the sample was heated to the boiling point by a heat gun. The hot solution became completely clear and upon cooling the hydrogels formed. Successfully formed gels were prepared in triplicates.

Method B

For preparing hydrogels by irradiation, the composition of 0.3 wt% F_2-PAP-DKP-Lys (**57**) and 0.1 wt% alginate was chosen. At lower concentrations, no gel formation was observed (by this method), while higher concentrations prevented gel-to-sol formation by irradiation or required higher irradiation times. Two solutions were prepared in bi-distilled water: 0.2 wt% of alginate and 0.6 wt% of **57**. The solution with the gelator was then irradiated at 523 nm for 1 h under periodical mixing.

To a 1.5 mL-vial (screw top) was added 250 µL of hydrogelator solution. Subsequently, 250 µL of the alginate solution were added and mixed by repetitive pipetting. Then, the mixture was irradiated at 410 nm for 30 min and the viscous mixture turned into stable hydrogel. The hydrogels stabilized further by equilibration overnight. Successfully formed gels were prepared in triplicates.

Melting temperatures were determined in triplicates. Gels were prepared as previously described. The vials were then mounted upside-down in a slowly stirred water bath (60 rpm) on a magnetic hotplate stirrer equipped with a thermometer. The water bath was heated (1.5 °C/min) until the gel started to melt and formed a sol.

Light induced gel-to-sol transitions were observed for specific concentrations after irradiation. For investigation of gel-to-sol transitions, the hydrogels were prepared in vials and subsequently irradiated with a 523 nm, 10 W LED diode to isomerize the azobenzene unit of hydrogelator **57** from its *E* to *Z* isomer.

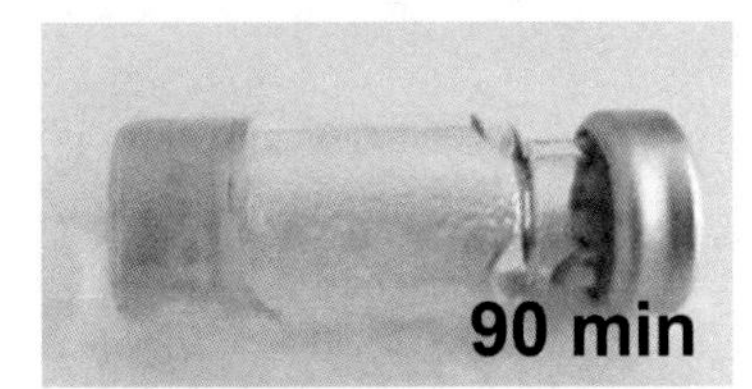

Figure 137: Light induced gel-to-sol transition of 0.2 wt% **57** and 0.2 wt% alginate after 60, 90, 105 and 120 minutes of irradiation at 523 nm.

Light induced sol to gel transitions were observed for specific concentrations after irradiation. For investigation of sol-to-gel transitions, the hydrogels which were first characterized by gel-to-sol transitions were subsequently irradiated with a 410 nm, 10 W LED diode.

Table 23: Hydrogelation experiments.

Composition		Description	T_m [°C]	$t_{gel\text{-}to\text{-}sol}$ [min]	$t_{sol\text{-}to\text{-}gel}$ [min]
Gelator **57**	Alginate				
0.6 wt%	**0.6 wt%**	**clear, orange gel**	**>100**	**-**	[1]
0.5 wt%	0.5 wt%	clear, orange gel	[1]	[1]	[1]
0.4 wt%	0.4 wt%	clear, orange gel	[1]	[1]	[1]
0.3 wt%	0.3 wt%	clear, yellow gel	[1]	-	[1]
0.2 wt%	**0.2 wt%**	**clear, yellow gel**	**97±3**	**120**	**no gel**
0.1 wt%	0.1 wt%	clear, yellow liquid	-	-	-
0.1 wt%	0.2 wt%	clear, yellow liquid	-	-	-
0.3 wt%	0.2 wt%	clear, yellow gel	[1]	[1]	[1]
0.4 wt%	0.2 wt%	clear, yellow gel	[1]	[1]	[1]
0.2 wt%	0.1 wt%	clear, yellow gel	[1]	120	no gel
0.3 wt%	**0.1 wt%**	**clear, yellow gel**	**100[2]**	**135**	**30**
0.3 wt%	**0.1 wt%**	**clear, yellow gel[3]**	**91±6[3]**	**120[3]**	**30[3]**
0.4 wt%	0.1 wt%	clear, yellow gel	[1]	210	30
0.5 wt%	0.1 wt%	clear, orange gel	[1]	[1]	[1]
0.2 wt%	0.05 wt%	clear, yellow gel	[1]	105	no gel
0.3 wt%	0.05 wt%	clear, yellow gel	[1]	105	no gel
0.4 wt%	0.05 wt%	clear, yellow gel	[1]	[1]	[1]
0.6 wt%	0 wt%	clear, orange liquid	-	-	-
0 wt%	0.6 wt%	colorless liquid	-	-	-

[1] not determined; [2] mechanical stimulation required; [3] prepared according to method B

Pure compound **57** was investigated by Johannes Karcher. Compound **57** formed stable and homogenous gels in the range between 4.0 – 7.0 wt% in PBS buffer.[165] At a lower concentration of 3.0 wt%, gelation was slow (overnight)

and the resulting gel of low mechanical stability (sensitive upon shaking). Therefore, the behavior of the gel upon addition of sodium alginate was investigated. >20–fold increased stability was observed when gelator **57** was combined with sodium alginate. The lowest critical gelation concentration was determined at 0.2 wt% gelator/ 0.2 wt% alginate with T_m = 97 °C and all formed gels were completely.

5.7.3 Gelation in biological medium

The following composite gelation experiments were performed by Simon Ludwig in his bachelor thesis under the supervision of Anna-Lena Leistner and repeated by Anna-Lena Leistner.[212]

The gelation properties of **57** were investigated under biological conditions with DMEM as fluid component according to preparation method B. The composition of 0.6 wt% F_2-PAP-DKP-Lys (**57**) and 1.5 wt% alginate was chosen. At lower concentrations, no gel formation was observed. Two solutions were prepared in DMEM: 3.0 wt% of alginate and 1.2 wt% of **57**. The solution with the gelator was then irradiated at 523 nm for 1 h under periodical mixing.

To a 1.5 mL-vial (screw top) was added 250 µL of hydrogelator solution. Subsequently, 250 µL of the alginate solution were added and mixed by repetitive pipetting. After 15 min a highly viscous liquid was formed. Then, the mixture was irradiated at 410 nm for 30 min and the viscous mixture turned into stable hydrogel. The hydrogels stabilized further by equilibration overnight. Successfully formed gels were prepared in triplicates.

Table 24: Hydrogelation experiments under biological conditions in DMEM.

Composition		Description	T_m [°C]
Gelator **57**	Alginate		
0.3 wt%	0.1 wt%	red liquid	-
0.3 wt%	0.6 wt%	red liquid	-
0.3 wt%	0.9 wt%	red liquid	-

0.6 wt%	**1.5 wt%**	**opaque, red gel**	**75±11**

The gelation of pure F_2-PAP-DKP-Lys (**57**) in Medium199 was investigated and the samples were prepared according to Method A.

Table 25: Hydrogelation experiments under biological conditions in Medium199.

Composition (x mg of **57** + 500 µL M199)	Approx. concentration	Description
10	2 wt%	opaque gel, precipitate formed upon irradiation at 523 nm
15	3 wt%	slightly opaque gel
20	4 wt%	slightly opaque gel

5.7.4 Diffusion of the gelator 57 after addition of Ca^{2+} ions

The following diffusion experiments were performed by Simon Ludwig in his bachelor thesis under the supervision of Anna-Lena Leistner.[212]

Sodium alginate forms gels by addition of divalent ions, for example Ca^{2+}. Therefore, the influence of Ca^{2+} ions on the composite hydrogelator **57** / alginate gels was assessed. For this purpose, 500 µL of a hydrogel sample (0.3 wt% **57** and 0.1 wt% alginate) was prepared in a 1.5 mL crimp top vial according to Method A.

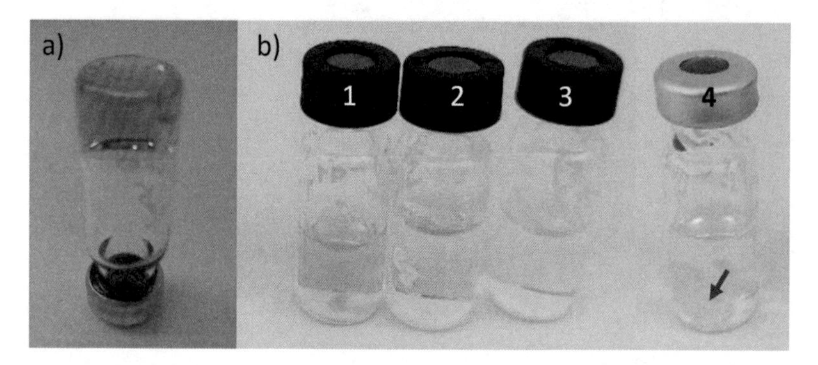

Figure 138: a) Hydrogel (0.3 wt% **57** and 0.1 wt% alginate). b) vial 1-3: Supernatant after 1-3 h, vial 4 contains shrunken gel and clear solution.

On top of these gels was added 500 µL of a 10 wt% solution of $CaCl_2$ in water and equilibrated for 1 h. Subsequently, the supernatant was removed and fresh $CaCl_2$ solution was added. The procedure was repeated three times in total.

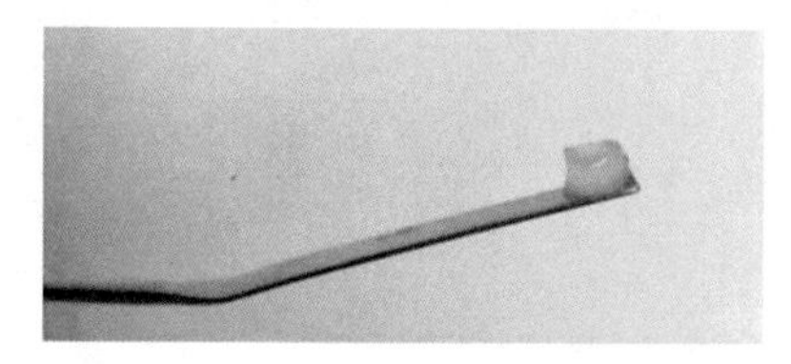

Figure 139: Product of the $CaCl_2$ experiment. The gel shrank to 1/8 of its former size while keeping its original shape.

5.7.5 Microscopy images

Scanning Electron Microscopy (SEM):

For the Scanning Electron Microscopy images, samples of hydrogelator **57** and alginate depicted in Table 26 were prepared with diH_2O after method A. The resulting material was freeze dried by lyophilization. The samples were placed on the measurement plate, additionally fixed with conductive silver and the coated with a thin layer of platinum (10 nm). Volker Zibat operated the microscope.

Table 26: Sample compositions measured under the scanning electron microscope.

Composition		Comment
Gelator **57**	Alginate	
0.6 wt%	0.6 wt%	
0.3 wt%	0.3 wt%	
0.3 wt%	0.2 wt%	
0.3 wt%	0.1 wt%	
0.3 wt%	0.1 wt%	After light induced gel-to-sol transition
0.3 wt%	0.05 wt%	
0.6 wt%	1.5 wt%	Prepared in DMEM

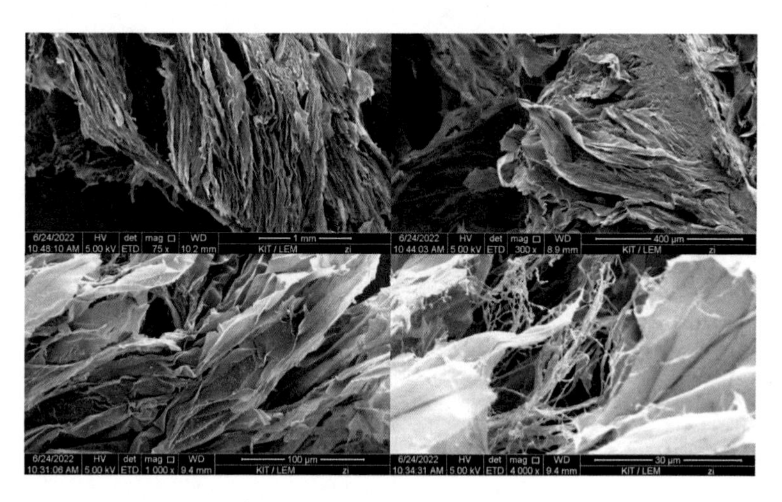

Figure 140: Scanning electron microscopy (SEM) pictures of 0.6 wt% gelator **57** and 0.6 wt% alginate.

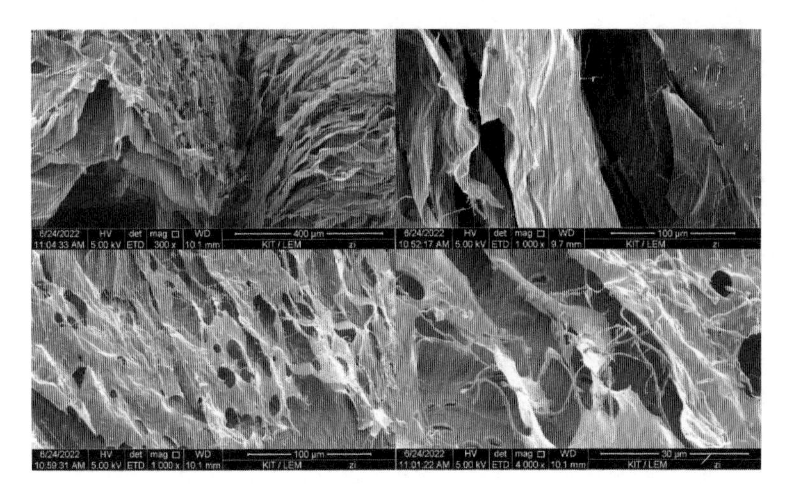

Figure 141: Scanning electron microscopy (SEM) pictures of 0.3 wt% gelator **57** and 0.3 wt% alginate.

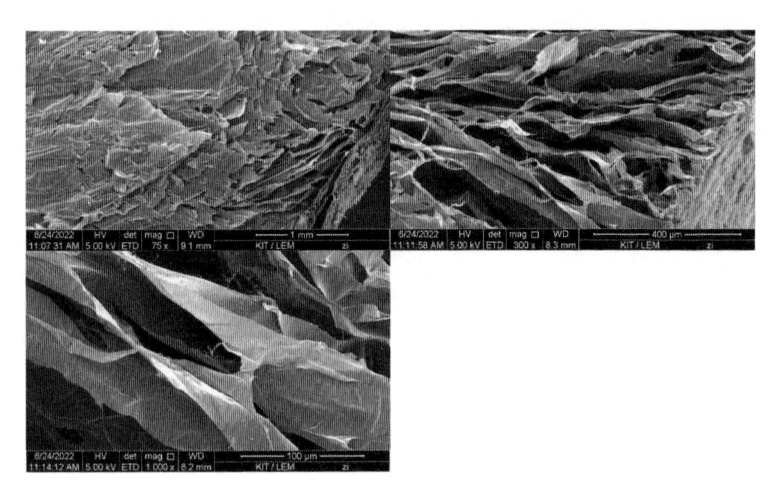

Figure 142: Scanning electron microscopy (SEM) pictures of 0.3 wt% gelator **57** and 0.2 wt% alginate.

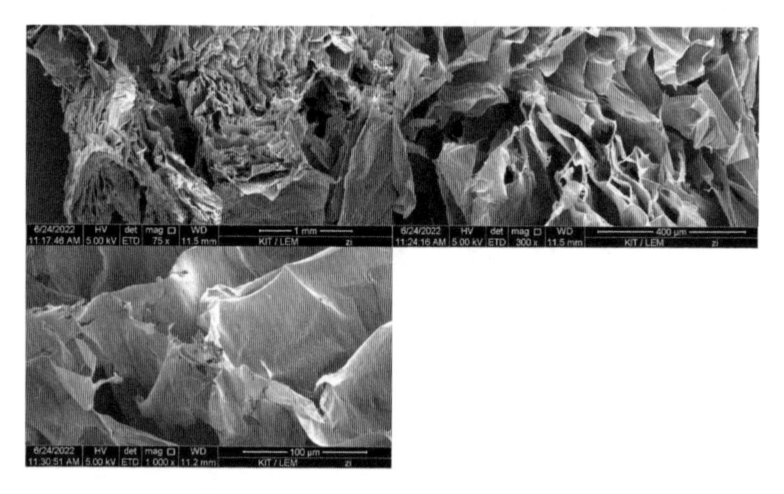

Figure 143: Scanning electron microscopy (SEM) pictures of 0.3 wt% gelator **57** and 0.1 wt% alginate.

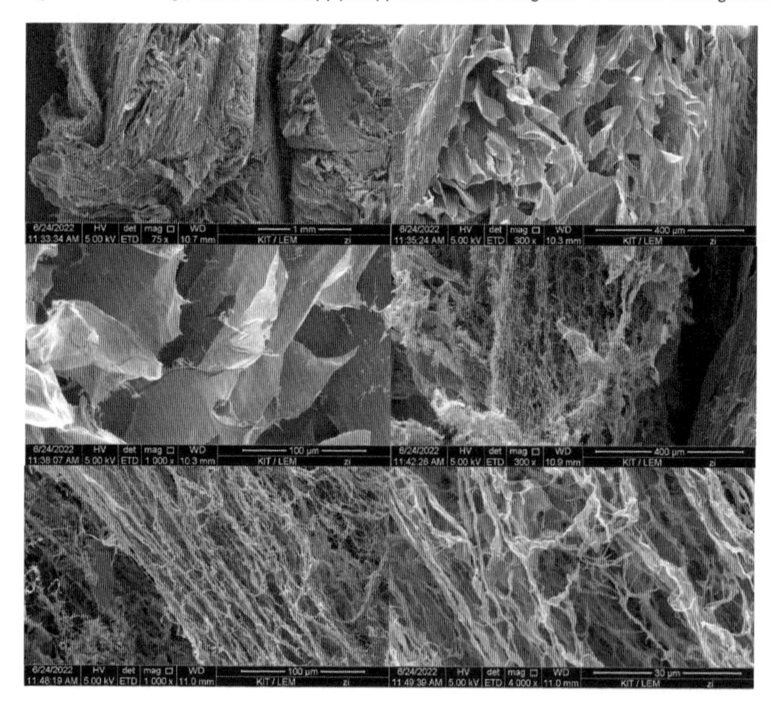

Figure 144: Scanning electron microscopy (SEM) pictures of 0.3 wt% gelator **57** and 0.1 wt% alginate after gel-to-sol transition.

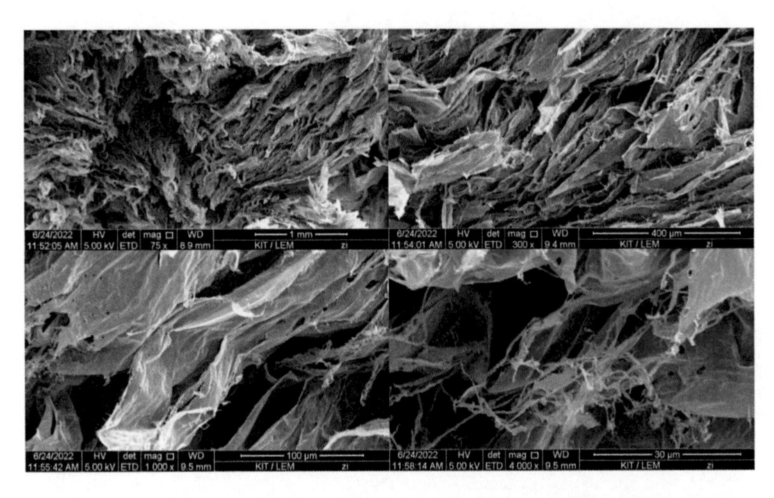

Figure 145: Scanning electron microscopy (SEM) pictures of 0.3 wt% gelator **57** and 0.05 wt% alginate.

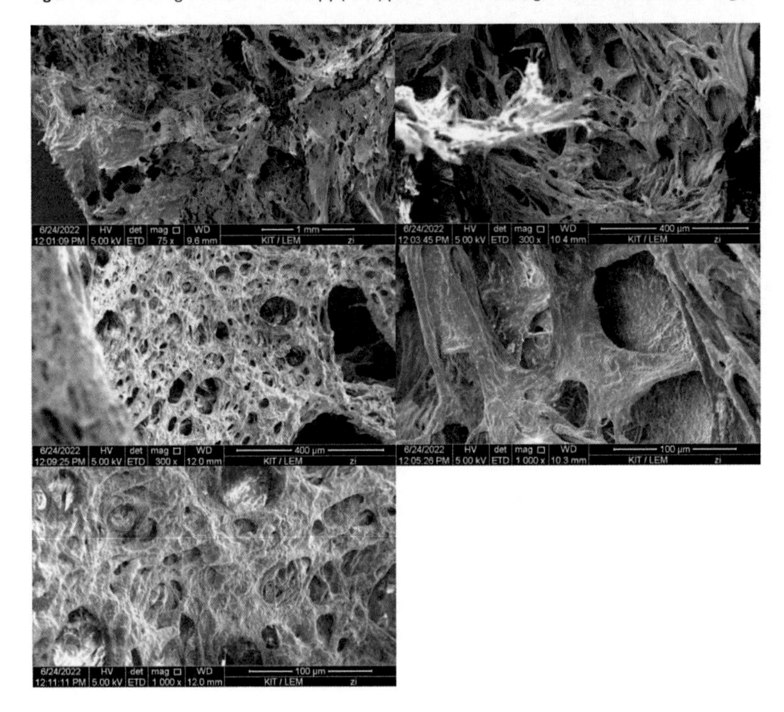

Figure 146: Scanning electron microscopy (SEM) pictures of 0.6 wt% gelator **57** and 1.5 wt% alginate prepared in DMEM.

5.7.6 Rheological experiments

Oscillatory shear experiments were conducted by Maxi Hoffmann on a strain-controlled ARES-G2 (TA Instruments) rheometer with a parallel plate geometry (25 mm diameter and 1 mm gap). A volume of 2.5 mL of a sample consisting of 0.3 wt% hydrogelator **57** and 0.1 wt% alginate (prepared as described in in 5.7.2, Method A) was poured onto the lower plate. The temperature was maintained at 25 °C by a Peltier element. A strain sweep was conducted by varying the strain amplitude in the range $\gamma = 0.01 - 91$ % at an angular frequency $\omega = 6.28$ rad s^{-1} to determine the linear viscoelastic (LVE) regime. A frequency sweep was recorded in the range $\omega = 0.10$ - 100 rad s^{-1} at $\gamma = 0.1$ % within the LVE regime.

The sample consisting of 0.3 wt% hydrogelator **57** and 0.1 wt% alginate was irradiated after the initial measurements. Here, no metal cooling block could be used. To avoid changes of the material`s properties by evaporation, the gel was covered by a glass plate during irradiation (Figure 147). The gel was first irradiated with a 523 nm, 10 W LED for 30 min, then a frequency sweep was carried out. Subsequently, the gel was irradiated with a 410 nm, 10 W LED for 160 sec and equilibrated in the dark for 20 min before measuring again a frequency sweep.

Figure 147: Irradiation setup: Gel on the lower plate is covered by a glass plate.

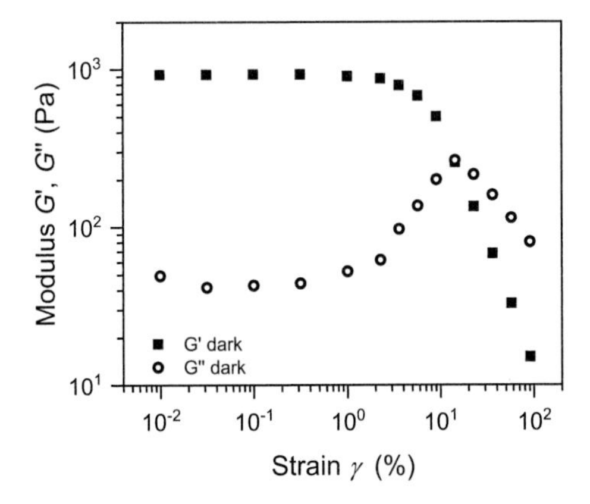

Figure 148: Strain sweep of a sample consisting of 0.3 wt% hydrogelator **57** and 0.1 wt% alginate at an angular frequency ω = 6.28 rad s^{-1} to determine the LVE.

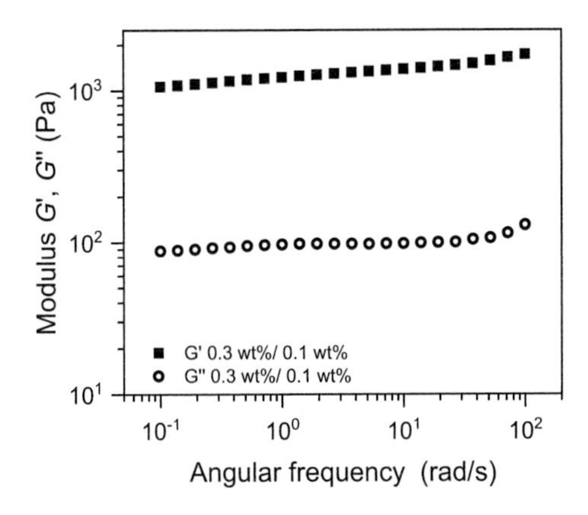

Figure 149: Frequency sweep of a sample consisting of 0.3 wt% hydrogelator **57** and 0.1 wt% alginate in water.

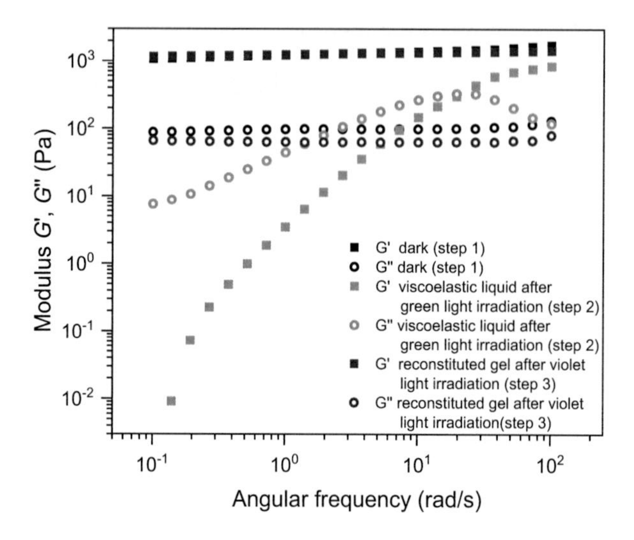

Figure 150: Frequency sweep of a sample consisting of 0.3 wt% hydrogelator **57** and 0.1 wt% alginate. The sample was irradiated directly on the lower plate.

Oscillatory shear experiments were conducted on a strain-controlled ARES-G2 (TA Instruments) rheometer with a parallel plate geometry (25 mm diameter and 1 mm gap). A volume of 2.5 mL of a sample consisting of 0.6 wt% hydrogelator **57** and 0.6 wt% alginate (prepared as described in 5.7.2, Method A) was poured onto the lower plate. The temperature was maintained at 25 °C by a Peltier element. A strain sweep was conducted by varying the strain amplitude in the range γ = 0.01 – 18 % at an angular frequency ω = 6.28 rad s^{-1} to determine the linear viscoelastic (LVE) regime. A frequency sweep was recorded in the range ω = 0.10 - 100 rad s^{-1} at γ = 0.1 % within the LVE regime.

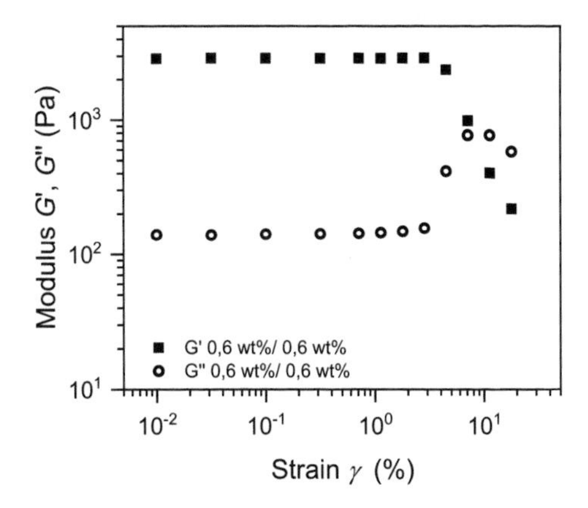

Figure 151: Strain sweep of a sample consisting of 0.6 wt% hydrogelator **57** and 0.6 wt% alginate at an angular frequency ω = 6.28 rad s^{-1} to determine the LVE.

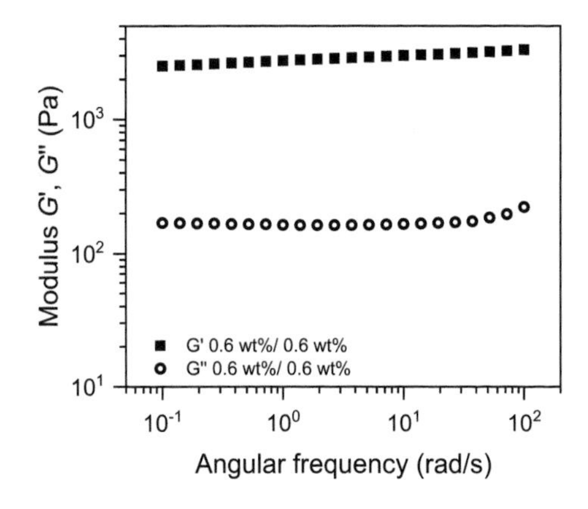

Figure 152: Frequency sweep of a sample consisting of 0.6 wt% hydrogelator **57** and 0.6 wt% alginate in water.

Oscillatory shear experiments were conducted on a strain-controlled ARES-G2 (TA Instruments) rheometer with a parallel plate geometry (25 mm diameter and 1 mm gap). A volume of 2.5 mL of a sample consisting of 0.6 wt% hydrogelator **57** and 1.5 wt% alginate in DMEM (prepared as described in 5.7.2, Method B) was poured onto the lower plate. The temperature was maintained at 25 °C by a Peltier element. A frequency sweep was recorded in the range ω = 0.14 - 100 rad s^{-1} at γ = 0.1 % within the LVE regime.

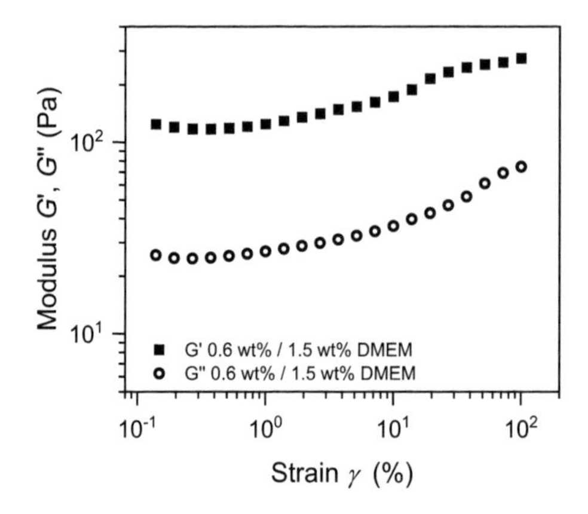

Figure 153: Frequency sweep of a sample consisting of 0.6 wt% hydrogelator **57** and 1.5 wt% alginate in DMEM.

A volume of 2.5 mL of a sample consisting of 2.0 wt% hydrogelator **57** (prepared as described in in 5.7.2, Method A) was poured onto the lower plate. The temperature was maintained at 25 °C by a Peltier element. A strain sweep was conducted by varying the strain amplitude in the range γ = 0.01 – 2.94 % at an angular frequency ω = 6.28 rad s^{-1} to determine the linear viscoelastic (LVE) regime. A frequency sweep was recorded in the range ω = 0.10 - 100 rad s^{-1} at γ = 0.19 % within the LVE regime.

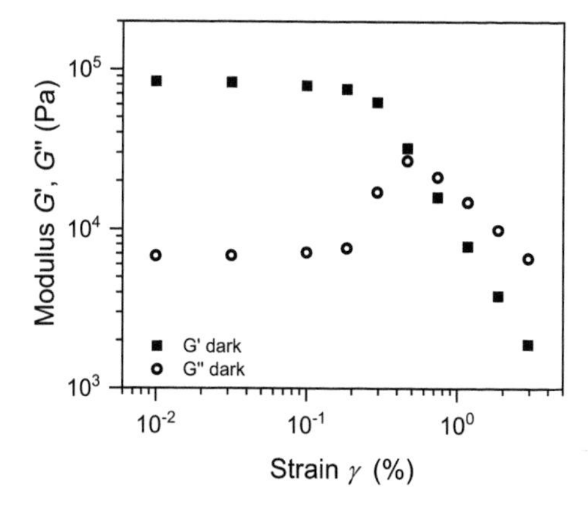

Figure 154: A strain sweep of a sample consisting of 2.0 wt% hydrogelator **57** was conducted by varying the strain amplitude in the range $\boldsymbol{\gamma}$ = 0.01 – 3 % at an angular frequency $\boldsymbol{\omega}$ = 6.28 rad s^{-1} to determine the LVE.

5.8 A new red light triggered LMWG – Cl_4-PAP-DKP-Lys

5.8.1 Synthesis

(*S*)-2-amino-3-(4-nitrophenyl)propanoic acid (102)

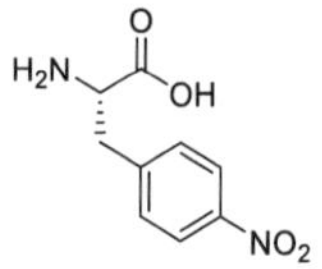

To stirred concentrated sulfuric acid (40 mL) was added *S*-phenylalanine (**70**, 30.0 g, 182 mmol, 1.00 eq) at 50 °C. After the *S*-phenylalanine was dissolved, the mixture was cooled to 0 °C and concentrated nitric acid (29.2 g, 21.0 mL, 463 mmol, 2.55 eq) was added dropwise over 2 h. The reaction mixture was stirred for another 30 min at 0 °C and was allowed to warm to room temperature and stirred for 1 h. The reaction mixture was poured onto ice/water (200 mL), and ammonia (26~28% in water) was added to adjust the pH to 7~8. Then the solution was filtrated and the precipitate was washed with cold water (3× 20 mL) and dried to give the crude product. The crude product was recrystallized from water (250 mL) to give **102** as white crystals (23.1 g, 110 mmol, 61%).

^{1}H NMR (400 MHz, MeOD): δ = 8.21 (d, J = 8.6 Hz, 2H), 7.55 (d, J = 8.4 Hz, 2H), 3.85 (dd, J = 8.2, 4.9 Hz, 1H), 3.40 (dd, J = 14.5, 4.9 Hz, 1H), 3.18 (dd, J = 14.4, 8.2 Hz, 1H) ppm. **^{1}H NMR (400 MHz, D_2O with KOH):** δ = 8.17 (d, J = 8.6 Hz, 1H), 7.44 (d, J = 8.6 Hz, 1H), 3.55 (t, J = 5.6 Hz, 1H), 3.07 (dd, J = 13.4, 5.9 Hz, 1H), 2.97 (dd, J = 13.4, 7.2 Hz, 1H) ppm. **^{13}C NMR (101 MHz, D_2O with KOH):** δ = 181.8, 146.8, 146.4, 130.2, 123.6, 57.3, 40.8 ppm. **TLC:** R_f = 0.45 (developed in 79% DCM, 20% MeOH, 1% Et_3N). **HRMS (EI+):** *m/z* calcd. for $C_9H_{10}N_2O_4$ [M] = 210.0635 Da, found 210.0637 Da. **IR (ATR, $\tilde{\nu}$)** = 3291 (w), 3272 (w), 3259 (w), 3245 (w), 3234 (w), 3210 (w), 3200 (w), 3193 (w), 3109 (w), 3084 (w), 3065 (w), 3053 (w), 3043 (w), 2997 (w), 2982 (w), 2944 (w), 2901 (w), 2885 (w), 2877 (w), 2776 (w), 2752 (w), 2738 (w), 2725 (w), 2646 (w), 1697 (w), 1643 (w), 1611 (s), 1568 (s), 1534 (vs), 1514 (vs), 1494 (m), 1442 (m), 1417 (s), 1344 (vs), 1312 (s), 1293 (m), 1242 (w), 1207 (w), 1191 (w), 1176 (w), 1140 (w), 1105 (m), 1071 (m), 1013 (w), 946 (w), 877 (m), 863 (s), 844 (w), 815 (w), 768 (m), 744 (s), 717 (m), 697 (vs), 653 (s), 630 (m), 615 (m), 567 (w), 524 (vs), 492 (s), 459 (w), 443 (w), 416 (m), 395 (w), 378 (w) cm^{-1}.

(*S*)-2-acetamido-3-(4-nitrophenyl)propanoic acid (110)

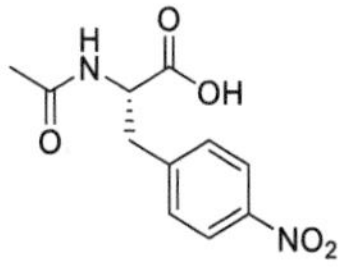

(2*S*)-2-azaniumyl-3-(4-nitrophenyl)propanoate (**102**) (20.0 g, 95.2 mmol, 1.00 eq) was dissolved in methanol (80.0 mL) and acetyl acetate (26.2 g, 24.3 mL, 257 mmol, 2.70 eq) was added. The reaction was heated to reflux for 6 h. The reaction was cooled to room temperature and the solvent was removed under reduced pressure. The crude was recrystallized from EtOH to give the product **110** as beige crystals (16.5 g, 65.4 mmol, 69%).

^{1}H NMR (400 MHz, DMSO-*d*$_6$) δ = 12.75 (s, 1H), 8.25 (d, *J* = 8.2 Hz, 1H), 8.15 (d, *J* = 8.7 Hz, 2H), 7.51 (d, *J* = 8.7 Hz, 2H), 4.49 (ddd, *J* = 9.7, 8.2, 4.9 Hz, 1H), 3.19 (dd, *J* = 13.8, 5.0 Hz, 1H), 2.98 (dd, *J* = 13.8, 9.7 Hz, 1H), 1.77 (s, 3H) ppm. **^{13}C NMR (101 MHz, DMSO-*d*$_6$)** δ = 172.7, 169.2, 146.3, 146.1, 130.4, 123.2, 52.8, 36.5, 22.3 ppm. **HRMS(EI+):** *m/z* calcd. for $C_{11}H_{12}N_2O_5$ [M] = 252.0746 Da, found 252.0739 Da. **IR (ATR, ṽ):** = 3322 (w), 3114 (w), 3081 (w), 2979 (w), 2949 (w), 2932 (w), 2859 (w), 1742 (s), 1730 (s), 1599 (s), 1567 (s), 1553 (s), 1524 (vs), 1493 (w), 1442 (m), 1405 (m), 1384 (w), 1370 (w), 1344 (vs), 1310 (m), 1285 (w), 1256 (w), 1222 (m), 1198 (s), 1184 (vs), 1126 (m), 1105 (m), 1044 (w), 1030 (w), 1014 (w), 990 (w), 969 (w), 929 (w), 877 (m), 858 (s), 846 (vs), 823 (w), 790 (w), 766 (w), 749 (w), 735 (m), 724 (m), 696 (vs), 637 (m), 594 (s), 581 (s), 538 (w), 520 (s), 500 (m), 398 (m) cm^{-1}.

(*S*)-2-acetamido-3-(4-aminophenyl)propanoic acid (111)

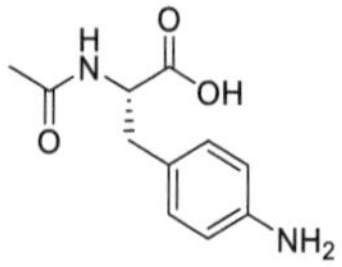

(2*S*)-2-acetamido-3-(4-nitrophenyl)propanoic acid (**110**) (16.0 g, 63.4 mmol, 1.00 eq) was added to a Schlenk flask and dissolved in methanol (667 mL) and Pd/C (160 mg, 1.50 mmol, 0.0237 eq) was added. The atmosphere was changed 3× with H_2 and the mixture was stirred overnight. The reaction mixture was filtered through Celite® and the solvent was removed under reduced pressure to give the pure product **111** in quantitative yield as a beige solid (14.1 g, 63.2 mmol, quant.).

^{1}H NMR (400 MHz, DMSO-*d*$_6$) δ = 8.05 (d, *J* = 8.0 Hz, 1H), 6.85 (d, *J* = 8.3 Hz, 2H), 6.45 (d, *J* = 8.3 Hz, 2H), 4.26 (ddd, *J* = 9.3, 8.0, 5.0 Hz, 1H), 2.83 (dd, *J* = 13.8, 5.0 Hz, 1H), 2.63 (dd, *J* = 13.9, 9.2 Hz, 1H), 1.78 (s, 3H) ppm. **^{13}C NMR (101 MHz, DMSO-*d*$_6$)** δ = 173.5, 169.2, 147.0, 129.5, 124.5, 113.8, 54.1, 36.2, 22.4 ppm. **HRMS(EI+):** *m/z* calcd. for $C_{11}H_{14}N_2O_3$ [M] = 222.1004 Da, found 222.1001 Da. **IR (ATR, ṽ):** = 3291 (w), 3081 (vw), 2918 (w), 2850 (w), 2588 (w), 1728 (vw), 1623 (vs), 1591 (vs), 1545 (vs), 1511 (vs), 1432 (m), 1395 (vs), 1368 (s), 1326 (m), 1292 (s), 1208 (m), 1183 (m), 1123 (m), 1075 (w), 1055 (w), 1040 (w), 1020 (w), 1010 (w), 960 (w), 911 (w), 861 (w), 849 (w), 817 (m), 802 (m), 756 (w), 710 (m), 679 (m), 649 (m), 640 (m), 613 (s), 596 (m), 554 (vs), 527 (vs), 493 (s), 483 (vs), 433 (s), 418 (s), 387 (vs), 377 (s) cm^{-1}.

(*S*,*E*)-2-acetamido-3-(4-(phenyldiazenyl)phenyl)propanoic acid (71)

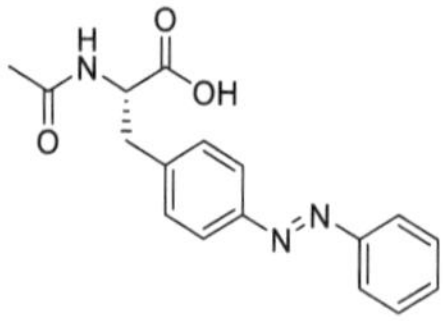

Nitrosobenzene (1.58 g, 14.7 mmol, 1.10 eq) was suspended in glacial AcOH (28 mL) and (2*S*)-2-acetamido-3-(4-aminophenyl)propanoic acid (**111**) (3.50 g,

13.4 mmol, 1.00 eq) was added. The resulting mixture was stirred for 1 d at room temperature. The solution was diluted with toluene (100 mL) and concentrated under reduced pressure. The crude product was purified by silica gel column chromatography (gradient: 3-4% MeOH in DCM with 1% FA; R_f = 0.2 in 1% FA, 4% MeOH in DCM). The pure product **71** was obtained as an orange solid (3.11 g, 9.97 mmol, 75%).

^{1}H NMR (400 MHz, DMSO-*d*$_6$) δ = 12.77 (s, 1H), 8.25 (d, *J* = 8.1 Hz, 1H), 7.94 – 7.86 (m, 2H), 7.82 (d, *J* = 8.2 Hz, 2H), 7.64 – 7.51 (m, 3H), 7.45 (d, *J* = 8.2 Hz, 2H), 4.54 – 4.44 (m, 1H), 3.15 (dd, *J* = 13.8, 4.9 Hz, 1H), 2.95 (dd, *J* = 13.8, 9.6 Hz, 1H), 1.79 (s, 3H) ppm. **^{13}C NMR (101 MHz, DMSO-*d*$_6$)** δ = 173.0, 169.2, 152.0, 150.7, 141.8, 131.4, 130.2, 129.5, 122.5, 122.4, 53.2 ppm. **HRMS (ESI):** *m/z* calcd. for $C_{17}H_{17}N_3O_3$ [M$^+$] = 311.1270 Da, found 311.2343 Da. **IR (ATR, $\tilde{\nu}$):** = 3298 (w), 3078 (vw), 3037 (vw), 2979 (vw), 2929 (w), 2868 (w), 2803 (w), 2619 (w), 2448 (w), 1909 (vw), 1706 (vs), 1643 (w), 1611 (s), 1602 (s), 1548 (vs), 1500 (m), 1483 (w), 1468 (w), 1445 (m), 1418 (m), 1377 (m), 1346 (w), 1300 (s), 1285 (m), 1254 (vs), 1224 (m), 1203 (m), 1184 (w), 1156 (m), 1113 (m), 1103 (m), 1064 (m), 1044 (m), 1011 (m), 980 (m), 965 (s), 929 (m), 853 (vs), 841 (m), 824 (s), 792 (w), 765 (vs), 728 (m), 707 (m), 686 (vs), 664 (vs), 635 (m), 616 (w), 598 (vs), 562 (vs), 531 (vs), 506 (s), 486 (s), 456 (w), 441 (m), 425 (w), 411 (w), 402 (w) cm^{-1}. **UV-Vis (MeCN):** λ_{max} = 323, 440 nm.

(*S*,*E*)-2-acetamido-3-(3,5-dichloro-4-((2,6-dichlorophenyl)diazenyl)phenyl)propanoic acid (72)

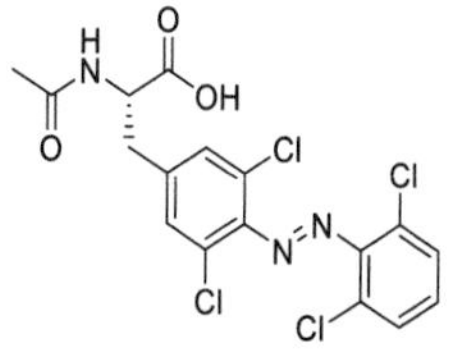

(*S*,*E*)-2-acetamido-3-(4-(phenyldiazenyl)phenyl)propanoic acid (**71**) (3.90 g, 12.5 mmol, 1.00 eq), NCS (8.36 g, 62.6 mmol, 5.00 eq) and Pd(OAc)$_2$ (281 mg,

1.25 mmol, 0.100 eq) were dissolved in AcOH (125 mL) under an argon atmosphere in a sealable reaction tube. The reaction was heated to 140 °C for 2.5 h during which time the color turned dark red. After cooling to room temperature, the mixture was transferred to a separatory funnel and DCM and brine were added. The organic phase was washed with brine (2×), dried over Na_2SO_4 and the solvent was removed under reduced pressure. After purification by column chromatography (10% MeOH in DCM to 1% FA 10% MeOH in DCM, second column chromatography: cH/EtOAc 1:1 1%FA; R_f = 0.2), the product **72** was obtained as a dark red solid (3.01 g, 6.70 mmol, 54%).

^{1}H NMR (400 MHz, DMSO-*d*$_6$) δ = 12.77 (s, 1H), 8.29 (d, J = 8.2 Hz, 1H), 7.70 (d, J = 8.1 Hz, 2H), 7.57 (s, 2H), 7.51 (dd, J = 8.6, 7.7 Hz, 1H), 4.58 – 4.47 (m, 1H), 3.15 (dd, J = 13.8, 5.0 Hz, 1H), 2.94 (dd, J = 13.8, 9.5 Hz, 1H), 1.82 (s, 3H) ppm. **^{13}C NMR (101 MHz, DMSO-*d*$_6$)** δ = 184.1, 163.0, 146.7, 144.7, 142.1, 131.1, 130.6, 129.9, 125.9, 125.9, 52.7, 35.7, 22.3 ppm. **HRMS (ESI):** *m/z* calcd. for $C_{17}H_{13}Cl_4N_3O_3$ [M+Na$^+$] =471.9574 Da, found 471.9567 Da. **IR (ATR, ṽ):** = 2970 (w), 2958 (w), 2897 (m), 2877 (m), 2850 (m), 2701 (w), 1727 (s), 1652 (w), 1591 (m), 1572 (w), 1560 (m), 1548 (s), 1487 (m), 1434 (vs), 1402 (s), 1374 (m), 1336 (w), 1272 (w), 1254 (w), 1200 (vs), 1159 (m), 1137 (m), 1102 (w), 1082 (w), 1057 (m), 983 (w), 967 (w), 933 (w), 902 (w), 875 (m), 857 (w), 837 (w), 805 (vs), 772 (vs), 739 (s), 727 (s), 667 (w), 640 (m), 595 (m), 569 (w), 551 (w), 535 (w), 516 (m), 494 (m), 460 (w), 452 (w), 433 (w), 422 (w), 411 (w), 388 (w), 375 (m) cm^{-1}.

(*E*)-2-amino-3-(3,5-dichloro-4-((2,6-dichlorophenyl)diazenyl)phenyl)propanoic acid (73)

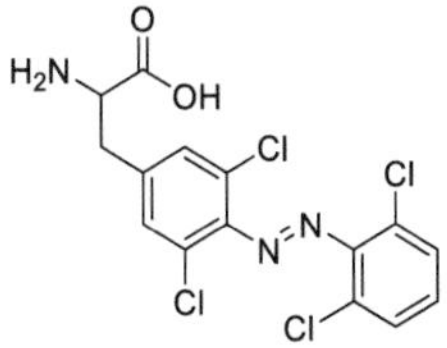

(*S*,*E*)-2-acetamido-3-(3,5-dichloro-4-((2,6-dichlorophenyl)diazenyl)phenyl)propanoic acid (**72**) (3.00 g, 6.68 mmol, 1.00 eq) was suspended in 6 M HCl and heated to 110 °C for 24 h. Upon cooling to room temperature, the product precipitated. It was filtered off, washed with diluted HCl and dried to give the racemic product (chiral RP-HPLC chromatography; 1100 Series from Agilent Technologies with a G1322A degasser, a G1211A pump, a G1313A autosampler, a G1316A column oven, and a G1315B diode array system using a Daicel Chiralpak® AD-H (4.6 × 250 mm, 5 µm particle size) column. The separations were performed with a 20 min isocratic mixture of HPLC-grade hexane/isopropanol (85/15), flow rate 1 mL/min, wavelength 256 nm or 330 nm for detection: signals at 7.4 min 50% peak area, 12.5 min 50% peak area) as a dark orange solid (2.11 g, 5.18 mmol, 78%).

^{1}H NMR (400 MHz, DMSO-*d*$_6$) δ = 14.06 (s, 1H), 8.52 (s, 3H), 7.72 (d, J = 8.1 Hz, 2H), 7.67 (s, 2H), 7.53 (t, J = 8.1 Hz, 2H), 4.35 (t, J = 6.4 Hz, 1H), 3.32 (dd, J = 14.2, 5.8 Hz, 1H), 3.21 (dd, J = 14.2, 7.1 Hz, 1H) ppm. **^{13}C NMR (101 MHz, DMSO-*d*$_6$)** δ = 170.0, 146.7, 145.3, 139.3, 131.2, 131.1, 130.0, 126.1, 125.9, 52.5, 34.4 ppm. **HRMS (FAB):** *m/z* calcd. for $C_{15}H_{12}Cl_4N_3O_2$ [M+H] = 405.9678 Da, found 405.9679 Da. **IR (ATR, ṽ):** = 2975 (m), 2953 (m), 2897 (s), 2859 (s), 2711 (w), 2653 (w), 2609 (w), 2582 (w), 2557 (w), 2548 (w), 2537 (w), 1728 (vs), 1591 (m), 1574 (w), 1560 (m), 1551 (m), 1502 (s), 1465 (w), 1434 (vs), 1405 (m), 1397 (m), 1360 (w), 1258 (w), 1218 (s), 1201 (vs), 1157 (w), 1137 (w), 1112 (w), 1084 (w), 1057 (m), 892 (w), 880 (m), 853 (m), 837 (m), 807 (vs), 772 (vs), 741 (s), 724 (m), 683 (w), 667 (w), 643 (w), 611 (w), 595 (w), 571 (w), 552 (w), 535 (w), 516 (m), 494 (w), 476 (w), 469 (w), 455 (w), 436 (w), 415 (w), 398 (w), 387 (w), 375 (m) cm^{-1}.

(*E*)-2-((*tert*-butoxycarbonyl)amino)-3-(3,5-dichloro-4-((2,6-dichloro-phenyl)diazenyl)phenyl)propanoic acid (74)

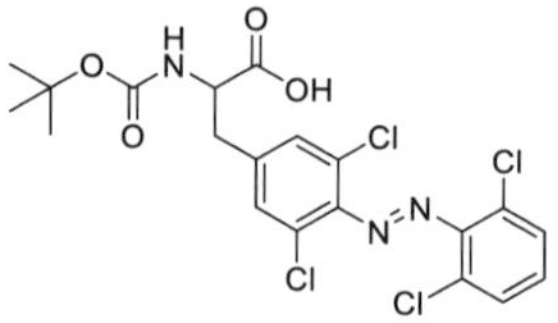

73 (2.10 g, 5.16 mmol, 1.00 eq) was suspended in 1,4-dioxane (31.5 mL) and $NaHCO_3$ (2.17 g, 25.8 mmol, 5.00 eq) in water (25 mL) was added, followed by Boc_2O (1.24 g, 5.67 mmol, 1.10 eq). The mixture was stirred at room temperature for 20 h. EtOAc was added, and the mixture was transferred to a separatory funnel. The organic phase was washed with $KHSO_4$, followed by brine and dried over Na_2SO_4. The crude was purified by column chromatography (DCM/MeOH/FA 97:2:1; R_f = 0.26). The product was obtained as a dark red foam (2.18 g, 5.16 mmol, 83%).

^{1}H NMR (400 MHz, DMSO-*d*$_6$) δ = 12.94 (s, 1H), 7.70 (d, J = 8.1 Hz, 2H), 7.59 (s, 2H), 7.54 – 7.48 (m, 1H), 7.23 (d, J = 8.7 Hz, 1H), 4.23 (ddd, J = 10.7, 8.5, 4.4 Hz, 1H), 3.16 (dd, J = 13.6, 4.6 Hz, 1H), 2.90 (dd, J = 13.7, 10.8 Hz, 1H), 1.32 (s, 9H) ppm. **^{13}C NMR (101 MHz, DMSO-*d*$_6$)** δ = 172.9, 155.4, 146.8, 144.6, 142.5, 131.0, 130.7, 129.9, 125.9, 125.8, 78.2, 54.2, 28.1 ppm. **HRMS (FAB):** *m/z* calcd. for $C_{20}H_{19}Cl_4N_3O_4$ [M+H] = 505.0124 Da, found 505.0126 Da. **IR (ATR, $\tilde{\nu}$):** = 2976 (w), 2929 (w), 2859 (w), 1710 (s), 1594 (w), 1562 (w), 1547 (w), 1506 (m), 1452 (w), 1435 (s), 1394 (s), 1367 (s), 1341 (m), 1254 (s), 1222 (m), 1200 (s), 1157 (vs), 1119 (s), 1081 (m), 1055 (s), 1026 (m), 966 (w), 888 (m), 868 (vs), 849 (s), 802 (vs), 773 (vs), 738 (s), 710 (m), 694 (m), 684 (m), 673 (m), 649 (m), 612 (s), 591 (s), 578 (m), 551 (m), 537 (m), 516 (s), 506 (m), 480 (m), 472 (m), 458 (m), 436 (s), 425 (m), 409 (m), 399 (s), 390 (m) cm^{-1}. **UV-Vis (MeCN):** λ_{max} = 295, 443 nm.

Methyl *N*⁶-(*tert*-butoxycarbonyl)-*N*²-(2-((*tert*-butoxycarbonyl)amino)-3-(3,5-dichloro-4-((*E*)-(2,6-dichlorophenyl)diazenyl)phenyl)propanoyl)-L-lysinate (75)

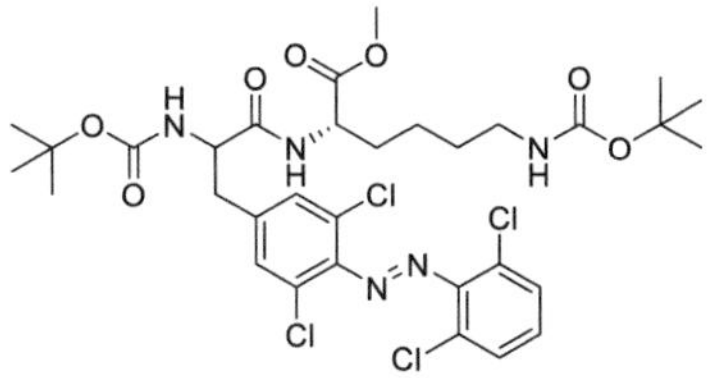

74 (2.17 g, 4.06 mmol, 1.00 eq), HBTU (1.62 g, 4.27 mmol, 1.05 eq) and DIPEA (1.31 g, 1.77 mL, 10.2 mmol, 2.50 eq) were dissolved in anhydrous *N,N*-dimethylformamide (6 mL) and stirred for 10 min at room temperature under an argon atmosphere. H-Lys(Boc)-OMe·HCl (1.27 g, 4.27 mmol, 1.05 eq) and DIPEA (1.31 g, 1.77 mL, 10.2 mmol, 2.50 eq) were added and the resulting mixture was stirred for 2 h under an argon atmosphere until the starting material was consumed. The reaction was quenched by sat. NH_4Cl solution and extracted with EtOAc. The organic phase was washed with sat. NH_4Cl (3×) and brine (1×) and was dried over Na_2SO_4. The crude was purified by column chromatography (cH/EtOAc 2:1 --> 1:1; R_f= 0.46) to yield 3.00 g of **75** (4.00 mmol, 98%) as a dark red solid.

¹H NMR (400 MHz, DMSO-*d*₆) *δ* = 8.39 (dd, *J* = 34.2, 7.7 Hz, 1H), 7.70 (dd, *J* = 8.0, 1.2 Hz, 2H), 7.63 (s, 2H), 7.51 (t, *J* = 8.1 Hz, 1H), 7.01 (dd, *J* = 9.0, 2.8 Hz, 1H), 6.76 (q, *J* = 6.3 Hz, 1H), 4.29 (ddt, *J* = 16.7, 8.2, 3.5 Hz, 2H), 3.64 (d, *J* = 8.2 Hz, 3H), 3.04 (ddd, *J* = 18.3, 13.3, 4.1 Hz, 1H), 2.90 (p, *J* = 6.8 Hz, 2H), 2.82 – 2.74 (m, 1H), 1.70 (dt, *J* = 8.2, 3.4 Hz, 1H), 1.60 (dtd, *J* = 14.0, 9.2, 5.8 Hz, 1H), 1.35 (d, *J* = 8.2 Hz, 9H), 1.30 (d, *J* = 2.5 Hz, 9H), 1.27 – 1.21 (m, 4H) ppm. **¹³C NMR (101 MHz, DMSO-*d*₆)** *δ* = 172.5, 171.2, 170.9, 155.5, 155.2 (d, *J* = 8.3 Hz), 146.8, 144.5, 142.6, 131.0, 130.7, 129.9, 125.9 (d, *J* = 3.6 Hz), 125.8, 78.2 (d, *J* = 5.3 Hz), 77.3, 54.9, 52.1 – 51.5 (m), 37.2, 36.7, 31.0, 30.7, 29.0 (d, *J* = 7.3 Hz), 28.2 (d, *J* = 2.7 Hz), 28.0, 26.3, 22.5 (d, *J* = 6.8 Hz) ppm. **HRMS (FAB):** *m/z* calcd. for $C_{32}H_{42}Cl_4N_5O_7$ [M+H] =748.1833 Da, found 748.1835 Da. **IR (ATR, ṽ):** = 3306 (w), 3298 (w), 3067 (vw), 2975 (w), 2952 (w), 2932 (w), 2864 (vw), 1742 (w), 1679 (s), 1657 (vs), 1592 (w), 1511 (s), 1453 (m), 1435 (s), 1391 (m), 1366 (s),

1268 (m), 1248 (s), 1204 (m), 1163 (vs), 1084 (w), 1048 (m), 1018 (m), 943 (w), 904 (w), 853 (w), 802 (m), 776 (s), 737 (m), 710 (w), 664 (w), 647 (m), 633 (m), 619 (m), 611 (m), 594 (m), 572 (m), 555 (w), 541 (m), 534 (m), 520 (m), 504 (w), 492 (w), 475 (w), 462 (w), 435 (w), 425 (w), 411 (w), 399 (w), 378 (w) cm^{-1}. **UV-Vis (MeCN):** λ_{max} = 295, 452 nm.

Methyl (2-amino-3-(3,5-dichloro-4-((*E*)-(2,6-dichlorophenyl)diazenyl)-phenyl)propanoyl)-*L*-lysinate (76) PAP-Cl_4-Lys-OMe

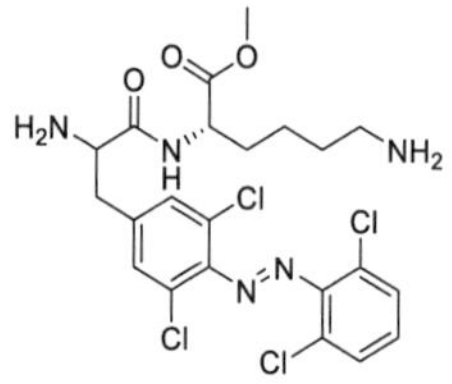

Compound **75** (1.30 g, 1.73 mmol, 1.00 eq) was dissolved in DCM (17 mL). 2,2,2-trifluoroacetic acid (17 mL) and 1% (v/v) triisopyropylsilane were added. The mixture was stirred for 1 h at room temperature. Toluene was added and the solvent was removed under reduced pressure to yield the product as the TFA salt (dark red solid, 1.15 g, 1.73 mmol, quant.). The product was used without further purification in the next step.

^{1}H NMR (400 MHz, DMSO-*d*$_6$) δ = 9.07 (dd, *J* = 15.1, 7.6 Hz, 1H), 8.35 (t, *J* = 6.6 Hz, 3H), 7.86 (s, 3H), 7.71 (d, *J* = 8.1 Hz, 2H), 7.61 (s, 2H), 7.55 – 7.49 (m, 1H), 4.39 – 4.28 (m, 1H), 4.23 (s, 1H), 3.67 (d, *J* = 3.8 Hz, 3H), 3.24 (ddd, *J* = 14.0, 5.1, 2.3 Hz, 1H), 3.13 – 3.01 (m, 1H), 2.83 – 2.71 (m, 2H), 1.84 – 1.60 (m, 2H), 1.56 (q, *J* = 7.5 Hz, 2H), 1.45 – 1.25 (m, 2H) ppm. **^{13}C NMR (101 MHz, DMSO-*d*$_6$)** δ = 171.9, 167.9 (d, *J* = 3.8 Hz), 146.7, 139.0 (d, *J* = 23.4 Hz), 131.2, 131.1 (d, *J* = 8.8 Hz), 130.0, 126.2, 125.9, 118.1, 53.1, 52.8, 52.2 (d, *J* = 3.2 Hz), 52.0, 38.5 (d, *J* = 2.5 Hz), 36.2, 35.8, 30.7, 30.4, 26.6, 22.2 (d, *J* = 5.0 Hz), 19.2 ppm. **HRMS (FAB):** *m/z* calcd. for $C_{22}H_{26}Cl_4N_5O_3$ [M+H] = 548.0784 Da, found 548.0785 Da. **IR (ATR, ṽ):** = 3067 (w), 2958 (w), 2927 (w), 2861 (w), 2649 (vw), 1735 (w), 1664 (vs), 1594 (m), 1561 (m), 1551 (m), 1534 (m), 1473 (w), 1460 (w), 1435 (m), 1401 (w), 1357 (w), 1307 (w), 1290 (w), 1258 (w), 1196 (vs), 1177 (vs), 1132 (vs), 1014 (w), 1003 (w), 990 (w), 941 (w), 908 (w), 884 (w), 839 (m), 798 (vs), 776 (s), 738 (w), 722 (vs), 705 (m), 636 (w), 628 (w), 596 (w), 560 (w), 551 (w), 538 (w), 517 (w), 483 (w), 467 (w), 433 (w), 409 (w), 390 (w), 375 (w) cm^{-1}. **UV-Vis (MeCN):** λ_{max} = 293, 444 nm.

(3*S*,6*S*)-3-(4-aminobutyl)-6-(3,5-dichloro-4-((*E*)-(2,6-dichlorophenyl)diazenyl)benzyl)piperazine-2,5-dione and (3*S*,6*R*)-3-(4-aminobutyl)-6-(3,5-dichloro-4-((*E*)-(2,6-dichlorophenyl)diazenyl)benzyl)-piperazine-2,5-dione (58), PAP-Cl₄-DKP-Lys

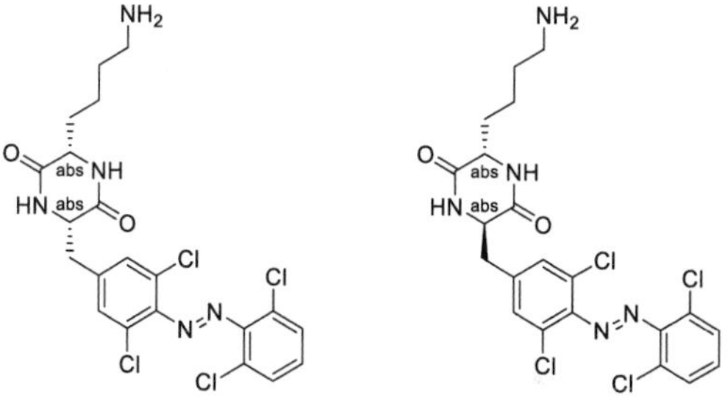

PAP-Cl_4-Lys-OMe **76** (1.13 g, 1.70 mmol, 1.00 eq) was dissolved in butan-2-ol (70.0 mL). Subsequently, AcOH (377 mg, 359 µL, 6.27 mmol, 3.69 eq), 4-methylmorpholine (220 mg, 242 µL, 2.18 mmol, 1.28 eq) and DIPEA (319 mg, 429 µL, 2.46 mmol, 1.45 eq) were added. The mixture was heated to 120 °C for 2 h, then cooled down. Ca. 80% of the solvent was removed under reduced pressure. The precipitate was filtered off and washed 3x with cold MeCN. The product was obtained as 7:3 diastereomeric mixture as a dark red-brown solid (464 mg, 0.897 mmol, 53%). The isomers were separated by preparative HPLC (30 min gradient of 35-65% MeCN in H_2O, 0.1% TFA; **product 1** (**P1**) at 13-14 min, **product 2** (**P2**) at 16-17 min) and obtained as TFA salts after lyophilization. The orange solids were dissolved in H_2O and ion exchange resin (Amberlite® IRA-900, Cl-form) was added to remove excess of TFA, which inhibits gelation processes. The resin mixture was stirred for 2 d, then the resin was removed by filtration and the solution was again lyophilized.

P1: ^{1}H NMR (400 MHz, DMSO-d_6) δ 8.31 (d, J = 19.9 Hz, 2H), 7.70 (d, J = 8.1 Hz, 2H), 7.60 – 7.43 (m, 3H), 4.33 (t, J = 5.1 Hz, 1H), 3.81 (t, 1H), 3.19 (dd, J = 13.6, 4.7 Hz, 1H), 3.03 (dd, J = 13.6, 5.1 Hz, 1H), 2.61 (t, J = 7.7 Hz, 2H), 1.52 – 1.28 (m, 4H), 1.12 – 0.89 (m, 2H) ppm. **^{13}C NMR (101 MHz, DMSO-d_6)** δ 167.3, 166.4, 146.7, 144.9, 141.0, 131.4, 131.1, 129.9, 126.0, 125.8, 54.7, 53.5, 38.6, 36.9, 31.9, 27.0, 20.4 ppm. **HRMS (FAB):** *m/z* calcd. for $C_{21}H_{22}{}^{35}Cl_2{}^{37}Cl_2N_5O_2$ [M+H] = 518.0307 Da, found 518.0309 Da. **IR (ATR, ṽ):** = 3201 (w), 3189 (w), 3050 (w), 3017 (w), 2959 (w), 2935 (w), 2918 (w), 2894 (w), 2874 (w), 2529 (vw), 2167 (vw), 1680 (vs), 1594 (w), 1574 (w), 1560 (w), 1550 (w), 1509 (w), 1499 (w), 1456 (m), 1435 (m), 1402 (w), 1330 (w), 1288 (vw), 1252 (vw), 1201 (w), 1179 (w), 1130 (w), 1102 (w), 1060 (vw), 1033 (vw), 931 (vw), 914 (vw), 836 (w), 807 (m), 799 (m), 778 (s), 739 (w), 722 (w), 676 (w), 657 (w), 611 (vw), 595 (w), 518 (w), 510 (w), 492 (w), 462 (w), 442 (w), 433 (w), 418 (w), 409 (w) cm^{-1}.

P2: ^{1}H NMR (400 MHz, DMSO-d_6) δ 8.31 (d, J = 14.7 Hz, 2H), 7.94 (s, 3H), 7.71 (d, J = 8.2 Hz, 2H), 7.57 (s, 2H), 7.52 (t, J = 8.1 Hz, 1H), 4.31 (t, J = 5.5 Hz, 1H), 3.65 (t, J = 5.2 Hz, 1H), 3.20 (dd, J = 13.6, 5.3 Hz, 1H), 3.04 (dd, J = 13.6, 5.2 Hz, 1H), 2.73 (d, J = 8.1 Hz, 2H), 1.66 (ddt, J = 25.0, 13.7, 6.4 Hz, 2H), 1.53 (p, J = 7.5 Hz, 2H), 1.45 – 1.21 (m, 2H) ppm. **^{13}C NMR (101 MHz, DMSO-d_6)** δ 167.9, 166.8, 146.7, 144.8, 141.0, 131.3, 131.2, 129.9, 125.9, 125.8, 54.6, 53.5, 38.6, 36.7, 32.1, 26.7, 20.7 ppm. **HRMS (FAB):** *m/z* calcd. for $C_{21}H_{22}{}^{35}Cl_2{}^{37}Cl_2N_5O_2$ [M+H] = 518.0307 Da, found 518.0308 Da. **IR (ATR, ṽ):** 3204 (w), 3186 (w), 3047 (w), 2958 (w), 2885 (w), 1664 (vs), 1594 (w), 1574 (w), 1560 (w), 1550 (w), 1513 (vw), 1449 (m), 1434 (m), 1401 (w), 1334 (w), 1288 (vw), 1252 (vw), 1201 (w), 1179 (w), 1130 (w), 1102 (w), 1060 (vw), 1030 (vw), 963 (vw), 939 (vw), 911 (vw), 857 (w), 834 (w), 807 (m), 778 (m), 739 (w), 721 (w), 691 (vw), 616 (vw), 595 (vw), 516 (vw), 452 (w), 414 (vw), 394 (vw) cm^{-1}.

5.8.2 Photophysical properties of **58**

The diastereomers **P1** and **P2** of compound **58** were investigated separately. First, the isosbestic point was determined, which was then used for quantifying the PSS by HPLC measurements.

UV/Vis isomerization experiments

A 1.50 mM stock solution was prepared in 10% H_2O in MeCN and diluted to reach a final concentration of 150 µM. The cuvette with the sample was irradiated with light of different wavelengths (410 nm, 430 nm, 523 nm, 660 nm with 630 nm cut-off filter) directly before the measurement.

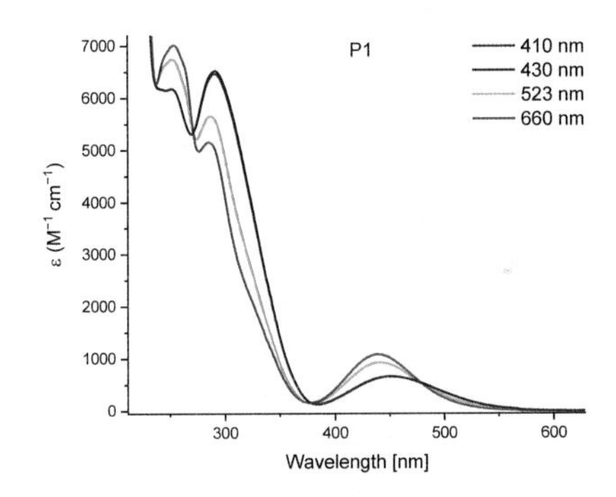

Figure 155: Molar absorptivity of compound **58 P1** (150 µM, 10% H_2O/MeCN) before and after irradiation. The shown curves after irradiation correspond to the photostationary states and distinct isosbestic points are displayed at 272 nm, 379 nm, and 479 nm (the point at 272 nm was used for quantification by HPLC).

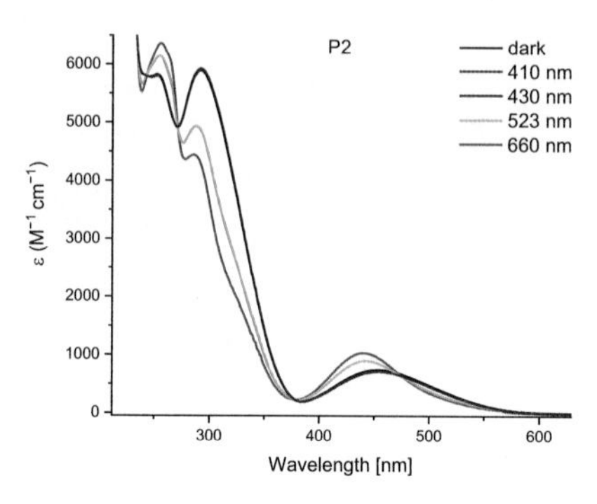

Figure 156: Molar absorptivity of compound **58 P2** (150 µM, 10% H_2O/MeCN) before and after irradiation. The shown curves after irradiation correspond to the photostationary states and distinct isosbestic points are displayed 272 nm, 379 nm, and 475 nm (the point at 272 nm was used for quantification by HPLC).

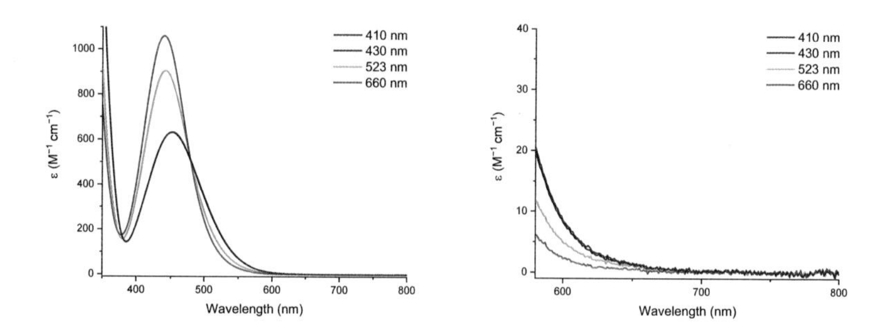

Figure 157: Molar absorptivity of compound **58 P1** (1.50 mM, 10% H_2O/MeCN) after irradiation. The shown curves correspond to the photostationary states; absorption >600 nm is visible.

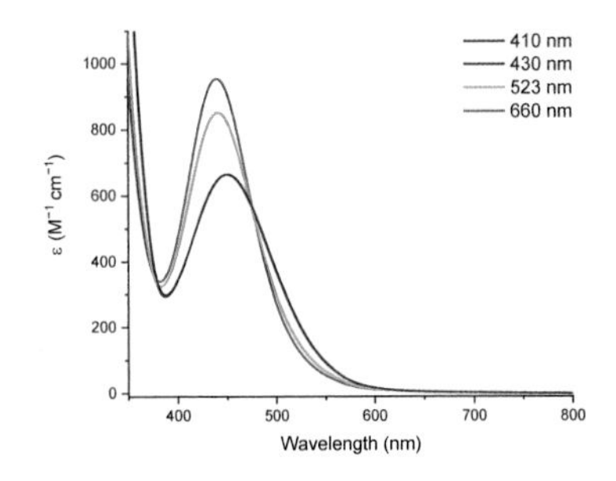

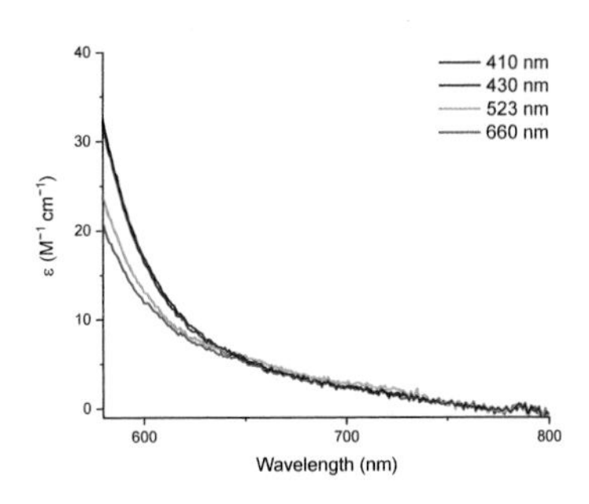

Figure 158: Molar absorptivity of compound **58 P2** (1.50 mM, 10% H_2O/MeCN) after irradiation. The shown curves correspond to the photostationary states; absorption >600 nm is visible.

For HPLC analysis, compound **58 P1** and **P2** were dissolved in DMSO (500 μM) and subsequently irradiated (410 nm, 430 nm, 455 nm, 523. nm, 660 nm) until no further change in the isomer ratio was detected by HPLC at the isosbestic point of 272 nm.

Table 27: PSS of **58** determined by isosbestic point at 272 nm in the 500 μM solution in DMSO by HPLC. Values agree with NMR analysis and are depicted as % of *Z*-isomer. * a 630 nm cut off filter was used.

Compound	410 nm	430 nm	455 nm	523 nm	660 nm*
P1	12	14	19	53	85
P2	12	14	19	55	85

Thermal stability

The thermal stability was determined for compound **58 P1** and **P2.** Samples at a concentration of 500 μM were prepared in AcOH and MeCN/H_2O 1:1 and irradiated at 660 nm to yield a high PSS of the *Z*-isomer. The solutions were kept at 60 °C (AcOH additionally at 25 °C) and the isomer ratio was determined by HPLC in intervals. The obtained data was processed by calculating the $\ln(X_0/X_t)$, where X is the percentage of the respective *Z*-isomer and linear fitting (Equation (1)) of the obtained values. The calculated slope corresponds to the degradation rate constant *k* which is used to calculate the half-life $t_{1/2}$.

(1)
$$x_t = x_0 \cdot e^{-k \cdot t} \leftrightarrow ln\left(\frac{x_0}{x_t}\right) = k \cdot t$$

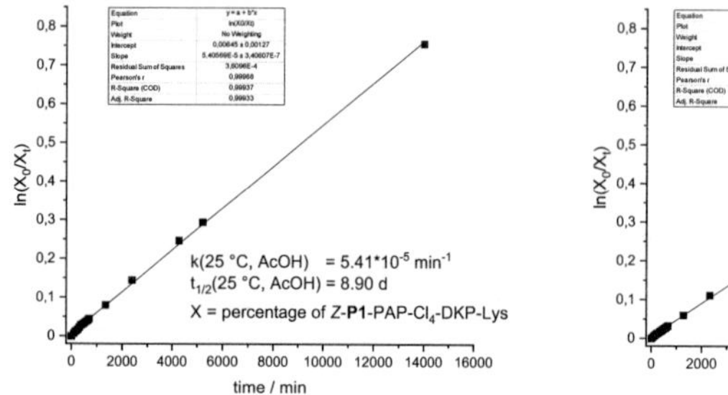

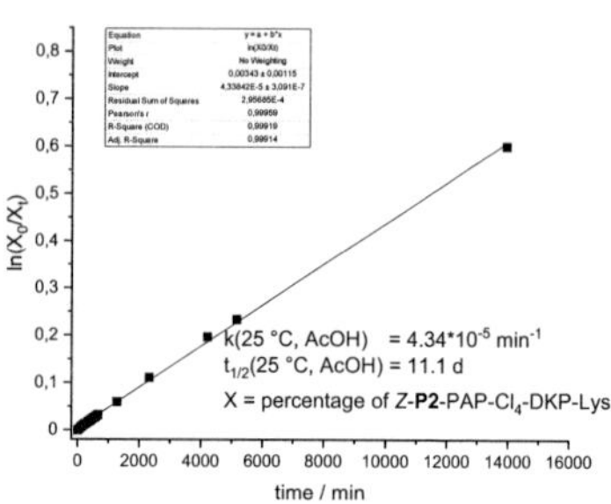

Figure 159: Linear fit of the decay of the *Z*-isomer of compound **58 P1** and **P2** at 25 °C in AcOH for first-order kinetics.

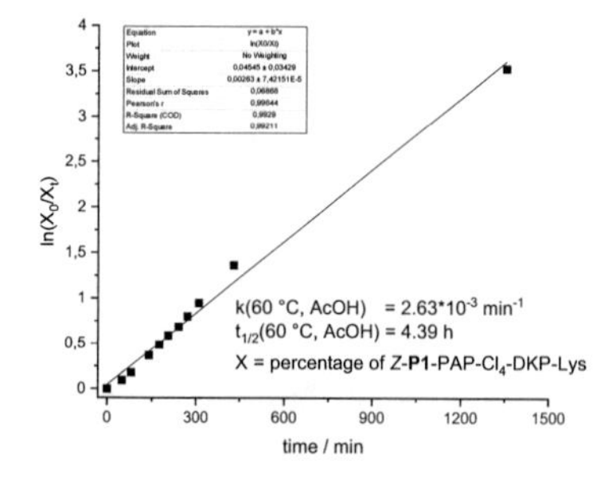

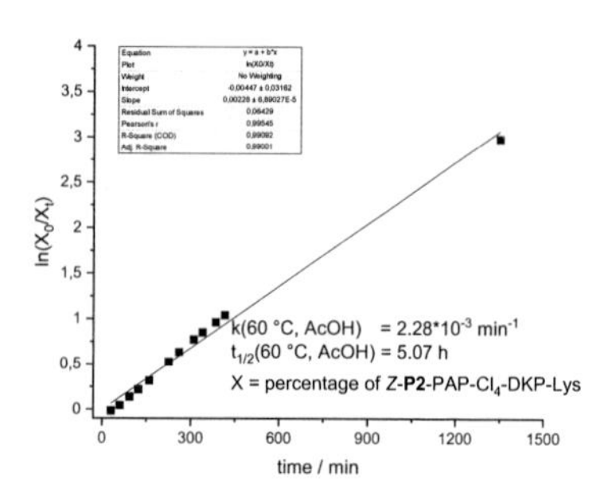

Figure 160: Linear fit of the decay of the *Z*-isomer of compound **58 P1** and **P2** at 60 °C in AcOH for first-order kinetics.

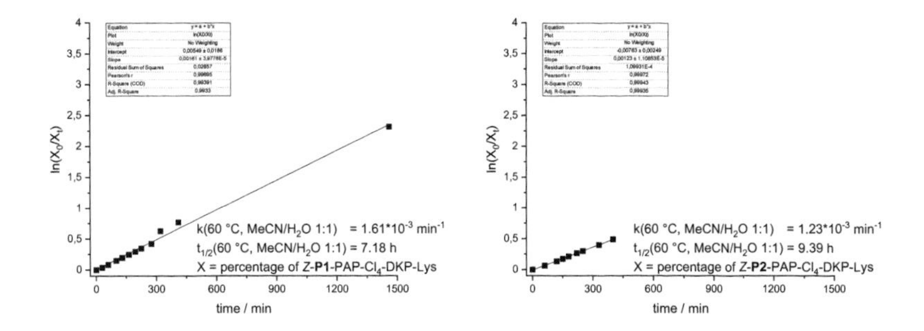

Figure 161: Linear fit of the decay of the *Z*-isomer of compound **58 P1** and **P2** at 60 °C in MeCN/H_2O 1:1 for first-order kinetics.

5.8.3 Gelation properties of compound 58

Gel preparation: To a 1.5 mL-vial (crimp top, 12×32 mm) was added the photochromic material **58** as a powder and 500 µL of the aqueous solution. This suspension was sonicated for 2 min followed by heating to 80 °C in a vial block. After equilibration at 80 °C for 5 min, the sample was heated to the boiling point by a heat gun. The hot solution became completely clear and upon cooling the hydrogels formed. Successfully formed gels were prepared in triplicates.

Melting temperatures were determined in triplicates. Gels were prepared as previously described. The vials were then put upside-down in a slowly stirred water bath (60 rpm) on a magnetic hotplate stirrer equipped with a thermometer. The water bath was heated (1.5 °C/min) to 100 °C while the gel started to shrink and released liquid.

Table 28: Gelation properties of compound **58** in PBS buffer before HPLC purification. The samples contain a 7:3 mixture of **P1**:**P2**.

Composition of the solution (x mg of **58** + 500 µL PBS)	Approx. Concentration	Description *after 5 min cooling **after 1 h cooling	T_m (°C) [1]shrinking
10	2.0 wt%	Clear gel*	>100
5	1.0 wt%	Clear gel*	>100
3	0.6 wt%	Clear gel*	>100

2.5	0.5 wt%	Clear gel*	>100
2	0.4 wt%	Clear gel*	>100
1.5	0.3 wt%	Clear gel*	>100
1	0.2 wt%	Clear gel*	92[1]
0.5	0.1 wt%	Clear gel*	74[1]
0.25	0.05 wt%	Clear weak gel**	-

Table 29: Gelation properties of compound **58** in Ringer's solution before HPLC purification. The samples contain a 7:3 mixture of **P1**:**P2**. Only single measurements were done.

Composition of the solution (x mg of **58** + 500 µL Ringer's solution)	Approx. Concentration	Description *after 5 min cooling **after 1 h cooling	T_m (°C) [1]shrinking
1.5	0.3 wt%	Clear gel*	69[1]
1	0.2 wt%	Clear gel*	-
0.5	0.1 wt%	Clear weak gel**	-

So far, gelation was only observed from the precipitate collected after the reaction. HPLC purification gives the TFA salt of the product. However, the gel system is highly sensitive on salt concentrations and pH (see 5.5.3), therefore ion exchange resin (Amberlite® IRA-900, Cl-form) was used to remove excess of TFA. After this treatment gelation was observed, the results are summarized in Table 30.

Table 30: Gelation properties of compound **58** after HPLC purification and ion exchange. Measurements were done in triplicates; single measurements are marked in grey.

Composition (Ratio **P1**:**P2**, gel in 500 µL PBS)		Approx. Concentration	Description *after 5 min cooling	T_m (°C)
P1	**P2**			
7	3	0.2 wt%	Clear gel[1]*	>99[2]
7	3	0.1 wt%	Clear gel[1]*	64-X[2]
3	7	0.2 wt%	Clear gel[1]*	94-99[2]
3	7	0.1 wt%	Clear gel[1]*	80-84[2]
1	1	0.2 wt%	Clear gel[1]*	91[2]

1	1	0.1 wt%	Clear gel[1*]	64[2]
1	0	0.4 wt%	Clear gel[1*]	-
1	0	0.3 wt%	Clear gel[1*]	Stable at 100 °C
1	0	0.2 wt%	Clear gel[1*]	90[2]
0.5	0	0.1 wt%	Clear Gel[*]	
0	1	0.4 wt%	Clear gel[1*]	-
0	1	0.3 wt%	Clear gel[1*]	90[2]
0	1	0.2 wt%	Clear gel[1*]	-

[1] no complete dissolution; [2]shrinking and releasing liquid instead of melting, reversible by heating >100 °C

5.8.4 Rheological experiments

Oscillatory shear experiments were conducted by Maxi Hoffmann on a strain-controlled ARES-G2 (TA Instruments) rheometer with a parallel plate geometry (40 mm diameter and 1 mm gap). A volume of 8 mL of a sample consisting of 0.3 wt% and 1.0 wt% hydrogelator **58** in PBS was poured directly onto the lower plate (50 mm diameter) at ca. 95 °C. The sample was covered with parafilm and incubated overnight in darkness. During the measurement, the temperature was maintained at 25 °C by a Peltier element. A strain sweep was conducted by varying the strain amplitude in the range $\gamma = 10^{-4} - 5 \cdot 10^{-2}$ % at an angular frequency $\omega = 10$ rad s^{-1} to determine the linear viscoelastic (LVE) regime. A frequency sweep was recorded in the range $\omega = 0.10 - 100$ rad s^{-1} at $\gamma = 0.01$ % within the LVE regime.

Additionally, a temperature dependent frequency sweep was performed for the 0.3 wt% sample in the range of 25 °C to 90 °C with a heating rate of 5 °C/min.

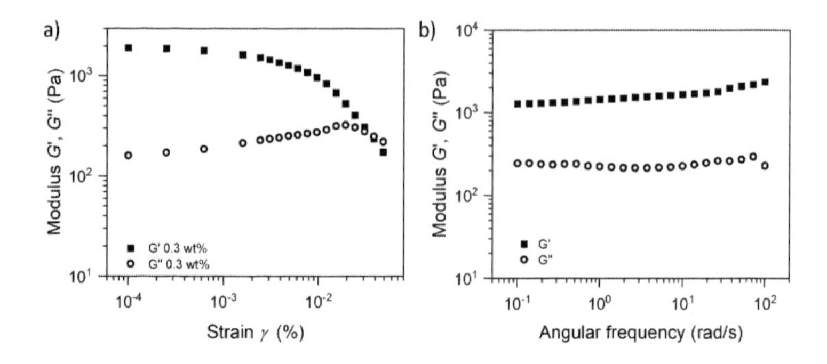

Figure 162: a) Strain sweep of a sample consisting of 0.3 wt% hydrogelator **58** in PBS at an angular frequency ω = 10 rad s^{-1}. b) frequency sweep of a sample consisting of 0.3 wt% hydrogelator **58** in PBS at γ = 0.01 % within the LVE regime.

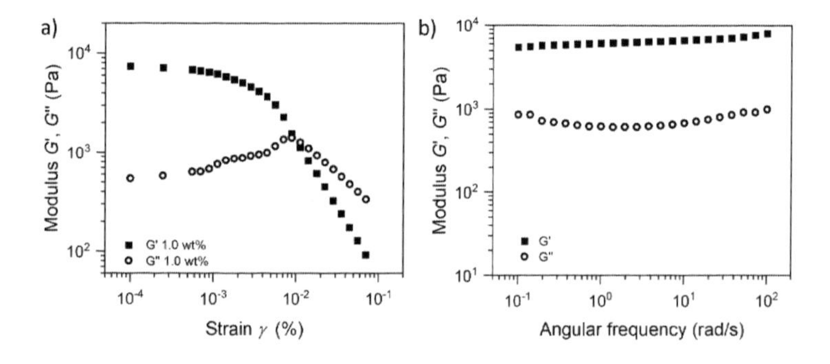

Figure 163: a) Strain sweep of a sample consisting of 1.0 wt% hydrogelator **58** in PBS at an angular frequency ω = 10 rad s^{-1}. b) frequency sweep of a sample consisting of 1.0 wt% hydrogelator **58** in PBS at γ = 0.01 % within the LVE regime.

5.8.5 SEM characterization

For the Scanning Electron Microscopy images, a sample of hydrogelator **58** before HPLC (P1:P2, 7:3) was prepared in PBS at concentrations of 1.0 wt%, 0.5 wt% and 0.1 wt%. The resulting material was freeze dried by lyophilization. The sample was placed on the measurement plate, additionally fixed with conductive silver, and then coated with a thin layer of platinum (6.5 nm).

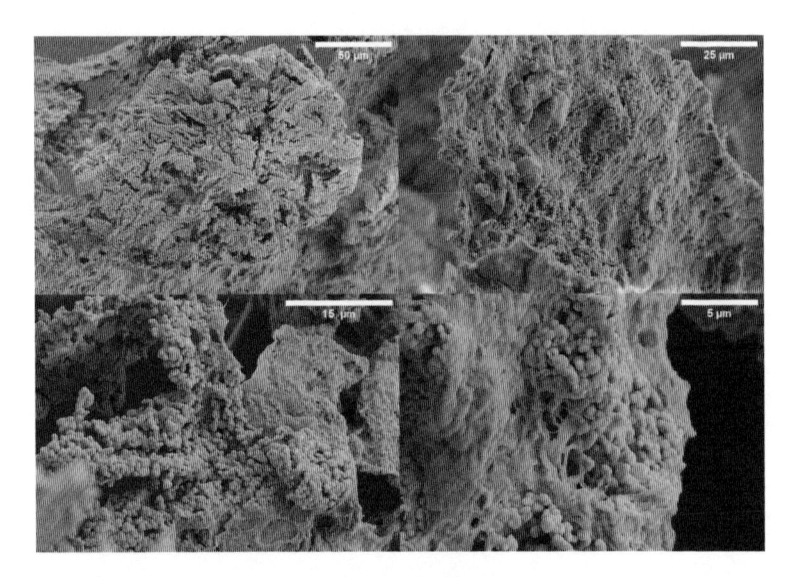

Figure 164: SEM images of freeze-dried gel (0.1 wt% Cl_4-PAP-DKP-Lys **58** in PBS).

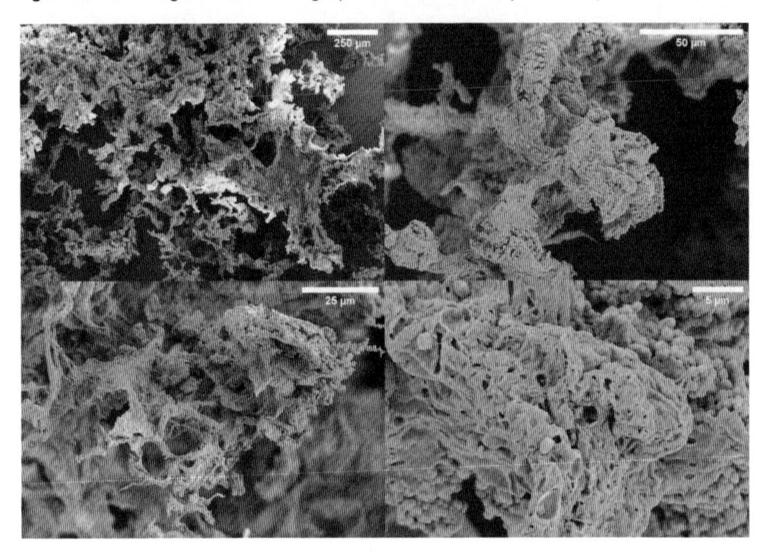

Figure 165: SEM images of freeze-dried gel (0.5 wt% Cl_4-PAP-DKP-Lys **58** in PBS).

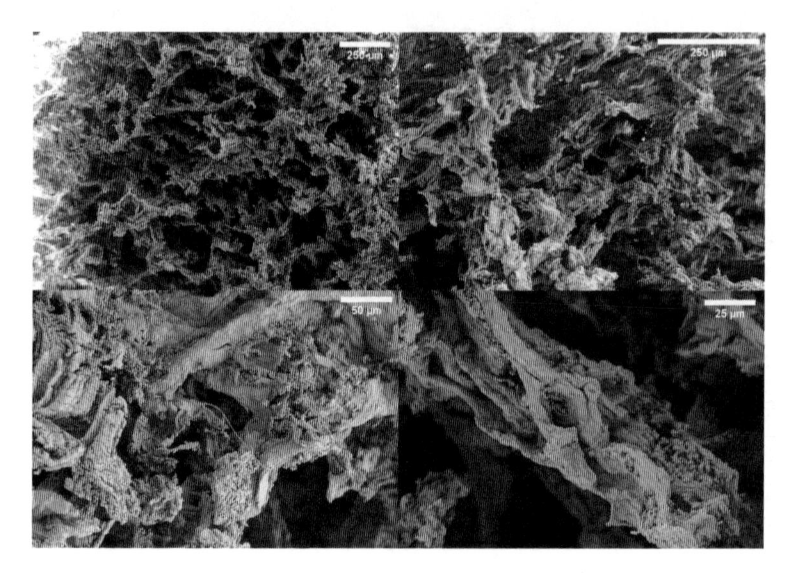

Figure 166: SEM images of freeze-dried gel (1.0 wt% Cl_4-PAP-DKP-Lys **58** in PBS).

5.8.6 Cell viability assays

Compatibility of the smart materials with the biological environment was assessed for the hydrogelator **58** using cell viability assays (MTT assays). A human cancerous (HeLa) cell line was treated with increasing concentrations of the gelator **58** (**P1** and **P2**) in order to determine the range of IC_{50} values. Treatment of the cells was also performed with a mixture obtained upon irradiation red light (660 nm), until the photostationary state was achieved (60 min). The experimental protocols are summarized below and the results, summarized in Table 31, were plotted in Figure 167 and Figure 168.

Protocol for the cell viability assay on cancerous HeLa cells

Hela cells were grown in DMEM (Dulbecco's Modified Eagle Medium), which was modified with 10% FCS (fetal calf serum) and 1% penicillin/streptomycin solution (10,000 units/mL of penicillin and 10,000 µg/mL of streptomycin) in a humid incubator at 37 °C with 5% CO_2. Cells were detached from the surfaces with Trypsin-EDTA (0.25%) from Gibco®. Cells were washed with PBS (Phosphate-Buffered Saline) from Gibco®. 96-well plates with a flat bottom were prepared by filling all wells on the outer border with 200 µL PBS and the remaining wells with 100 µL of a cell suspension (30.000 cells/mL) in DMEM. The prepared plate was incubated overnight to ensure cell attachment to the well-bottom and cell growth.

For the dilution-series of each compound, a stock solution of DMEM modified with 0.25% DMSO was prepared to ensure that all cells are treated with the same conditions. Consequently, the first sample of the dilution series was prepared by dissolving the substance in DMSO and adding a specific amount of this solution to a specific amount of non-modified DMEM so that a final concentration of 0.25% DMSO is reached.

To apply the substances to the 96-well plate, the DMEM was removed without disturbing the cells grown in the plate and adding subsequently 100 µL to each well. To ensure the same treatment to the control rows, the DMEM was removed from the wells and DMSO-modified DMEM (100 µL) was added to the corresponding wells. The 96-well plate was incubated for 48 h.

The positive control was treated with 5 µL of Triton™ X-100 detergent (10% solution (w/v)) per well for at least 5 min before adding 15 µL of MTT dye-solution (3-(4,5-dimethylthiazol-2-yl)-2,5-diphenyltetrazoliumbromid in water / CellTiter 96®Non-Radioactive Cell Proliferation Assay from Promega) to all sample wells and incubating for 3 h in the dark. 100 µL of stop solution was added after incubation to stop the reduction of MTT to formazan, therefore prevent from overreaction and solubilize the formazan crystals. After 24 h of solubilization in the incubator, the plate was read out with a plate reader (BioTek® EPOCH^2, Gen5 Data Analysis) by measuring the absorption of each well at 570 nm.

Absorbance data was averaged over the technical replicates, the positive control subtracted as background, then normalized to viable cell count from negative control cells (% control) as 100% using Microsoft Excel 2019 Version 1808

(Microsoft Corporation). Higher accuracy can be obtained by measuring in triplicates and will be performed before publication. Data were plotted against the log of agonist concentration (log_{10}([agonist]) (M)) with mean and SD in GraphPad Prism Version 9.1.1 for Windows, GraphPad Software, San Diego, California USA.

Table 31: Cell viability of Cl_4-PAP-DKP-Lys **58** against HeLa cell line in the dark state and after irradiation.

		P1 dark		**P1** 660 nm	
Concentration	Concentration	Cell viability		Cell viability	
	[M]	[%]	Stdev	[%]	Stdev
1 mM	1.00E-03	**-5**	**0.16**	-5	7.14
100 µM	1.00E-04	95	3.75	**41**	**10.33**
10 µM	1.00E-05	96	3.52	79	11.19
1 µM	1.00E-06	95	2.68	89	8.73
100 nM	1.00E-07	98	2.23	99	17.27
10 nM	1.00E-08	98	3.13	91	9.34
1 nM	1.00E-09	97	2.73	90	13.18
100 pM	1.00E-10	101	9.76	95	6.55

		P2 dark		**P2** 660 nm	
Concentration	Concentration	Cell viability		Cell viability	
	[M]	[%]	Stdev	[%]	Stdev
1 mM	1.00E-03	-5	3.86	-5	4.88
100 µM	1.00E-04	-6	3.71	26	6.64
10 µM	1.00E-05	**47**	**4.93**	**49**	**7.11**
1 µM	1.00E-06	71	5.83	77	7.76
100 nM	1.00E-07	84	5.56	81	10.12
10 nM	1.00E-08	94	3.74	92	8.31
1 nM	1.00E-09	89	3.77	93	6.84
100 pM	1.00E-10	96	1.75	88	12.16

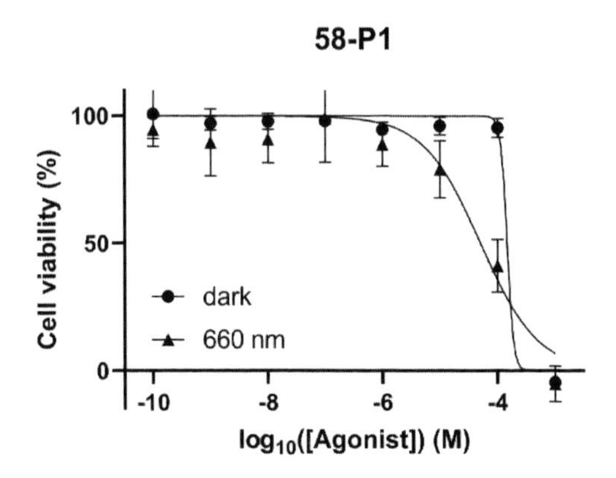

Figure 167: Cell viability assays of human cancerous HeLa cells treated with increasing concentrations of dark adapted **58-P1** (*E*-1) and **58** irradiated with red light until the photostationary state is achieved.

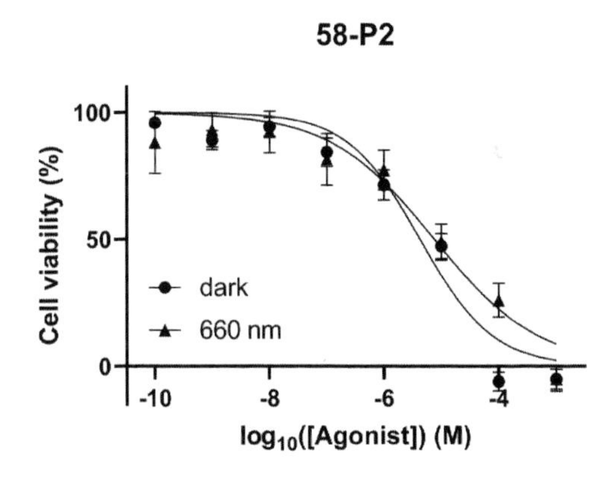

Figure 168: Cell viability assays of human cancerous HeLa cells treated with increasing concentrations of dark adapted **58-P2** (*E*-1) and **58** irradiated with red light until the photostationary state is achieved.

5.8.7 Chiral HPLC analysis

The stereochemical outcome of the synthesis of amino acid **73** has been investigated with chiral RP-HPLC chromatography on the HPLC chromatograph 1100 Series from AGILENT TECHNOLOGIES with a G1322A degasser, a G1211A pump, a

G1313A autosampler, a G1316A column oven, and a G1315B diode array system using a Daicel Chiralpak® AD-H (4.6 × 250 mm, 5 µm particle size) column. The separations were performed with a 20 min isocratic mixture of HPLC-grade hexane/isopropanol (85/15), flow rate 1 mL/min, slit = 4 nm, wavelength 256 nm or 330 nm for detection. The photochromic amino acid **73** was obtained as racemic mixture (Figure 169).

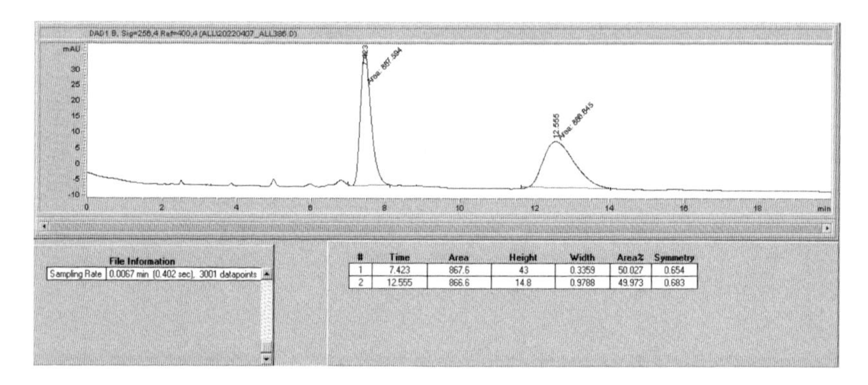

#	Time	Area	Height	Width	Area%	Symmetry
1	7.423	867.6	43	0.3359	50.027	0.654
2	12.555	866.6	14.8	0.9788	49.973	0.683

Figure 169: Chromatogram of chiral HPLC analysis of **73**, detection at 256 nm.

5.9 A new red light triggered LMWG – CL_2-F_2-PAP-DKP-Lys

5.9.1 Synthesis

The following syntheses towards compound **59** were performed by Mario Michael Most in his bachelor thesis under the supervision of Anna-Lena Leistner and are cited in verbatim.[200]

Methyl L-serinate HCl (83)

The synthesis was performed according to Vaswani *et al.*[213]

(2*S*)-2-Amino-3-hydroxypropanoic acid (50.0 g, 476 mmol, 1.00 eq) was dissolved in methanol (200 mL). Thionyl dichloride (65.1 g, 39.7 mL, 547 mmol, 1.15 eq) was added dropwise at 0 °C and the resulting mixture was allowed to warm to 21 °C and stirred at room temperature overnight. The solvent was removed under reduced pressure and the white solid residue was dissolved in ca 140 mL of MeOH. The product precipitated by dripping the solution into diethyl ether. The product **83** was obtained as a colorless solid (70.2 g, 451 mmol, 95%).

^{1}H NMR (400 MHz, DMSO-*d*$_6$): δ = 8.58 (s, 3H), 5.62 (s, 1H), 4.09 (t, *J* = 3.6 Hz, 1H), 3.82 (t, *J* = 3.0 Hz, 2H), 3.74 (s, 3H) ppm. **^{13}C NMR (101 MHz, DMSO-*d*$_6$)**: δ = 168.5, 59.4, 54.3, 52.7 ppm. **HRMS (EI):** *m/z* calcd. for $C_4H_{10}NO_3$ $[M+H]^+$ = 120.0655 Da, found 120.0655 Da. IR (ATR, ṽ) = 3343 (s), 3187 (vw), 3152 (vw), 3023 (m), 2966 (s), 2917 (vs), 2751 (w), 2732 (w), 2662 (w), 2635 (w), 2604 (w), 2551 (w), 2490 (vw), 1926 (vw), 1745 (vs), 1708 (w), 1592 (m), 1578 (w), 1509 (vs), 1472 (m), 1443 (m), 1431 (m), 1381 (w), 1344 (w), 1296 (m), 1249 (vs), 1242 (vs), 1187 (m), 1159 (m), 1128 (m), 1094 (vs), 1038 (vs), 979 (m), 966 (s), 899 (m), 844 (w), 795 (w), 674 (vw), 666 (vw), 659 (vw), 572 (vs), 564 (vs), 469 (m), 429 (w), 404 (w) cm^{-1}.

Methyl acetyl-L-serinate (84)

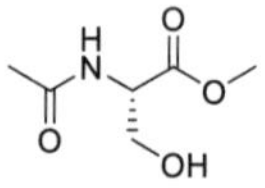

Methyl L-serinate hydrochloride (**83**, 10.0 g, 64.3 mmol, 1.00 eq) was dissolved in dry DCM (200 mL) under argon atmosphere. The solution was cooled to 0 °C and Et_3N (13.7 g, 18.8 mL, 135 mmol, 2.10 eq) was added. Subsequently, acetyl chloride (5.30 g, 4.82 mL, 67.5 mmol, 1.05 eq) was added dropwise at 0 °C. The resulting mixture was stirred for 1 h at room temperature until the TLC showed full conversion. Then, the solvent was removed under reduced pressure. EtOAc (150 mL) was added, the mixture was filtered, and the filtrate was evaporated under reduced pressure. Afterwards, the crude product was purified by silica gel column chromatography using DCM/MeOH 9:1. The product **84** was obtained as a slightly yellowish oil (8.99 g, 55.8 mmol, 87%).

^{1}H NMR (400 MHz, $CDCl_3$): δ = 7.14 (d, J = 7.9 Hz, 1H), 4.51 (dt, J = 7.7, 3.7 Hz, 1H), 4.26 (s, 1H), 3.86 (dd, J = 11.3, 4.0 Hz, 1H), 3.74 (dd, J = 11.3, 3.4 Hz, 1H), 3.66 (s, 3H), 1.95 (s, 3H) ppm. **^{13}C NMR (101 MHz, $CDCl_3$)**: δ = 171.3, 171.1, 62.4, 54.7, 52.5, 22.7 ppm. **TLC:** R_F = 0.38 (developed in DCM/MeOH 9:1). **HRMS (EI):** *m/z* calcd. for $C_6H_{12}NO_4$ $[M+H]^+$ = 162.0761 Da, found 162.0763 Da. **IR (ATR, $\tilde{\nu}$)** = 3292 (m), 3080 (w), 3003 (w), 2953 (w), 2887 (w), 2851 (w), 1737 (vs), 1647 (vs), 1534 (vs), 1435 (vs), 1374 (vs), 1346 (s), 1285 (vs), 1208 (vs), 1146 (vs), 1077 (vs), 1040 (vs), 980 (s), 895 (m), 856 (m), 793 (m), 759 (m), 664 (s), 594 (vs), 574 (vs), 528 (vs), 499 (vs), 452 (s), 443 (s), 433 (s), 408 (m), 377 (s) cm^{-1}.

Methyl (*R*)-2-acetamido-3-iodopropanoate (85)

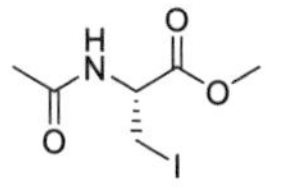

Triphenylphosphine (18.9 g, 72.1 mmol, 1.30 eq) was dissolved in dry DCM (230 mL) under argon atmosphere. Imidazole (4.91 g, 72.1 mmol, 1.30 eq) was added and dissolved (ca. 15 min) to gain a clear solution. Afterwards, the solution was cooled to 0 °C and molecular iodine (18.3 g, 72.1 mmol, 1.30 eq) was added slowly in darkness. The orange suspension was stirred for 10 min at 20 °C and then cooled to 0 °C again. Methyl acetyl-L-serinate **84** (8.99 g, 55.5 mmol, 1.00 eq) was dissolved in dry DCM (60 mL) and added dropwise within 1 h in darkness. The suspension became more red and white precipitation was observed. After the reaction mixture was stirred at room temperature for 2 h, the progress was controlled by TLC and the starting material **84** was fully converted. The mixture was filtered through a short column with silica gel and cH/EtOAc (1:1) as the eluent. Then, the solvent was removed under reduced pressure yielding a yellowish solid which was purified by silica gel column chromatography using DCM/MeOH 99:1. The product **85** (7.72 g, 22.0 mmol, 40%) was isolated as a slightly yellowish solid with a purity of ca. 78% determined by ^{1}H NMR spectroscopy (contaminant: Triphenylphosphine oxide and decomposition product methyl 2-acetamidoacrylate).

The product was stored in darkness and frozen to prevent elimination of HI to the decomposition product. Additionally, the reaction and purification were performed at the same day. Due to decomposition the ^{13}C-NMR was not measured.

^{1}H NMR (400 MHz, $CDCl_3$): δ = 6.27 (s, 1H), 4.79 (dt, J = 7.2, 3.7 Hz, 1H), 3.82 (s, 3H), 3.61 (dd, J = 3.7, 2.8 Hz, 2H), 2.08 (s, 3H) ppm. **TLC:** R_F = 0.20 (developed in DCM/MeOH 99:1). **HRMS (FAB):** *m/z* calcd. for $C_6H_{11}NO_3I$ [M+H] = 271.9778 Da, found 271.9777 Da. **IR (ATR, ṽ)** = 3361 (w), 3299 (s), 3216 (w), 3077 (w), 3048 (w), 2996 (w), 2951 (w), 2849 (w), 1723 (vs), 1638 (vs), 1545 (vs), 1438 (vs), 1411 (s), 1370 (vs), 1340 (vs), 1302 (vs), 1222 (vs), 1191 (vs), 1146 (vs), 1133 (vs), 1027 (vs), 1001 (vs), 984 (s), 925 (m), 904 (m), 868 (m), 849 (w), 807 (w), 779 (s), 707 (s), 681 (s), 609 (vs), 592 (vs), 526 (m), 475 (s), 455 (vs) cm^{-1}.

Decomposition product: methyl 2-acetamidoacrylate **86**

^{1}H NMR (400 MHz, $CDCl_3$): δ = 7.71 (s, 1H), 6.60 (s, 1H), 5.88 (d, J = 1.5 Hz, 1H), 3.85 (s, 3H), 2.13 (s, 3H) ppm.

***Tert*-butyl (4-bromo-2-fluorophenyl)-carbamate (112)**

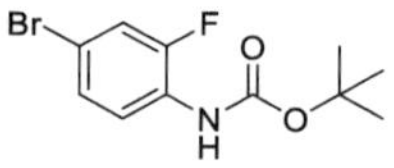

4-Bromo-2-fluoroaniline (7.00 g, 36.8 mmol, 1.00 eq) was dissolved in DMF (75 mL) and carbonyldiimidazole (17.9 g, 111 mmol, 3.00 eq) was added to gain a slightly yellowish suspension. The solution was slowly (over ca. 20 min) heated to 105 °C and stirred for 2 h at this temperature while the solution got clear again. Afterwards, the temperature was reduced to 95 °C and *tert*-butanol (8.19 g, 10.4 mL, 111 mmol, 3.00 eq) was added. The heat source was removed and the solution was stirred over night while cooling to 20 °C. After full conversion was observed by TLC, water (210 mL) was added which led to weak heat generation and clouding of the solution. The solution was stirred vigorously for 1 h and afterwards the product was extracted with DCM (3×200 mL). Following, the solvent was removed under reduced pressure and the crude product was purified by silica gel chromatography using cH/EtOAc 19:1. Compound **112** was yielded as a white, crystalline solid (5.57 g, 19.2 mmol, 52%).

^{1}H NMR (400 MHz, $CDCl_3$): δ = 8.02 (t, *J* = 7.9 Hz, 1H), 7.31 – 7.19 (m, 2H), 6.67 (s, 1H), 1.54 (s, 9H) ppm. **^{13}C NMR (101 MHz, $CDCl_3$)**: δ = 153.0, 152.3, 150.6, 127.8, 126.4, 121.1, 118.5, 118.3, 114.2, 81.6, 28.4 ppm. **^{19}F NMR** (376 MHz, $CDCl_3$): δ = –130.0 ppm. **TLC:** R_F = 0.20 (developed in cH/DCM 9:1) and 0.53 (developed in cH/EtOAc 19:1). **HRMS (EI):** *m/z* calcd. for $C_{11}H_{13}NO_2FBr$ [M+] = 289.0108 Da, found 289.0110 Da. **IR (ATR, $\tilde{\nu}$)** = 3376 (m), 3017 (w), 3000 (w), 2983 (m), 2931 (w), 2881 (w), 1711 (vs), 1609 (w), 1592 (m), 1523 (vs), 1509 (vs), 1485 (vs), 1459 (m), 1448 (m), 1405 (vs), 1385 (s), 1370 (s), 1364 (s), 1319 (s), 1268 (m), 1239 (vs), 1198 (m), 1152 (vs), 1119 (vs), 1064 (s), 1048 (vs), 1023 (vs), 945 (m), 899 (s), 873 (vs), 858 (vs), 812 (vs), 773 (vs), 761 (vs), 742 (s), 639 (s), 584 (vs), 569 (vs), 551 (vs), 523 (m), 456 (s), 431 (m), 402 (w), 384 (w) cm^{-1}.

Methyl (*S*)-2-acetamido-3-(4-((*tert*-butoxycarbonyl)amino)-3-fluorophenyl)propanoate (113)

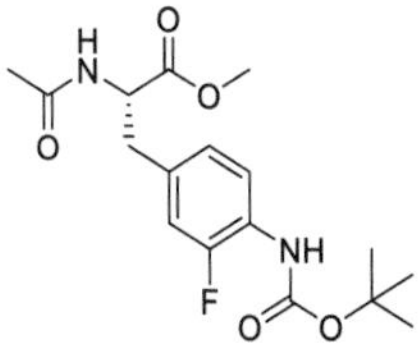

All substrates except methyl (*R*)-2-acetamido-3-iodopropanoate **85** due to its instability and the molecular iodine were dried overnight under high vacuum prior to the reaction. Additionally, compound **85** was dried for 45 min under the same conditions just before the reaction was started.

To zinc dust (991 mg, 15.2 mmol, 3.00 eq) under argon atmosphere was added dry DMF (9.90 mL) and afterwards molecular iodine (192 mg, 758 µmol, 0.15 eq). The suspension was stirred for 20 min until it turned clear. Methyl (*R*)-2-acetamido-3-iodopropanoate **85** was added, followed by molecular iodine (192 mg, 758 µmol, 0.15 eq) and the solution was stirred for 20 min until it cooled down to 20 °C again. Afterwards, $Pd_2(dba)_3$ (116 mg, 126 µmol, 0.025 eq), SPhos (104 mg, 253 µmol, 0.05 eq) and *tert*-butyl (4-bromo-2-fluorophenyl)-carbamate **112** (1.91 g, 6.57 mmol, 1.30 eq) were added and the dark suspension was stirred for 43 h under an argon atmosphere at room temperature. After the TLC showed full conversion, the suspension was filtered over Celite® with EtOAc as the eluent and concentrated under reduced pressure to gain a yellow oil. The crude product was purified by silica gel column chromatography using cH/EtOAc 1:1 to yield the product **113** as a colorless solid (1.17 g, 3.30 mmol, 65%).

^{1}H NMR (400 MHz, CDCl$_3$): δ = 7.99 (t, J = 8.2 Hz, 1H), 6.86 – 6.76 (m, 2H), 6.68 – 6.63 (m, 1H), 5.94 (d, J = 7.7 Hz, 1H), 4.84 (dt, J = 7.7, 5.5 Hz, 1H), 3.73 (s, 3H), 3.06 (qd, J = 14.0, 5.6 Hz, 2H), 1.99 (s, 3H), 1.51 (s, 9H) ppm. **^{13}C NMR (101 MHz, CDCl$_3$)**: δ = 172.0, 169.7, 153.1, 152.5, 150.7, 131.0, 126.0, 125.5, 120.1, 115.6, 81.2, 53.2, 52.6, 37.2, 28.4, 23.3 ppm. **^{19}F NMR** (376 MHz, CDCl$_3$): δ = –132.2 ppm. **TLC:** R_F = 0.30 (developed in cH/EtOAc 1:1). **HRMS (FAB):** *m/z* calcd. for $C_{17}H_{24}N_2O_5F$ [M+H] = 355.1664 Da, found 355.1666 Da. **IR (ATR, $\tilde{\nu}$)** = 3336 (m), 2986 (w), 2961 (vw), 2929 (vw), 1747 (vs), 1700 (vs), 1647 (vs), 1595 (w), 1521 (vs), 1504 (vs), 1463 (w), 1426 (s), 1395 (w), 1374 (s), 1316 (w), 1276 (s), 1248 (vs), 1232 (s), 1217 (vs), 1159 (vs), 1130 (vs), 1058 (s), 1031 (m), 1006 (w), 980 (w), 962 (w), 922 (w), 902 (w), 885 (w), 867 (w), 830 (w), 816 (m), 788 (w), 765 (m), 742 (m), 708 (w), 669 (m), 618 (s), 582 (vs), 567 (s), 517 (m), 465 (m), 455 (w), 433 (m), 416 (w) cm^{-1}.

Methyl (*S*)-2-acetamido-3-(4-amino-3-fluorophenyl)propanoate (90)

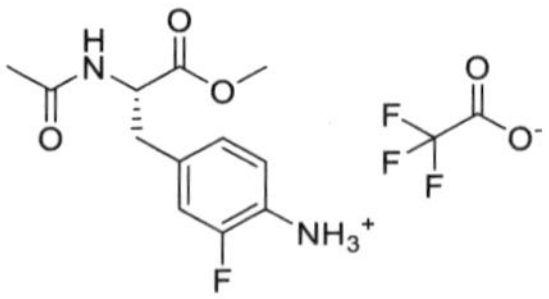

Methyl (*S*)-2-acetamido-3-(4-((*tert*-butoxycarbonyl)amino)-3-fluorophenyl)propanoate **113** (410 mg, 1.16 mmol, 1.00 eq) was dissolved in DCM (12.0 mL). Afterwards, TFA (12.0 mL) and 1% (v/v) TIPS (0.24 mL) were added, and the mixture was stirred for 2 h at room temperature. After full conversion was observed by TLC, toluene (ca. 50 mL) was added, and the solvent was removed under reduced pressure. The product **90** was obtained in quantitative yield (426 mg, 1.16 mmol) as TFA salt.

^{1}H NMR (400 MHz, DMSO-d_6): δ = 8.60 (s, 3H), 8.29 (d, J = 7.8 Hz, 1H), 7.00 (dd, J = 12.2, 1.8 Hz, 1H), 6.92 – 6.80 (m, 2H), 4.39 (ddd, J = 9.3, 7.8, 5.5 Hz, 1H), 3.59 (s, 3H), 2.91 (dd, J = 13.8, 5.5 Hz, 1H), 2.76 (dd, J = 13.8, 9.4 Hz, 1H), 1.78 (s, 3H) ppm. **^{13}C NMR (101 MHz, DMSO-d_6)**: δ = 172.1, 169.3, 158.3, 153.0, 150.6, 125.3, 118.8, 116.8, 116.0, 113.9, 53.6, 51.8, 35.8, 22.2 ppm. **^{19}F NMR (376 MHz, DMSO-d_6)**: δ = –75.0 (signal of TFA), –132.0 ppm. **TLC:** R_F = 0.08 (developed in cH/EtOAc 1:1). **HRMS (EI):** m/z calcd. for $C_{12}H_{15}N_2O_3F$ [M+] = 254.1061 Da, found 254.1063 Da. **IR (ATR, ṽ)** = 3293 (w), 3047 (w), 2931 (w), 2859 (w), 2594 (w), 2109 (vw), 1778 (w), 1727 (m), 1638 (vs), 1543 (s), 1513 (vs), 1436 (s), 1378 (m), 1285 (m), 1142 (vs), 965 (s), 915 (m), 884 (m), 839 (s), 816 (s), 798 (s), 724 (vs), 704 (s), 636 (m), 599 (s), 541 (s), 518 (s), 450 (s), 407 (m), 387 (m) cm^{-1}.

1-Fluoro-2-nitrosobenzene (91)

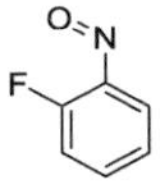

Oxone® (30.4 g, 99.0 mmol, 2.20 eq) was dissolved in diH_2O (300 mL) and combined with 2-fluoroaniline (5.00 g, 45.0 mmol, 1.00 eq) which was dissolved in DCM (180 mL). The mixture was stirred vigorously for 1 d at 20 °C. Afterwards, the organic layer was separated from the aqueous layer, which was additionally extracted with DCM (3×200 mL). All organic layers were combined and washed with 1 M HCl (3×200 mL), diH_2O (2×200 mL) and brine (200 mL). The organic layers were dried over Na_2SO_4 and concentrated under reduced pressure. Then, the crude product was purified by sublimation at 50 °C and 3 mbar to yield 1-fluoro-2-nitrosobenzene **91** as a slightly green to colorless solid (3.26 g, 26.0 mmol, 58%). The product was stored in darkness at –18 °C due to possible decomposition.

^{1}H NMR (400 MHz, $CDCl_3$): δ = 7.72 (dddd, J = 8.9, 7.1, 5.0, 1.8 Hz, 1H), 7.51 (ddd, J = 9.8, 8.4, 1.2 Hz, 1H), 7.14 (t, 1H), 6.49 (ddd, J = 8.4, 6.9, 1.8 Hz,

1H) ppm. **^{13}C NMR (101 MHz, CDCl$_3$)**: δ = 166.1, 163.4, 155.0, 138.2, 123.8, 118.8, 109.8 ppm. **^{19}F NMR (376 MHz, CDCl$_3$)**: δ = –129.5 ppm.

Methyl (*S,E*)-2-acetamido-3-(3-fluoro-4-((2-fluorophenyl)diazenyl)phenyl)propanoate (PAP-F$_2$-NAc-OMe) (92)

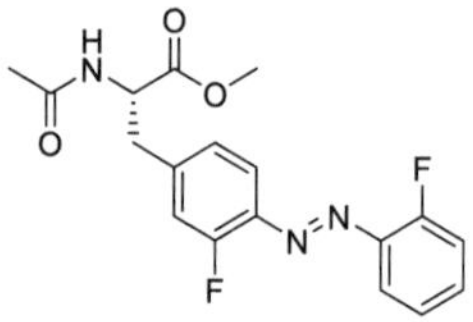

Methyl (*S*)-2-acetamido-3-(4-amino-3-fluorophenyl)propanoate **90** was suspended in glacial AcOH (17 mL) and 1-fluoro-2-nitrosobenzene **91** (519 mg, 4.15 mmol, 1.30 eq) was added. After the mixture was stirred for 42 h, it was concentrated under reduced pressure and the crude product was purified by silica gel column chromatography. First, cH/EtOAc 1:1 was used as the eluent until a yellow side product was eluted, followed by cH/EtOAc 1:2. The product PAP-F$_2$-NAc-OMe **92** (616 mg, 1.71 mmol, 53%) was obtained as an orange solid.

^{1}H NMR (400 MHz, CDCl$_3$) δ = 7.81 – 7.69 (m, 2H), 7.47 (dddd, *J* = 8.6, 6.9, 5.0, 1.8 Hz, 1H), 7.31 – 7.17 (m, 2H), 7.06 – 6.89 (m, 2H), 6.06 (d, *J* = 7.6 Hz, 1H), 4.93 (dt, *J* = 7.6, 5.7 Hz, 1H), 3.76 (s, 3H), 3.24 (dd, *J* = 13.8, 5.9 Hz, 1H), 3.15 (dd, *J* = 13.8, 5.5 Hz, 1H), 2.02 (s, 3H) ppm. **^{13}C NMR (101 MHz, CDCl$_3$)** δ = 171.8, 169.8, 161.7, 159.1, 142.2, 141.0, 140.0, 133.2, 125.5, 124.5, 118.0, 117.8, 117.3, 117.1, 53.1, 52.7, 37.8, 23.3 ppm. **^{19}F NMR (376 MHz, CDCl$_3$)**: δ = –123.7, –124.2 ppm. **TLC:** R_F = 0.40 (developed in cH/EtOAc 1:1). **HRMS (FAB):** *m/z* calcd. for $C_{18}H_{18}N_3O_3F_2$ [M+H] = 362.1311 Da, found 362.1309 Da. **IR (ATR, ṽ)** = 3323 (m), 3057 (w), 2955 (w), 1737 (vs), 1640 (s), 1613 (w), 1604 (m), 1584 (m), 1536 (vs), 1482 (m), 1446 (s), 1375 (m), 1339 (m), 1299 (m), 1259 (s), 1228 (vs), 1217 (vs), 1179 (m), 1150 (m), 1142 (m), 1119 (vs), 1037 (s), 994 (m), 952 (s), 878 (s), 850 (w), 840 (w), 827 (m), 795 (m), 765 (vs), 745 (m), 734 (m), 694 (s),

684 (s), 635 (m), 596 (vs), 560 (w), 496 (m), 480 (w), 463 (vs), 397 (s), 375 (m) cm^{-1}. **UV-VIS (MeCN):** λ_{max} = 228, 332, 443 nm.

Methyl (*S,E*)-2-acetamido-3-(3-chloro-4-((2-chloro-6-fluorophenyl)diazenyl)-5-fluorophenyl)propanoate (PAP-F_2Cl_2-NAc-OMe) (114)

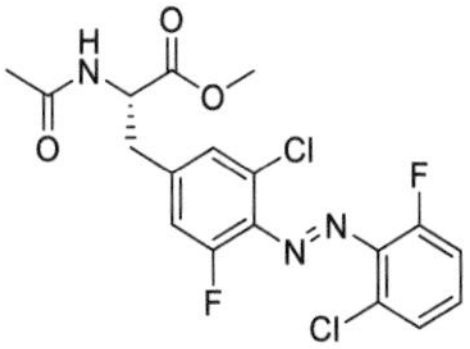

PAP-F_2-NAc-OMe **92** (351 mg, 971 µmol, 1.00 eq), NCS (389 mg, 2.91 mmol, 3.00 eq) and palladium diacetate (21.8 mg, 97.1 µmol, 0.10 eq) were dried under high vacuum for 1 h in a 20 mL vial. The components were dissolved in degassed AcOH (9.80 mL) and the reaction was heated to 140 °C for 30 min. During the whole procedure the vial was sealed. After cooling to room temperature, the mixture was transferred to a separatory funnel and DCM (ca. 50 mL) as well as brine solution (ca. 50 mL) were added. The organic layer was separated, washed with brine (2×50 mL) and dried over Na_2SO_4. The solvent was removed under reduced pressure and the residue was purified by silica gel column chromatography using cH/EtOAc 1:1 to obtain the product PAP-F_2Cl_2-NAc-OMe **114** (238 mg, 554 µmol, 57%) as an orange solid.

^{1}H NMR (400 MHz, $CDCl_3$) δ = 7.32 (ddd, *J* = 21.9, 15.7, 8.4 Hz, 2H), 7.14 (d, *J* = 9.4 Hz, 2H), 6.91 (d, *J* = 11.0 Hz, 1H), 6.12 (d, *J* = 7.4 Hz, 1H), 4.90 (q, *J* = 6.2 Hz, 1H), 3.78 (s, 3H), 3.21 (dd, *J* = 13.9, 5.9 Hz, 1H), 3.12 (dd, *J* = 13.8, 5.4 Hz, 1H), 2.04 (s, 3H) ppm. **^{13}C NMR (101 MHz, $CDCl_3$)** δ = 171.5, 169.9, 153.9, 151.3, 140.4, 139.6, 138.3, 132.6, 132.1, 130.8, 127.1, 126.3, 116.8, 116.0, 52.9, 37.5, 23.3 ppm. **^{19}F NMR (376 MHz, $CDCl_3$)**: δ = –123.0, –123.7 ppm. **TLC:** R_F = 0.25 (developed in cH/EtOAc 1:1). **HRMS (FAB):** *m/z* calcd. for $C_{18}H_{16}N_3O_3F_2Cl_2$ [M+H] = 430.0531 Da, found 430.0532 Da. **IR (ATR, ṽ)** = 3285 (m), 3077 (w), 2949 (w), 2929 (w), 1742 (vs), 1650 (vs), 1611 (s), 1592 (m), 1561 (s), 1533 (vs), 1449 (vs), 1436 (vs), 1371 (vs), 1351 (s), 1292 (m), 1264 (s), 1254 (vs), 1215 (s), 1191 (s), 1173 (vs), 1136 (s), 1122 (s), 1027 (w), 1013 (m), 984 (w), 967 (m), 933 (m), 914 (m), 891 (vs), 877 (vs), 866 (vs), 810 (w), 785 (vs), 765 (s), 744 (vs), 727 (vs), 714 (vs), 645 (w), 618 (s), 609 (s), 598 (vs), 572 (s), 531 (w), 513 (s), 482 (m), 445 (m), 429 (w), 399 (m) cm^{-1}. **UV-VIS (MeCN):** λ_{max} = 315, 449 nm.

(*S,E*)-2-Amino-3-(3-chloro-4-((2-chloro-6-fluorophenyl)diazenyl)-5-fluorophenyl)-propanoic acid (PAP-F_2Cl_2) (115)

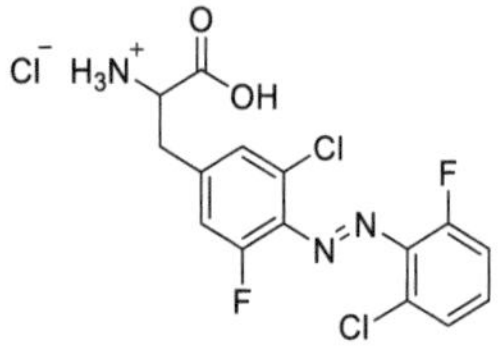

6 M HCl (50.0 mL) was added to compound PAP-F_2Cl_2-NAc-OMe **114** (366 mg, 850 µmol, 1.00 eq) and the mixture was heated to 110 °C for 2 d. The precipitate was separated by centrifugation to gain the product PAP-F_2Cl_2 **115** (179 mg, 436 µmol, 51%) as a brown solid.

The stereochemical outcome of the synthesis of PAP-F_2Cl_2 **115** was investigated with chiral RP-HPLC chromatography on the HPLC chromatograph 1100 Series from AGILENT TECHNOLOGIES with a G1322A degasser, a G1211A pump, a G1313A autosampler, a G1316A column oven, and a G1315B diode array system using a Daicel Chiralpak® AD-H (4.6 × 250 mm, 5 µm particle size) column. The separations were performed with a 20 min isocratic mixture of HPLC-grade hexane/isopropanol (85/15), flow rate 1 mL/min, slit = 4 nm, wavelength 450 nm for detection. The photochromic amino acid **115** was obtained as an enantiomeric mixture with the ratio of ca. 9:1 (Figure 170). The enantiomers were not separated but used as the mixture in the following reactions.

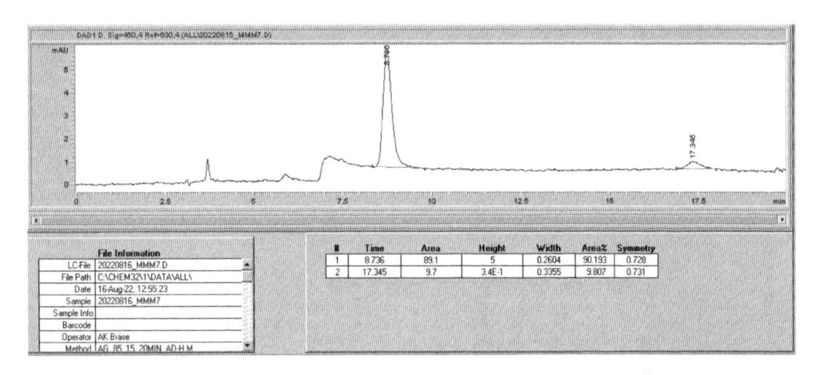

Figure 170: Stereochemical analysis of PAP-F_2Cl_2 **115** by chiral HPLC chromatography.

^{1}H NMR (400 MHz, DMSO-*d*$_6$) δ = 14.03 (s, 1H), 8.55 (s, 3H), 7.64 – 7.55 (m, 3H), 7.52 – 7.45 (m, 2H), 4.35 (s, 1H), 3.34 (dd, *J* = 14.2, 5.8 Hz, 1H), 3.23 (dd, *J* = 14.2, 7.1 Hz, 1H) ppm. **^{13}C NMR (101 MHz, DMSO-*d*$_6$)** δ = 170.0, 152.8, 150.2, 140.7, 138.5, 137.3, 132.4, 130.8, 130.3, 127.8, 126.7, 117.8, 116.5, 52.5, 34.7 ppm. **^{19}F NMR** (376 MHz, DMSO-*d*$_6$) δ = –124.7, –124.7 ppm. **HRMS (FAB):** *m/z* calcd. for $C_{15}H_{12}N_3O_2F_2Cl_2$ [M] = 374.0269 Da, found 374.0270 Da. **IR (ATR, ṽ)** = 2968 (m), 2893 (m), 1977 (vw), 1737 (vs), 1592 (m), 1572 (m), 1564 (m), 1487 (vs), 1451 (vs), 1419 (s), 1361 (w), 1340 (w), 1286 (w), 1258 (w), 1230 (s), 1210 (vs), 1152 (m), 1120 (w), 1098 (w), 1057 (m), 994 (w), 965 (w), 936 (w), 901 (vs), 871 (s), 840 (m), 782 (vs), 744 (vs), 711 (w), 674 (w), 652 (w), 611 (m), 565 (w), 511 (m), 482 (m), 445 (w), 426 (w), 397 (m), 375 (w) cm^{-1}.

(*S,E*)-2-((*Tert*-butoxycarbonyl)amino)-3-(3-chloro-4-((2-chloro-6-fluorophenyl)diazenyl)-5-fluorophenyl)propanoic acid (PAP-F_2Cl_2-NBoc) (116)

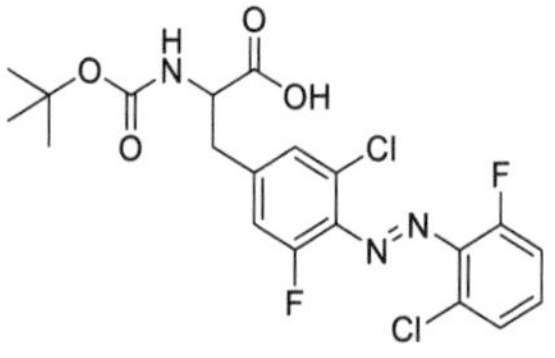

PAP-F_2Cl_2 **115** (179 mg, 436 µmol, 1.00 eq) was suspended in 1,4-dioxane (2.9 mL) and $NaHCO_3$ (201 mg, 2.39 mmol, 5.00 eq) dissolved in diH_2O (2.3 mL) was added, followed by Boc_2O (115 mg, 526 µmol, 1.10 eq). The mixture was stirred for 23 h at 20 °C. Afterwards, the mixture was transferred to a separation funnel and EtOAc (ca. 30 mL) was added. The organic layer was separated, washed with $KHSO_4$ (3×50 mL) solution, brine (50 mL) and dried over Na_2SO_4. After the solvent was removed under reduced pressure, the crude product was purified by silica gel column chromatography using DCM/MeOH/FA 97:2:1. The product PAP-F_2Cl_2-NBoc **116** (216 mg, 376 µmol, 86%) was isolated as an orange/red solid with a purity of ca. 83%, determined by ^{1}H NMR spectroscopy (contaminants: FA, DCM and 1,4-dioxane).

^{1}H NMR (400 MHz, DMSO-*d*$_6$) δ = 12.74 (s, 1H), 7.61 – 7.54 (m, 2H), 7.51 – 7.44 (m, 2H), 7.38 (dd, *J* = 11.8, 1.6 Hz, 1H), 7.22 (d, *J* = 8.6 Hz, 1H), 4.23 (ddd, *J* = 10.5, 8.6, 4.5 Hz, 1H), 3.23 – 3.10 (m, 1H), 2.91 (dd, *J* = 13.7, 10.8 Hz, 1H), 1.31 (s, 9H) ppm. **^{13}C NMR (101 MHz, DMSO-*d*$_6$)** δ = 172.9, 155.4, 152.8, 150.2, 144.1, 138.6, 136.7, 132.1, 130.6, 130.2, 127.4, 126.6, 117.2, 116.5, 78.2, 54.2, 36.0, 28.1 ppm. **^{19}F NMR (376 MHz, DMSO-*d*$_6$)** δ = –124.7, –124.7 ppm. **TLC:** R_F = 0.25 (developed in DCM/MeOH/FA 97:2:1). **HRMS (FAB):** *m/z* calcd. for $C_{20}H_{20}N_3O_4F_2Cl_2$ [M+H] = 474.0793 Da, found 474.0794 Da. **IR (ATR, ṽ)** = 3330 (w), 3078 (w), 2979 (w), 2929 (w), 2866 (w), 1711 (vs), 1686 (vs), 1609 (m), 1595 (m), 1565 (m), 1509 (s), 1453 (vs), 1421 (s), 1394 (s), 1368 (s), 1254 (vs), 1160 (vs), 1120 (m), 1057 (m), 1026 (w), 994 (w), 929 (w), 914 (m), 887 (s), 868 (s), 779 (vs), 742 (m), 670 (w), 635 (w), 613 (w), 588 (w), 577 (w), 555 (w), 544 (w), 479 (w), 466 (w), 429 (w) cm^{-1}.

Methyl *N*^{6}-(*tert*-butoxycarbonyl)-*N*^{2}-((*S*)-2-((*tert*-butoxycarbonyl)amino)-3-(3-chloro-4-((*E*)-(2-chloro-6-fluorophenyl)diazenyl)-5-fluorophenyl)propanoyl)-*L*-lysinate (PAP-F_2Cl_2-Lys-OMe-Boc) (117)

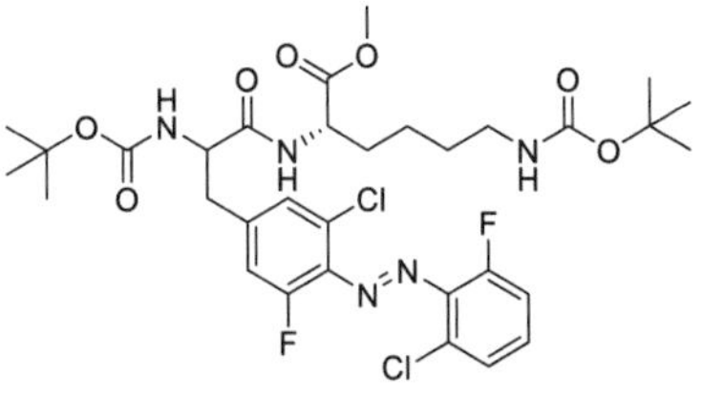

PAP-F_2Cl_2-NBoc **116** (179 mg, 376 µmol, 1.00 eq), HBTU (150 mg, 395 µmol, 1.05 eq) and DIPEA (122 mg, 164 µL, 941 µmol, 2.50 eq) were dissolved in anhydrous DMF (500 µL) under an argon atmosphere. After the mixture was stirred for 10 min at room temperature, H-Lys(Boc)-OMe·HCl (117 mg, 395 µmol, 1.05 eq) and DIPEA (122 mg, 164 µL, 941 µmol, 2.50 eq) were added. The resulting mixture was stirred for 1 h under an argon atmosphere until full conversion of the starting material was observed by TLC. EtOAc (ca. 50 mL) was added, and the organic phase was washed with aq. NH_4Cl solution (3×50 mL) and brine (50 mL). In the following, the organic layer was dried over Na_2SO_4 and concentrated under reduced pressure to gain a red solid. The residual solid was purified by silica gel column chromatography using cH/EtOAc 1:1 to yield the product PAP-F_2Cl_2-Lys-OMe-Boc **117** (217 mg, 283 µmol, 75%) as a sticky orange/red solid.

^{1}H NMR (400 MHz, DMSO-*d*$_6$) δ = 8.33 (d, *J* = 7.5 Hz, 1H), 7.64 – 7.44 (m, 5H), 7.42 – 7.35 (m, 1H), 7.02 (d, *J* = 8.8 Hz, 1H), 6.77 (d, *J* = 5.9 Hz, 1H), 4.27 (dtd, *J* = 16.2, 9.4, 8.1, 4.6 Hz, 2H), 3.62 (s, 3H), 3.32 (s, 6H), 3.07 (dd, *J* = 13.6, 4.3 Hz, 1H), 2.90 (q, *J* = 6.5 Hz, 2H), 1.67 (dt, *J* = 35.8, 7.8 Hz, 3H), 1.36 (s, 9H), 1.30 (s, 9H) ppm. **^{13}C NMR (101 MHz, DMSO-*d*$_6$)** δ = 172.4, 171.2, 155.6, 155.2, 153.0, 150.4, 144.2, 138.6, 136.6, 132.2, 130.7, 130.1, 127.4, 126.6, 117.4, 117.2, 116.6, 116.4, 78.2, 77.3, 54.8, 51.9, 37.0, 30.7, 29.1, 28.3, 28.0, 22.6 ppm. **^{19}F NMR (376 MHz, DMSO-*d*$_6$)** δ = –124.7, –124.8 ppm. **TLC:** R_F = 0.58 (developed in cH/EtOAc 1:1). **HRMS (FAB):** *m/z* calcd. for $C_{32}H_{42}N_5O_7F_2Cl_2$

[M+H] = 716.2424 Da, found 716.2426 Da. **IR (ATR, ṽ)** = 3327 (w), 3080 (vw), 2975 (w), 2932 (w), 2864 (vw), 1738 (w), 1681 (vs), 1653 (vs), 1609 (w), 1595 (w), 1564 (w), 1520 (vs), 1452 (s), 1390 (m), 1366 (s), 1323 (w), 1271 (s), 1249 (vs), 1211 (s), 1162 (vs), 1047 (m), 1014 (m), 996 (m), 933 (m), 888 (s), 850 (m), 779 (s), 742 (s), 636 (s), 613 (s), 560 (m), 483 (w), 462 (w), 429 (w), 399 (w) cm^{-1}.

Methyl ((*S*)-2-amino-3-(3-chloro-4-((*E*)-(2-chloro-6-fluorophenyl)diazenyl)-5-fluorophenyl)propanoyl)-*L*-lysinate (PAP-F_2Cl_2-Lys-OMe) (118)

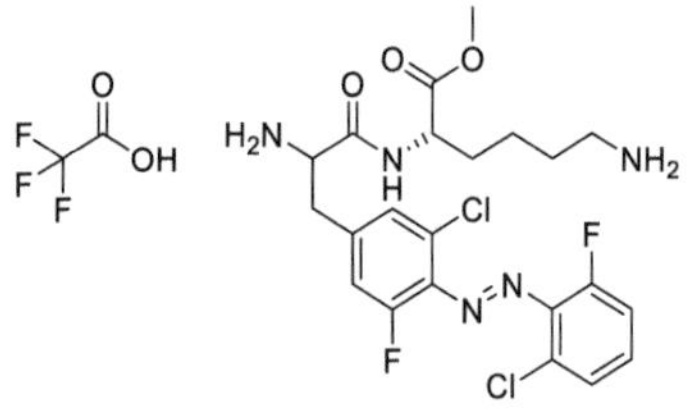

PAP-F_2Cl_2-Lys-OMe-Boc **117** (202 mg, 282 μmol, 1.00 eq) was dissolved in DCM (2.70 mL) and TFA (2.70 mL) was added. The mixture was stirred for 2 h at room temperature until full conversion was observed by TLC. Afterwards, toluene (ca. 40 mL) was added, and the solvent was removed under reduced pressure to yield the product PAP-F_2Cl_2-Lys-OMe **118** (146 mg, 282 μmol) quantitatively as the TFA salt. No further purification was done, and the material was directly used in the next reaction.

^{1}H NMR (400 MHz, DMSO-*d₆*) δ = 8.95 (d, *J* = 7.5 Hz, 1H), 8.27 (s, 3H), 7.74 (s, 3H), 7.64 – 7.55 (m, 2H), 7.54 – 7.46 (m, 2H), 7.37 (dd, *J* = 11.6, 1.7 Hz, 1H), 4.33 (td, *J* = 8.1, 5.4 Hz, 1H), 4.19 (s, 1H), 3.66 (s, 3H), 3.23 (dd, *J* = 14.0, 5.3 Hz, 1H), 3.09 (dd, *J* = 14.0, 8.0 Hz, 1H), 2.77 (q, *J* = 6.4 Hz, 2H), 1.78 (dq, *J* = 13.9, 6.5 Hz, 1H), 1.71 – 1.61 (m, 1H), 1.55 (p, *J* = 7.5 Hz, 2H), 1.37 (q, *J* = 9.8, 7.6 Hz, 2H) ppm. **^{13}C NMR (101 MHz, DMSO-d_6)** δ = 172.4, 171.2, 155.6, 155.2, 153.0, 150.4, 144.2, 138.6, 136.6, 132.2, 130.7, 130.1, 127.4, 126.6, 117.4, 117.2, 116.6, 116.4, 78.2, 77.3, 54.8, 51.9, 37.0, 30.7, 29.1, 28.3, 28.0, 22.6 ppm. **^{19}F NMR**

(376 MHz, DMSO-d_6) δ = –73.9 (signal of TFA), –124.3, –124.8 ppm. **HRMS (ESI):** *m/z* calcd. for $C_{22}H_{26}N_5O_3F_2Cl_2$ [M+H] = 516.1375 Da, found 516.1375 Da The IR was not recorded due to toluene and TFA residues.

(*3S,6S*)-3-(4-Aminobutyl)-6-(3-chloro-4-((*E*)-(2-chloro-6-fluorophenyl)di-azenyl)-5-fluorobenzyl)piperazine-2,5-dione (PAP-F_2Cl_2-DKP-Lys) (59)

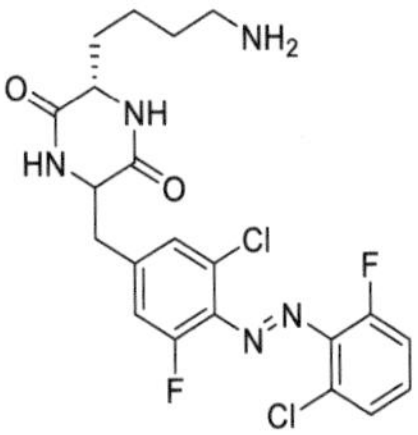

PAP-F_2Cl_2-Lys-OMe **118** (146 mg, 283 µmol, 1.00 eq) was dissolved in 2-butanol (5.50 mL). Subsequently, AcOH (62.6 mg, 59.6 µL, 1.04 mmol, 3.69 eq), *N*-methylmorpholine (36.6 mg, 40.2 µL, 362 µmol, 1.28 eq) and DIPEA (52.9 mg, 71.3 µL, 410 µmol, 1.45 eq) were added. The mixture was heated to 120 °C for 2 h. After it cooled down to room temperature, the solvent was removed under reduced pressure. To the residual red solid MeCN (ca. 50 mL) was added and the suspension was dispersed using an ultrasonic bath. Then, the solid was filtered off and dried to yield the crude product PAP-F_2Cl_2-DKP-Lys **59** as a red solid.

^{1}H NMR (400 MHz, DMSO-d_6) δ = 8.35 – 8.18 (m, 2H), 7.64 – 7.41 (m, 3H), 7.41 (s, 1H), 7.28 (dd, J = 11.9, 1.6 Hz, 1H), 4.38 – 4.25 (m, 1H), 3.82 (d, J = 5.5 Hz, 1H), 3.55 (t, J = 4.6 Hz, 1H), 3.20 (dd, J = 13.6, 4.8 Hz, 1H), 3.03 (dd, J = 13.6, 5.1 Hz, 1H), 2.66 – 2.58 (m, 1H), 1.42 – 1.33 (m, 2H), 1.15 (d, J = 6.7 Hz, 4H), 1.07 – 0.88 (m, 2H) ppm. **^{19}F NMR (376 MHz, DMSO-d_6)** δ = –124.6, –124.8 ppm. **HRMS (FAB):** *m/z* calcd. for $C_{21}H_{22}N_5O_2F_2Cl_2$ [M+H] = 484.1113 Da, found 484.1111 Da. **IR (ATR, ṽ)** = 3187 (w), 3051 (w), 2952 (w), 1663 (vs), 1612 (m), 1595 (m), 1567 (w), 1452 (s), 1424 (m), 1333 (w), 1255 (w), 1201 (s), 1180 (s), 1130 (vs), 999 (w), 933 (w), 891 (m), 836 (s), 799 (m), 781 (s), 741 (m), 722 (s),

673 (w), 643 (w), 612 (w), 599 (w), 569 (w), 550 (w), 534 (w), 517 (w), 484 (w), 446 (m), 438 (m), 428 (m), 419 (m), 405 (m), 387 (w) cm^{-1}. **UV-VIS (MeCN):** λ_{max} = 315, 473 nm.

For analytics, the crude was purified by preparative HPLC with the following settings: 15 mL/min, 30 min gradient 30-60% MeCN in bidest. H_2O with 0.1% TFA, detection at 310 nm, retention at 19.5 min. After lyophilization, the hydrogelator **59** and an ion exchange resin were suspended in diH_2O and stirred for 2 d. Subsequently, the sample was lyophilized to yield the pure product as an orange solid (29.0 mg, 59.9 µmol, 16%).

^{1}H NMR (400 MHz, DMSO-*d*$_6$) δ = 8.28 (q, *J* = 12.3 Hz, 2H), 7.65 – 7.53 (m, 2H), 7.51 – 7.43 (m, 1H), 7.42 (s, 1H), 7.34 – 7.25 (m, 1H), 4.33 (d, *J* = 5.1 Hz, 1H), 3.81 (d, *J* = 5.6 Hz, 1H), 3.19 (dd, *J* = 13.6, 4.8 Hz, 1H), 3.05 (dd, *J* = 13.6, 5.1 Hz, 1H), 2.58 (t, *J* = 7.7 Hz, 2H), 1.40 (pq, *J* = 10.3, 5.4 Hz, 4H), 1.10 – 0.85 (m, 2H) ppm. **^{13}C NMR (101 MHz, DMSO-*d*$_6$)** δ = 167.3, 166.4, 153.0, 150.1, 142.5, 138.6, 138.3, 136.8, 132.2, 130.5, 128.0, 126.6, 118.0, 116.5, 54.7, 53.4, 38.4, 37.4, 31.9, 26.5, 20.4 ppm. **^{19}F NMR (376 MHz, DMSO-*d*$_6$)** δ = –73.5 (signal of TFA), –118.4*, –119.2*, –124.6, –124.8 ppm (* product switched partially to *Z* isomer, determined by HPLC-MS analysis).

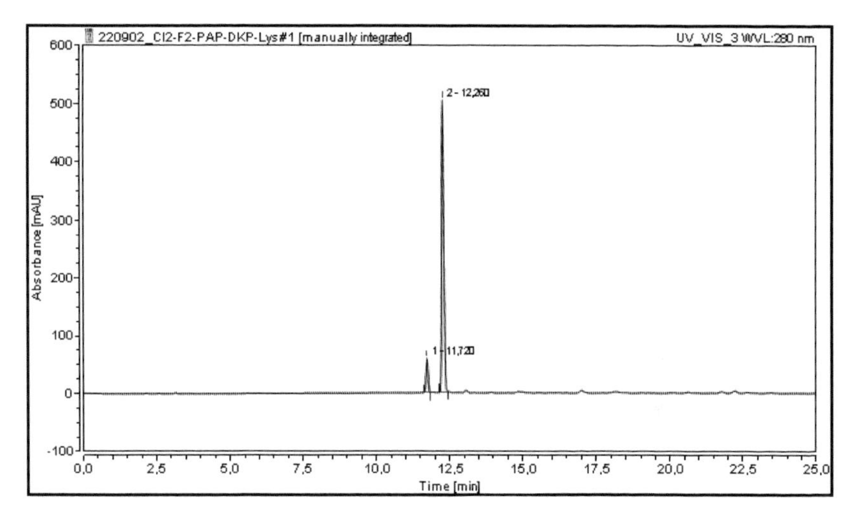

Figure 171: Chromatogram at 280 nm after a HPLC analysis (20 min gradient of 5-95% MeCN in H_2O, 0.1% TFA) of the compound **59**. The signal at 11.7 min was assigned to the *Z*-isomer.

No.	Retention Time	Area	Height	Relative Area	Relative Height
	min	mAU*min	mAU	%	%
1	11.720	4.721	59.520	9.74	10.57
2	12.260	43.723	503.705	90.26	89.43
Total:		**48.443**	**563.225**	**100.00**	**100.00**

Tert-butyl (*S*)-(6-chloro-8-fluoro-2-oxo-1,2,3,4-tetrahydroquinolin-3-yl)carbamate (89)

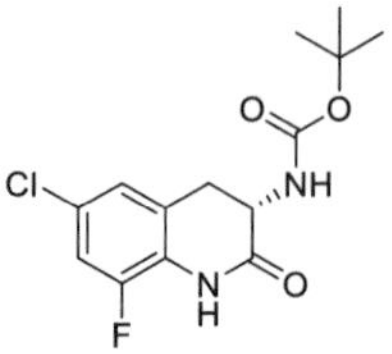

Dry *N,N*-dimethylformamide (36 mL) was added to zinc dust (3.58 g, 54.7 mmol, 3.00 equiv) under argon followed by molecular iodine (694 mg, 2.73 mmol, 0.150 eq) and the mixture was stirred until the solution turned clear again. Methyl (2*R*)-3-iodo-2-[(2-methylpropan-2-yl)oxycarbonylamino]propanoate (6.00 g, 18.2 mmol, 1.00 eq) was added followed by molecular iodine (694 mg, 2.73 mmol, 0.150 eq) and the solution was stirred for 15 min until it had cooled down to room temperature again. $Pd_2(dba)_3$ (417 mg, 456 µmol, 0.0250 eq), S-Phos (374 mg, 911 µmol, 0.0500 eq) and 4-bromo-2-chloro-6-fluoroaniline (5.32 g, 23.7 mmol, 1.30 eq) were added and the reaction mixture was stirred under argon for 3 d at room temperature. The crude product was filtrated over celite, concentrated under reduced pressure and purified by silica column chromatography (cH:EtOAc 5:1 with 1% Et_3N) to yield the product **89** (2.51 g, 7.99 mmol, 44%) as light brown solid.

^{1}H NMR (400 MHz, $CDCl_3$) δ = 7.71 (s, 1H), 7.05 (ddd, *J* = 9.8, 2.1, 1.0 Hz, 1H), 7.00 (d, *J* = 1.9 Hz, 1H), 5.58 (s, 1H), 4.36 (dt, *J* = 13.4, 5.4 Hz, 1H), 3.51 (dd, *J* = 15.7, 6.2 Hz, 1H), 2.86 (t, *J* = 14.9 Hz, 1H), 1.47 (s, 9H) ppm. **^{13}C NMR (101 MHz, $CDCl_3$)** δ = 168.6, 156.0, 151.1, 148.6, 128.8 (d, *J* = 9.7 Hz), 126.5 (d, *J* = 1.9 Hz), 124.6 (d, *J* = 3.2 Hz), 123.7, 115.6 (d, *J* = 21.5 Hz), 80.8, 52.8, 50.2, 32.7, 28.8 ppm. **^{19}F NMR (376 MHz, DMSO-*d*$_6$)** δ = –131.6 ppm. **TLC:** R_F = 0.21 (developed in cH/EtOAc 5:1 with 1% Et_3N).

5.9.2 X-ray structural analysis

Crystal structures were measured and solved by Dr. Martin Nieger at the University of Helsinki.

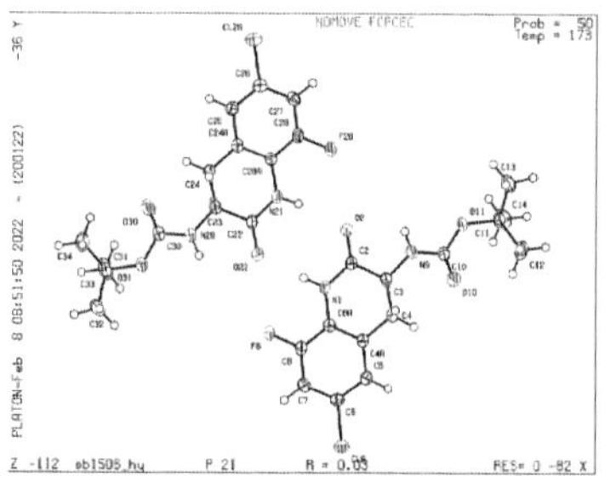

SB1506_HY

Compound 89
tert-butyl (*S*)-(6-chloro-8-fluoro-2-oxo-1,2,3,4-tetrahydroquinolin-3-yl)carbamate

absolute configuration (3S, 23S) determined crystallographically

Crystal data

$C_{14}H_{16}ClFN_2O_3$	$F(000) = 656$
$M_r = 314.74$	$D_x = 1.399$ Mg m^{-3}
Monoclinic, $P2_1$ (no.4)	Mo $K\alpha$ radiation, $\lambda = 0.71073$ Å
$a = 5.9620$ (2) Å	Cell parameters from 9962 reflections
$b = 18.7987$ (5) Å	$\theta = 2.6–27.4°$
$c = 13.3487$ (4) Å	$\mu = 0.28$ mm^{-1}
$\beta = 92.860$ (1)°	$T = 173$ K
$V = 1494.23$ (8) Å^3	Blocks, colourless
$Z = 4$	0.40 × 0.30 × 0.20 mm

Data collection

Bruker D8 VENTURE diffractometer with PhotonII CPAD detector	6726 reflections with $I > 2\sigma(I)$
Radiation source: INCOATEC microfocus sealed tube	$R_{int} = 0.050$
rotation in ϕ and ω, 1°, shutterless	$\theta_{max} = 27.5°$, $\theta_{min} = 2.2°$

scans	
Absorption correction: multi-scan *SADABS* (Sheldrick, 2014)	$h = -7 \rightarrow 7$
T_{min} = 0.667, T_{max} = 0.942	$k = -24 \rightarrow 24$
32070 measured reflections	$l = -17 \rightarrow 17$
6879 independent reflections	

Refinement

Refinement on F^2	**Secondary atom site location: difference Fourier map**
Least-squares matrix: full	**Hydrogen site location: difference Fourier map**
$R[F^2 > 2\sigma(F^2)] = 0.034$	**H atoms treated by a mixture of independent and constrained refinement**
$wR(F^2) = 0.087$	**$w = 1/[\sigma^2(F_o^2) + (0.048P)^2 + 0.2651P]$ where $P = (F_o^2 + 2F_c^2)/3$**
$S = 1.05$	**$(\Delta/\sigma)_{max} = 0.001$**
6879 reflections	**$\Delta\rangle_{max} = 0.22$ e Å^{-3}**
391 parameters	**$\Delta\rangle_{min} = -0.17$ e Å^{-3}**
5 restraints	**Absolute structure: Flack x determined using 3134 quotients [(I+)-(I-)]/[(I+)+(I-)] (Parsons, Flack and Wagner, Acta Cryst. B69 (2013) 249-259).**
Primary atom site location: dual	**Absolute structure parameter: -0.02 (3)**

5.10 New AIE-gen – Naphthalimide-(F_2)-PAP-DKP-Lys conjugate

5.10.1 Synthesis of PAP-DKP-Lys-Naphthalimide

6-Morpholino-1*H*,3*H*-benzo[*de*]isochromene-1,3-dione (119)

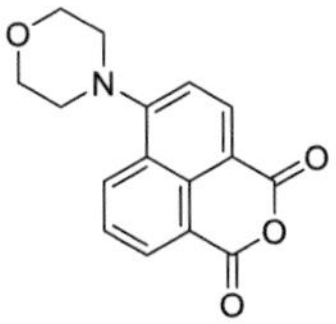

8-Bromo-3-oxatricyclo[7.3.1.05,13]trideca-1(12),5(13),6,8,10-pentaene-2,4-dione (2.00 g, 7.22 mmol, 1.00 eq) and morpholine (1.00 g, 11.5 mmol, 1.59 eq) were mixed in 2-methoxyethanol (15.0 mL). The resulting mixture was refluxed for 24 h under an argon atmosphere, then cooled down to 20 °C and further stirred overnight. The resulting yellow solid was filtered and washed with small amounts of MeOH to afford the product 6-Morpholino-1*H*,3*H*-benzo[de]isochromene-1,3-dione (1.63 g, 5.75 mmol, 80%).

^{1}H NMR (400 MHz, $CDCl_3$): δ = 8.55 (d, J = 7.2 Hz, 1H), 8.47 (dd, J = 11.6, 8.3 Hz, 2H), 7.73 (t, J = 7.9 Hz, 1H), 7.24 (d, J = 8.1 Hz, 1H), 4.06 – 3.99 (m, 4H), 3.39 – 3.19 (m, 4H) ppm. **^{13}C NMR (101 MHz, $CDCl_3$):** δ = 161.2, 160.5, 157.0, 135.0, 133.4, 132.3, 131.8, 126.3, 126.2, 119.6, 115.4, 112.4, 66.9, 53.4 ppm. **TLC:** R_f = 0.45 (developed in 79% DCM, 20% MeOH, 1% Et_3N). **HRMS (FAB):** *m/z* calcd. for $C_{16}H_{14}NO_4$ $[M+H]^+$ = 284.0917 Da, found 284.0918 Da (Δ = 0.2 ppm). **IR (ATR, $\tilde{\nu}$)** = 2953 (m), 2925 (w), 2901 (w), 2870 (w), 2853 (m), 1756 (s), 1742 (vs), 1717 (vs), 1582 (s), 1570 (vs), 1517 (s), 1470 (w), 1453 (m), 1434 (s), 1397 (s), 1374 (s), 1344 (m), 1324 (s), 1299 (vs), 1262 (m), 1235 (vs), 1224 (vs), 1210 (vs), 1171 (s), 1140 (w), 1115 (vs), 1088 (vs), 1068 (m), 1007 (vs), 990 (vs), 960 (s), 926 (m), 915 (vs), 863 (vs), 849 (vs), 834 (vs), 785 (vs), 756 (vs), 737 (s), 728 (vs), 676 (m), 664 (s), 652 (s), 608 (m), 588 (s), 572 (s), 557 (m), 534 (m), 523 (w), 492 (s), 476 (m), 455 (s), 428 (w), 412 (vs), 401 (m), 388 (vs) cm^{-1}.

2-(4-((2*S*,5*S*)-3,6-dioxo-5-(4-((*E*)-phenyldiazenyl)benzyl)piperazin-2-yl)butyl)-6-morpholino-1*H*-benzo[*de*]isoquinoline-1,3(2*H*)-dione (93)

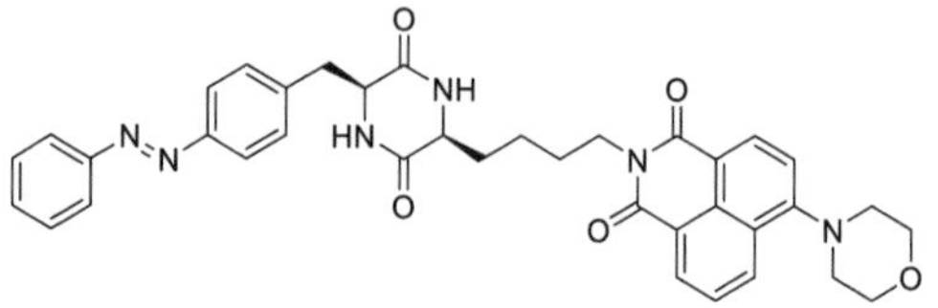

6-Morpholino-1*H*,3*H*-benzo[*de*]isochromene-1,3-dione (71.5 mg, 252 µmol, 1.00 eq) and PAP-DKP-Lys **52** (144 mg, 378 µmol, 1.50 eq) were suspended in DMF (3.5 mL). To this mixture was added Et_3N (102 mg, 140 µL, 1.01 mmol, 4.00 eq), and the reaction was heated to 110 °C for 12 h. The reaction mixture was cooled to 20 °C and turned to a thick suspension. Water (25 mL) was added to precipitate the desired product and to remove the unreacted PAP-DKP-Lys. The resultant mixture was sonicated for a while to ensure the formation of a suspension and then transferred into a 50 mL tube and centrifuged. The supernatant was decanted, and this procedure was repeated. Acetonitrile was added to the residue and the suspension was sonicated to remove traces of unreacted naphthalic anhydride. This suspension was further centrifuged, and the supernatant was decanted. Finally, the residue was purified by preparative HPLC with the following settings: 15 mL/min, 40 min gradient 65-95% MeCN in bidest. H_2O with 0.1% TFA, detection at 330 nm, retention at 19.5 min. After lyophilization, the pure product was obtained as a yellow solid (69.0 mg, 107 µmol, 42%).

^{1}H NMR (400 MHz, DMSO-d_6): δ = 8.45 (d, J = 8.4 Hz, 1H), 8.30 (s, 1H), 8.28 (dd, J = 24.3, 9.1 Hz, 2H), 8.12 (s, 1H), 7.81 (d, J = 7.9 Hz, 2H), 7.76 (t, J = 7.8 Hz, 1H), 7.62 – 7.47 (m, 2H), 7.39 (d, J = 8.0 Hz, 2H), 7.30 (d, J = 8.1 Hz, 1H), 6.91 (d, J = 7.2 Hz, 3H), 4.30 (s, 1H), 3.92 (t, J = 4.5 Hz, 4H), 3.64 (t, J = 6.5 Hz, 1H), 3.49 – 3.40 (m, 2H), 3.27 (d, J = 3.5 Hz, 1H), 3.23 (t, J = 4.6 Hz, 4H), 2.95 (dd, J = 13.4, 5.1 Hz, 1H), 1.22 – 1.07 (m, 3H), 0.94 – 0.67 (m, 3H) ppm. **^{13}C NMR (101 MHz, DMSO-d_6):** δ = 166.7, 165.8, 163.1, 162.6, 157.1, 155.3, 151.5, 150.8, 140.1, 131.9, 131.5, 130.5, 130.4, 130.3, 128.9, 128.5, 125.9, 125.2, 122.3, 121.8,

115.8, 114.9, 66.2, 55.2, 54.0, 53.1, 37.8, 33.0, 27.1, 21.3 ppm. **HRMS (FAB):** *m/z* calcd. for $C_{37}H_{37}N_6O_5$ $[M+H]^+$ = 645.2820 Da, found 645.2822 Da (Δ = 0.25 ppm). **IR (ATR, ṽ)** = 3191 (vw), 3043 (w), 2979 (w), 2962 (w), 2928 (vw), 2894 (w), 2877 (w), 2839 (vw), 2813 (vw), 2800 (vw), 1698 (w), 1673 (vs), 1656 (vs), 1615 (w), 1588 (m), 1577 (w), 1511 (w), 1459 (m), 1425 (w), 1392 (m), 1383 (w), 1357 (m), 1351 (m), 1333 (m), 1319 (w), 1302 (w), 1258 (w), 1241 (m), 1224 (w), 1203 (w), 1191 (w), 1171 (w), 1156 (w), 1146 (w), 1115 (s), 1088 (w), 1067 (w), 1048 (w), 1020 (w), 1009 (w), 972 (w), 925 (w), 905 (w), 882 (w), 851 (m), 839 (m), 827 (m), 799 (w), 782 (s), 771 (s), 759 (m), 690 (m), 666 (w), 656 (w), 640 (w), 608 (w), 592 (w), 569 (m), 537 (w), 514 (w), 503 (w), 476 (w), 463 (w), 448 (m), 435 (s), 416 (w), 405 (w), 375 (w) cm^{-1}. **UV-Vis (DMSO):** λ_{max} = 330 nm, 399 nm.

5.10.2 Photophysical properties of PAP-DKP-Lys-Naphthalimide

Photostationary states determined by ^{1}H NMR measurements

Photostationary states were determined by ^{1}H NMR measurements for compound **93** equilibrated under the indicated light wavelength (λ_{max} of the respective LED light diode). For analysis, non-overlapping NMR signals were assigned to the *E* and *Z* isomer, respectively. To assign the signals of the *Z* isomer, the dark spectrum was taken as comparison.

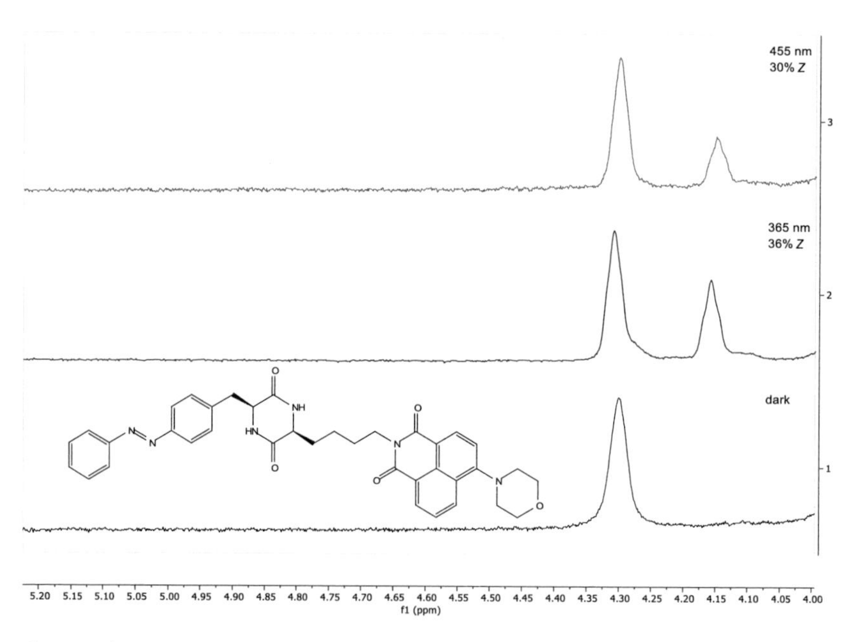

Figure 172: ^{1}H-NMR-spectrum (400 MHz, DMSO-d_6) of compound **93** in darkness and PSS after irradiation at 365 nm (36% *Z*-isomer) or 455 nm (30% *Z*-isomer).

5.10.3 Fluorescence of PAP-DKP-Lys-Naphthalimide

Fluorescence spectra of compound **93** (400 nm-700 nm) were measured (20 °C) in THF, $CHCl_3$, DCM, TCE, DMF, HFIP at a concentration of 50 μM and in MeOH, Acetone, ^{i}PrOH, MeCN due to low solubility at saturation <50 μM. The solutions were kept in the dark prior to measurement.

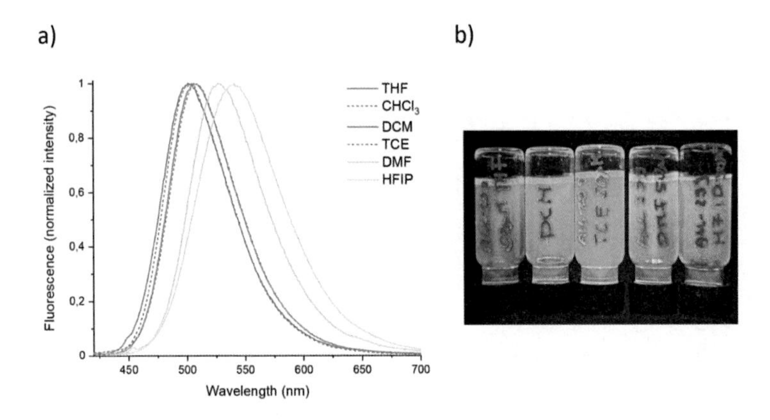

Figure 173: Compound **93** was dissolved (50 µM) in various solvents and is characterized by solvatochromism. a) Normalized fluorescence spectra, excitation at 400 nm. b) Photograph of solutions in THF, DCM, TCE, DMF, HFIP irradiated by a handheld UV-lamp.

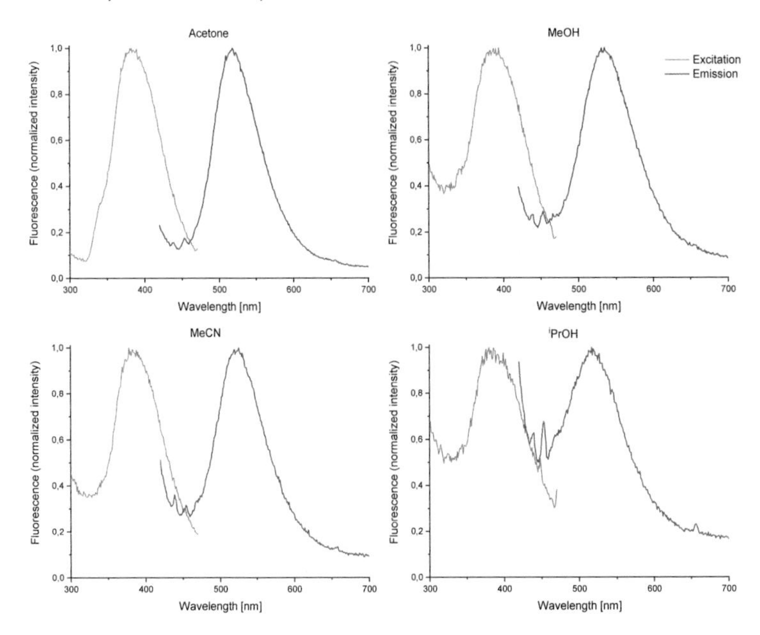

Figure 174: Normalized excitation (black) and emission spectrum (red, excited at 400 nm) of compound **93** in Acetone, MeOH, MeCN, iPrOH.

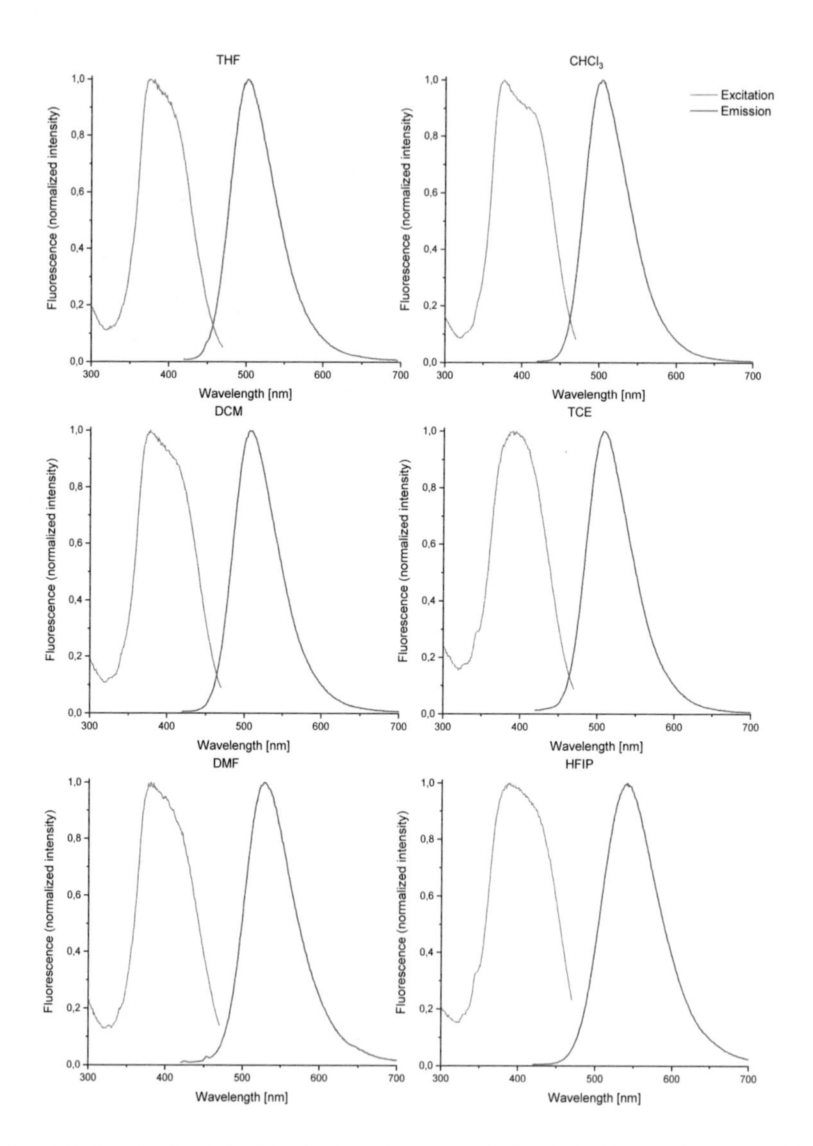

Figure 175: Normalized excitation (black) and emission spectrum (red, excited at 400 nm) of compound **93** in THF, $CHCl_3$, DCM, TCE, DMF, HFIP.

Fluorescence spectra after irradiation

In order to investigate the behavior upon isomerization, samples were irradiated, and the fluorescence was measured. Samples were prepared in triplicates at a concentration of 50 µM (**93**) in TCE and first measured in the dark, further measurements were done after each irradiation step. Subsequently, samples were irradiated at 365 nm (10 min), followed by 455 nm (10 min) and again at (365 nm).

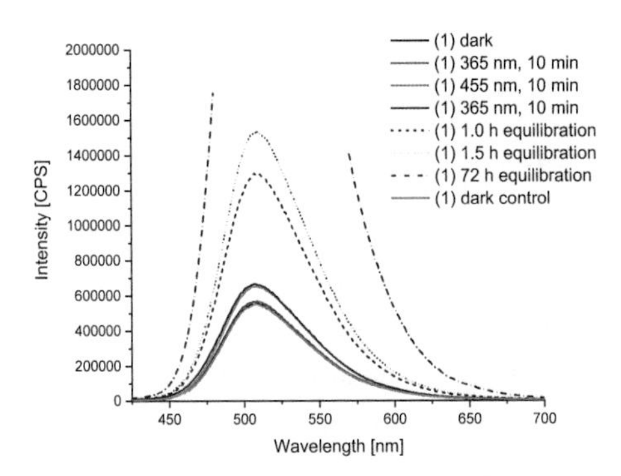

Figure 176: Measurement number 1: Emission spectra (excitation at 400 nm) of compound **93** in TCE. Upon irradiation the fluorescence is increasing over time.

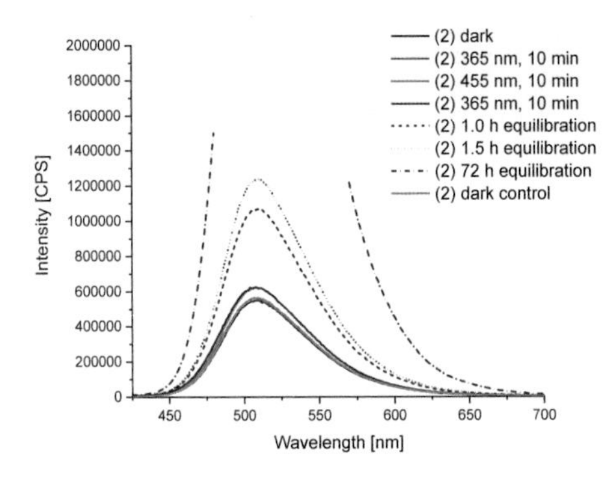

Figure 177: Measurement number 2: Emission spectra (excitation at 400 nm) of compound **93** in TCE. Upon irradiation the fluorescence is increasing over time.

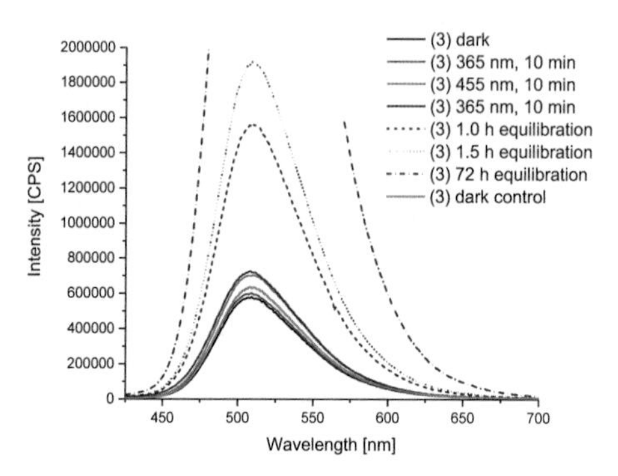

Figure 178: Measurement number 3: Emission spectra (excitation at 400 nm) of compound **93** in TCE. Upon irradiation the fluorescence is increasing over time.

Based on the low solubility of **93**, ultrasonic treatment was applied to the solutions at some point of the investigation. Here a new phenomenon – a hypsochromic shift of the fluorescence emission – was observed. This effect follows a distinct pattern and was occurring after ultrasonic treatment of solutions in TCE with an increase in fluorescence emission. Therefore, an experiment with a distinct pattern of sonication and irradiation was performed in independent triplicates by dissolving each time fresh compound **93** in TCE to obtain a 50 µM solution. Emission spectra were measured for samples treated in the following order 1) dark equilibrated, 2) irradiation induced fluorescence increase, 3) ultrasound, 4) irradiation at 365 nm, 5) 50 min equilibration.

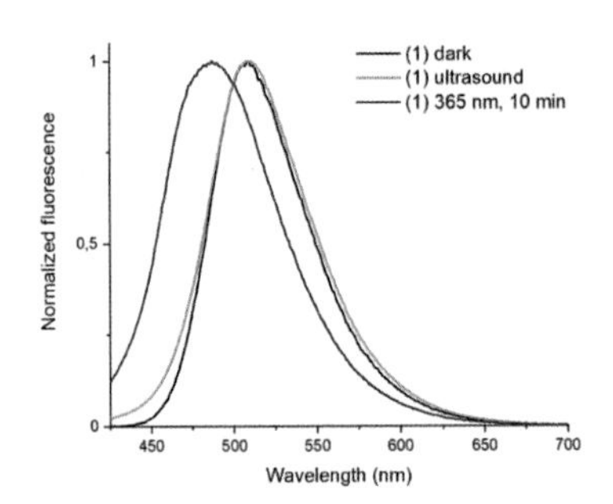

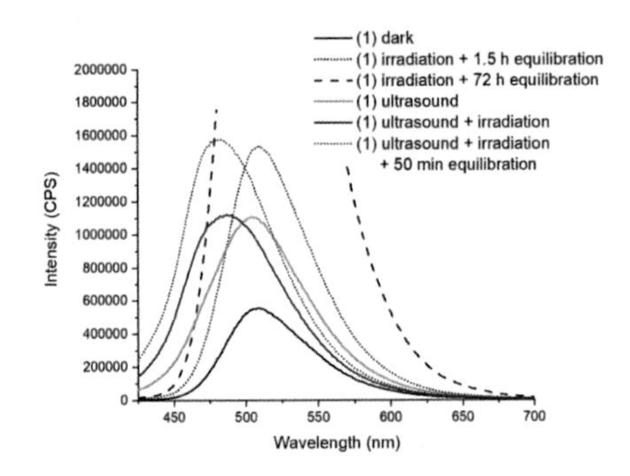

Figure 179: Demonstration of the hypsochromic shift described above. Left: normalized fluorescence emission of the relevant curves, dark state (Em_{max} = 509 nm), ultrasonic treatment after fluorescence increase and irradiation at 365 nm to the endpoint (Em_{max} = 481 nm). Right: measured fluorescence intensity of measurement 1; Excitation at 400 nm.

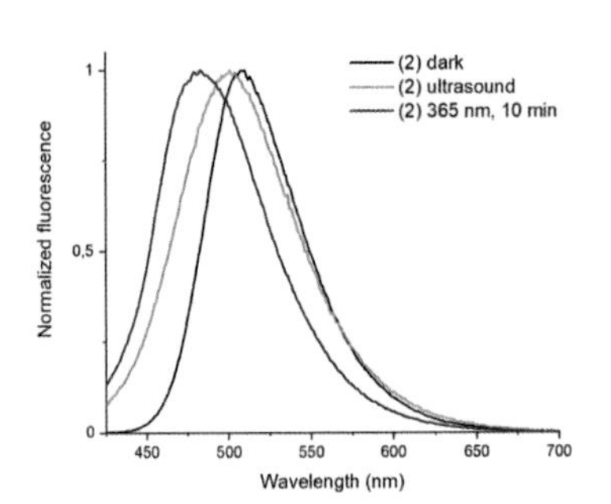

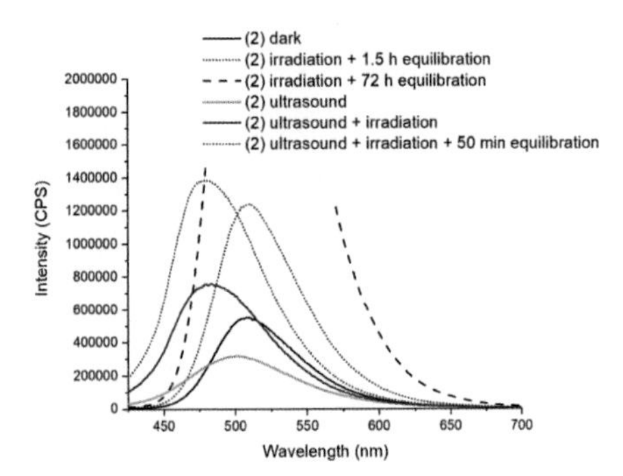

Figure 180: Demonstration of the hypsochromic shift described above. Left: normalized fluorescence emission of the relevant curves, dark state (Em_{max} = 509 nm), ultrasonic treatment after fluorescence increase and irradiation at 365 nm to the endpoint (Em_{max} = 480 nm). Right: measured fluorescence intensity of measurement 2; Excitation at 400 nm.

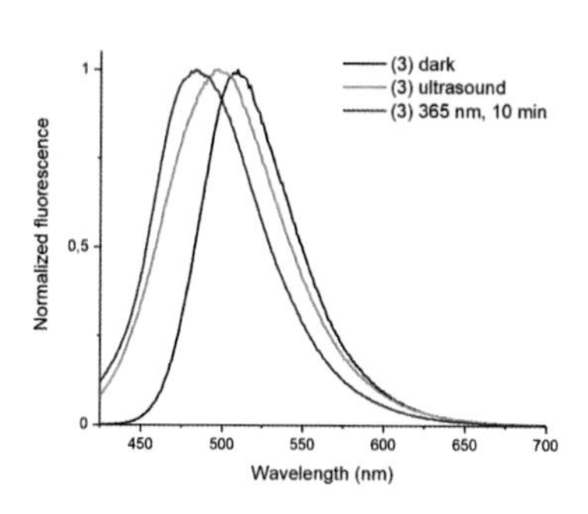

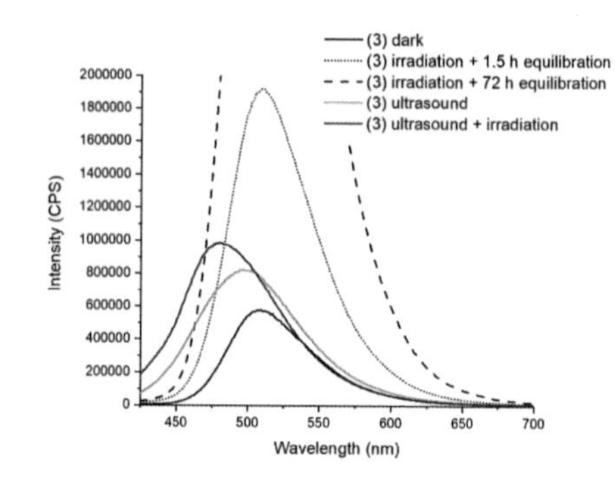

Figure 181: Demonstration of the hypsochromic shift described above. Left: normalized fluorescence emission of the relevant curves, dark state (Em_{max} = 509 nm), ultrasonic treatment after fluorescence increase and irradiation at 365 nm to the endpoint (Em_{max} = 480 nm). Right: measured fluorescence intensity of measurement 3; Excitation at 400 nm.

Furthermore, a control experiment where the samples were kept in the dark was performed. Three independent samples were dissolved in TCE at a concentration of 50 µM and stored in the dark. Here, no hypsochromic shift was observed and additional ultrasonic treatment after a week of equilibration in the dark did also not affect the emission maximum.

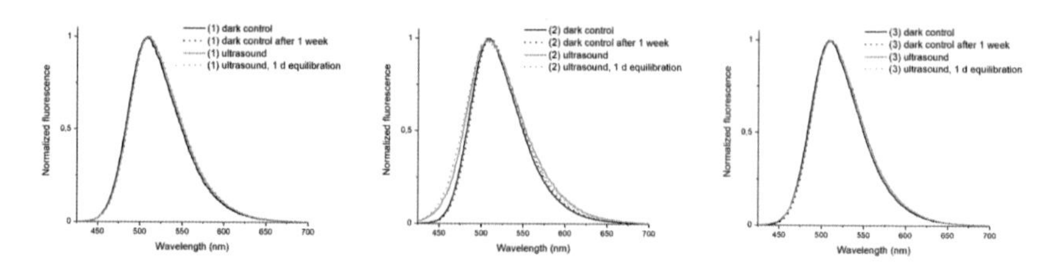

Figure 182: Control experiment; compound **93** was dissolved in TCE (50 µM) and kept in darkness. After one week equilibration in the dark, the samples were sonicated for 20 min and again fluorescence was measured. Additionally, fluorescence was measured one day after ultrasound treatment. Sample 2 displays a broadened curve shape for the ultrasound treated measurements in comparison to the dark curves. Though, the maximum is not shifted, and total exclusion of light cannot be assured.

The influence of the solvent on the hypsochromically shifted fluorescence emission phenomenon was investigated. Compound **93** was dissolved at a concentration of 50 µM in HFIP, DMF, DMSO, THF, DCM, $CHCl_3$ and again in TCE as a control. The solutions were freshly prepared, fluorescence was measured and then the solution was treated directly for 30 min with ultrasonic waves. Fluorescence was measured, then the sample was equilibrated for 60 min in darkness. Fluorescence was measured and in the following the solution was again treated with ultrasonic waves. Fluorescence was measured again, and the solution was then irradiated with light at 365 nm. Subsequently, a final fluorescence measurement was done. For selected samples further equilibration and ultrasound periods with subsequent fluorescence measurements were performed (see Figure 183).

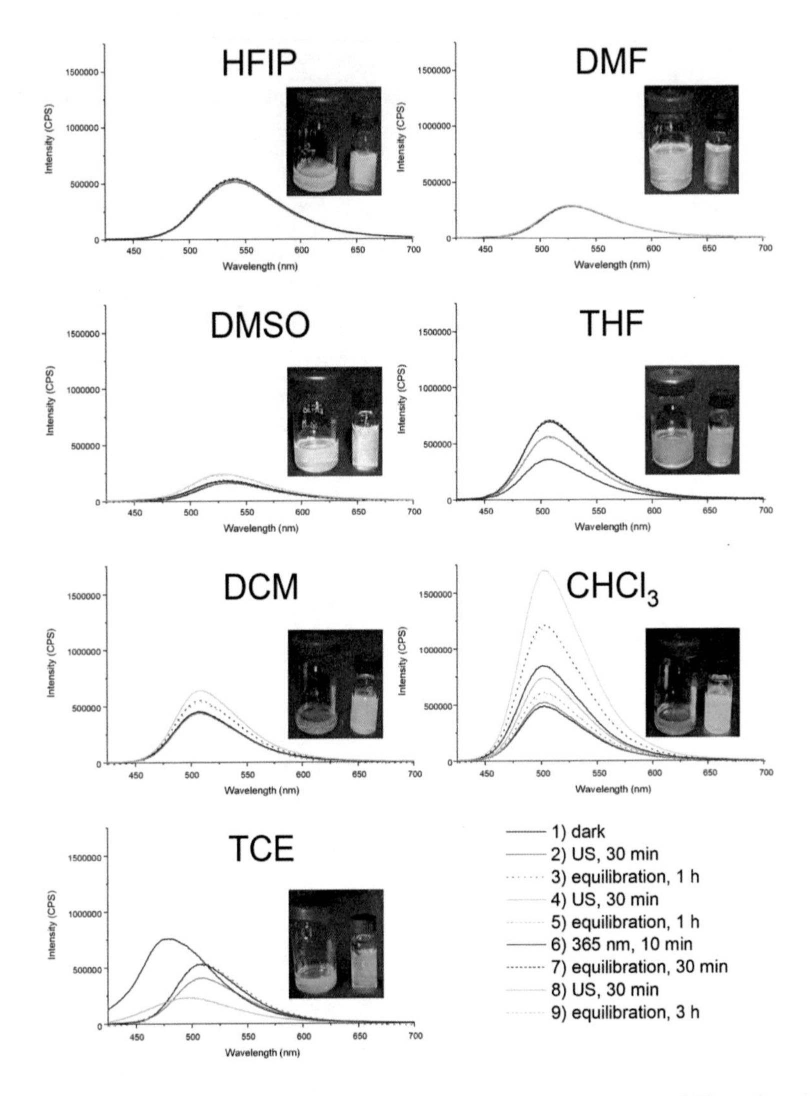

Figure 183: Transfer of the protocol for the hypsochromic shift of the emission maximum (TCE) to other solvents. The solvents are sorted by polarity from polar (top) to less polar (bottom).

5.10.4 Scanning Electron Microscopy (PAP-DKP-Lys-Naphthalimide)

The effect of hypsochromically shifted emission was only observed in the solvent TCE. To correlate the effects to macromolecular structures, scanning electron microscopy images were recorded. Therefore, compound **93** was dissolved at a concentration of 50 µM in TCE and previously cleaned silicon wafers were prepared by adding 10 µL of the solution by an Eppendorf pipette to the wafer. The droplets dried under atmospheric pressure at 20 °C (1 h). Before microscopic analysis, the wafers were coated with a thin layer of platinum (6-7 nm). Volker Zibat operated the microscope.Sample **a)** was taken directly after dissolution of the compound **93 (dark)**, sample **b)** was taken after ultrasonic treatment and 3 h of equilibration **(fluorescence increase)**, sample **c)** was taken after following additional ultrasonic treatment **(slight hypsochromic shift)** and sample **d)** was irradiated at 365 nm for 10 min after the previous steps **(finished hypsochromic shift)**. Due to the drying period of 1 h sample **a)** had additional equilibration time, which can explain the slight network formation visible in Figure 184. Another sample **e)** was prepared after equilibration of the hypsochromically shifted fluorescence sample overnight.

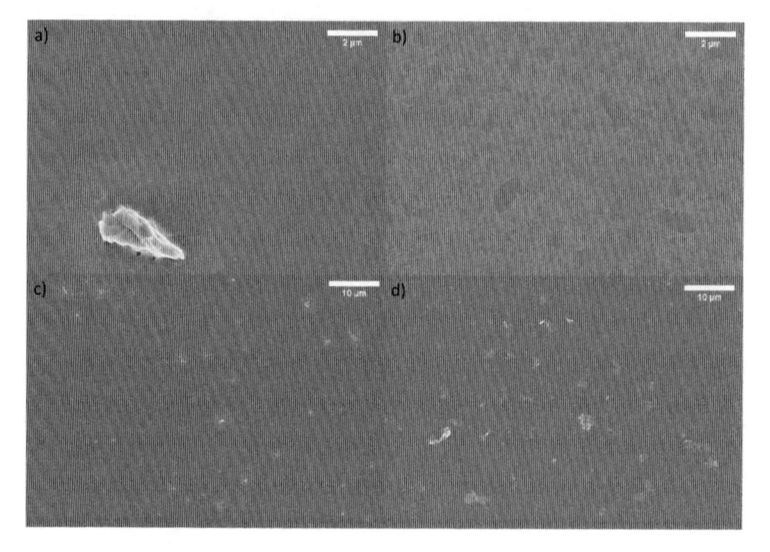

Figure 184: Scanning electron microscope images of sample **93** in TCE (50 µM). a) a network started to form during wafer preparation, b) a strong network was detected on the wafer, some small particles are enclosed, c) the network is still present and distinct particles are visible, d) the network disintegrates upon irradiation and small particles remain.

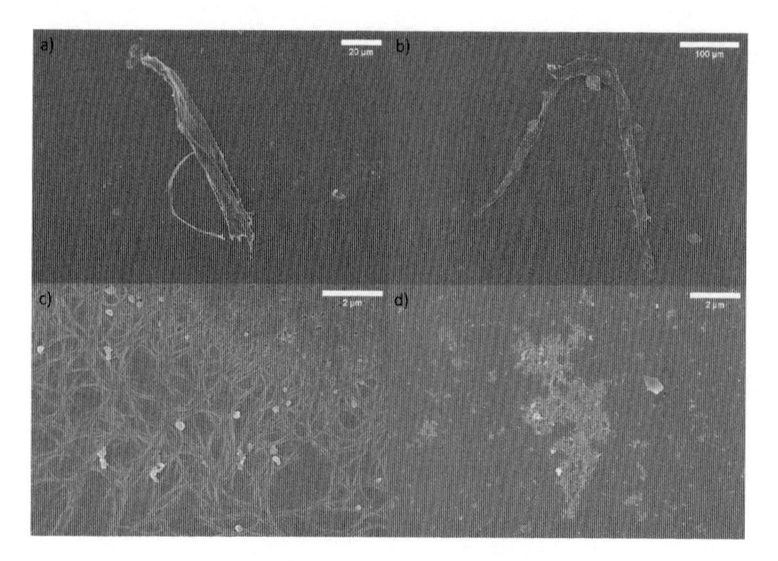

Figure 185: Scanning electron microscope images of sample **93** in TCE (50 µM). a) large structures start to assemble, b) large (>500 µm long) fibrillar structures were detected on the wafer, c) particles enclosed to a cross-linked network, d) the network disintegrates upon irradiation, rudimentary elements of the network are visible.

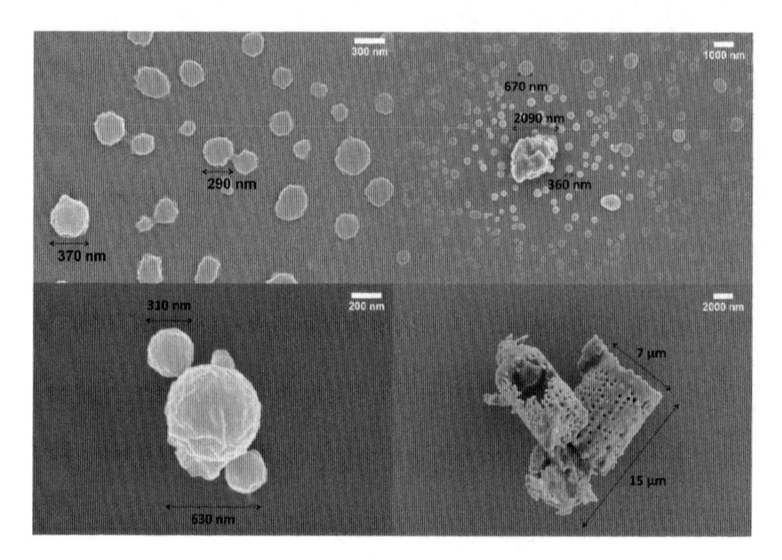

Figure 186: In sample e) the fragments and small particles (visible in Figure 185 d)) assembled to spheric nanoparticles, which tend to agglomerate.

5.10.5 Dynamic Light Scattering (DLS) measurement

To verify the observed particles in the microscope images, DLS measurements were performed. Compound **93** was dissolved at a concentration of 50 µM and the preparation sequence towards sample **e)** was repeated. Three independent measurements were performed, each with at least four technical replicates.

Hydrodynamic radii were determined with a ZETASIZER Nano-S ZEN1600 DLS spectrometer (Malvern Instruments Limited, Worcestershire, UK). The measurements were performed with the following parameters: as material, polystyrene latex (refractive index (RI) = 1.590) was chosen, dispersant was TCE at 25 °C with a viscosity of 1.70 cP and an RI of 1.493. Evaluation was done with Zetasizer Software 8.02 © 2002-2021 Malvern Panalytical and the analysis model was based on multiple narrow modes (high resolution).

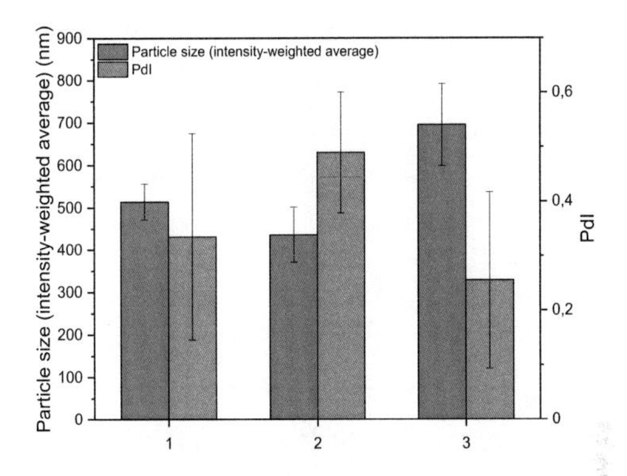

Figure 187: Results of the DLS measurement. Three independent measurements (number of technical replicates n≥4 respectively) were performed. The particle size is in good agreement with the microscopic observations.

5.10.6 Synthesis of F_2-PAP-DKP-Lys-Naphthalimide

2-(4-((2*S*,5*S*)-5-(4-((*E*)-(2,6-difluorophenyl)diazenyl)benzyl)-3,6-dioxopiperazin-2-yl)butyl)-6-morpholino-1*H*-benzo[*de*]isoquinoline-1,3(2*H*)-dione (94)

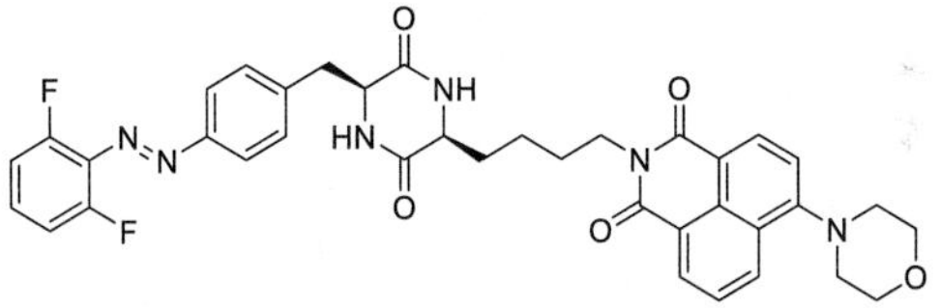

6-Morpholino-1*H*,3*H*-benzo[*de*]isochromene-1,3-dione (100 mg, 353 µmol, 1.00 eq) and F_2-PAP-DKP-Lys **57** (220 mg, 530 µmol, 1.50 eq) were suspended in DMF (5 mL). To this mixture was added Et_3N (102 mg, 140 µL, 1.01 mmol, 4.00 eq), and the reaction was heated to 110 °C for 12 h. The reaction mixture was cooled to 20 °C, whereupon it turned to a thick suspension. Water (40 mL) was added to precipitate the desired product and to remove the unreacted PAP-DKP-Lys. The resulting mixture was sonicated for a while to ensure the formation of a suspension and then transferred into a 50 mL tube and centrifuged. The supernatant was decanted, and this procedure was repeated. Acetonitrile was added to the residue and the suspension was sonicated to remove traces of unreacted naphthalic anhydride. This suspension was further centrifuged, and the supernatant was decanted. Finally, the orange-brown residue was purified by preparative HPLC with the following settings: 15 mL/min, 40 min gradient 65-95% MeCN in bidest. H_2O with 0.1% TFA, detection at 330 nm, retention at 20.5 min. After lyophilization, the pure product was obtained as a pale orange solid (75.0 mg, 110 µmol, 31%).

^{1}H NMR (400 MHz, DMSO-d_6): δ = 8.43 (d, J = 8.4 Hz, 1H), 8.30 (d, J = 6.9 Hz, 2H), 8.23 (d, J = 8.0 Hz, 1H), 8.14 (d, J = 2.3 Hz, 1H), 7.77 (d, J = 8.1 Hz, 2H), 7.76 – 7.71 (m, 1H), 7.39 (d, J = 8.1 Hz, 2H), 7.27 (d, J = 8.1 Hz, 1H), 6.97 (tt, J = 8.5, 6.0 Hz, 1H), 6.58 (t, J = 9.1 Hz, 2H), 4.30 (t, J = 5.0 Hz, 1H), 3.92 (t, J = 4.0 Hz, 4H), 3.65 – 3.58 (m, 3H), 3.29 (dd, J = 13.3, 3.6 Hz, 1H), 3.22 (t, J = 4.5 Hz, 4H), 2.95 (dd, J = 13.2, 5.0 Hz, 1H), 1.23 – 1.03 (m, 3H), 0.74 (dtd, J = 35.1, 12.2, 11.7, 6.0 Hz, 3H) ppm. **^{19}F NMR (376 MHz, DMSO-d_6):** δ = -122.15 ppm. **^{13}C NMR (101 MHz, DMSO-d_6):** δ = 166.8, 165.8, 163.1, 162.6, 155.7 (d, J = 3.8 Hz), 155.3, 153.2 (d, J = 4.4 Hz), 151.6, 141.2, 131.9, 131.6, 131.0 – 130.6 (m), 130.4 (d, J = 6.0 Hz), 130.1 – 129.8 (m), 128.9, 126.0, 125.3, 122.4, 115.8, 115.0, 112.2, 112.0, 66.3, 55.3, 54.0, 53.1, 38.0, 33.5, 27.1, 21.5 ppm. **HRMS (FAB):** *m/z* calcd. for $C_{37}H_{35}N_6O_5F_2$ $[M+H]^+$ = 681.2632 Da, found 681.2634 Da (Δ = 0.33 ppm). **IR (ATR, $\tilde{\nu}$)** = 3187 (w), 3047 (w), 2959 (w), 2932 (w), 2928 (w), 2917 (w), 2893 (w), 2860 (w), 1776 (vw), 1656 (vs), 1613 (m), 1588 (s), 1514 (w), 1468 (m), 1453 (s), 1429 (w), 1391 (m), 1360 (s), 1336 (m), 1302 (w), 1259 (w), 1239 (s), 1220 (w), 1207 (m), 1174 (m), 1153 (m), 1115 (m), 1089 (w), 1067 (w), 1017 (m), 925 (w), 902 (w), 849 (w), 822 (w), 782 (vs), 759 (s), 739 (w), 718 (w), 705 (w), 670 (w), 598 (vw), 540 (vw), 506 (vw), 433 (w), 418 (w), 407 (w) cm^{-1}. **UV-Vis (DMSO):** λ_{max} = 318 nm, 396 nm.

5.10.7 Fluorescence of F_2-PAP-DKP-Lys-Naphthalimide

Fluorescence spectra of compound **94** (400 nm-700 nm) were measured (20 °C) in $CHCl_3$, DCM, THF, TCE, Acetone, ^{i}PrOH, MeCN, DMF, MeOH and HFIP at a concentration of 50 µM. The solutions were kept in the dark prior to measurement.

Figure 188: Compound **94** was dissolved (50 µM) in various solvents and is characterized by solvatochromism. Photograph of solutions in $CHCl_3$, DCM, THF, TCE, Acetone, ^iPrOH, MeCN, DMF, MeOH and HFIP irradiated by a handheld UV-lamp.

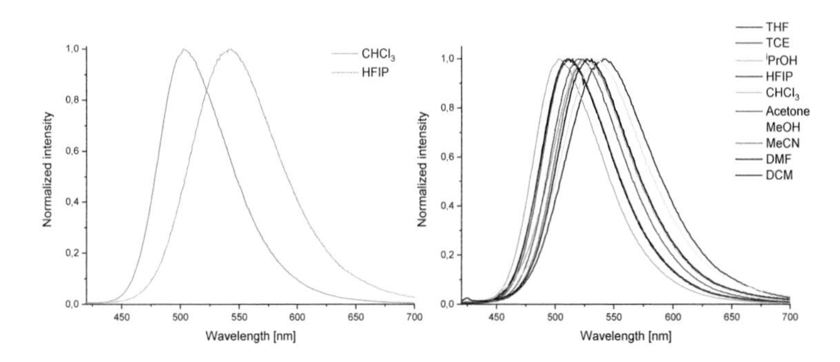

Figure 189: Normalized fluorescence spectra, excitation at 400 nm. Left: Fluorescence emission curve of **94** in $CHCl_3$ and HFIP, here Em_{max} has the highest difference. Right: Fluorescence emission curve of **94** in various solvents.

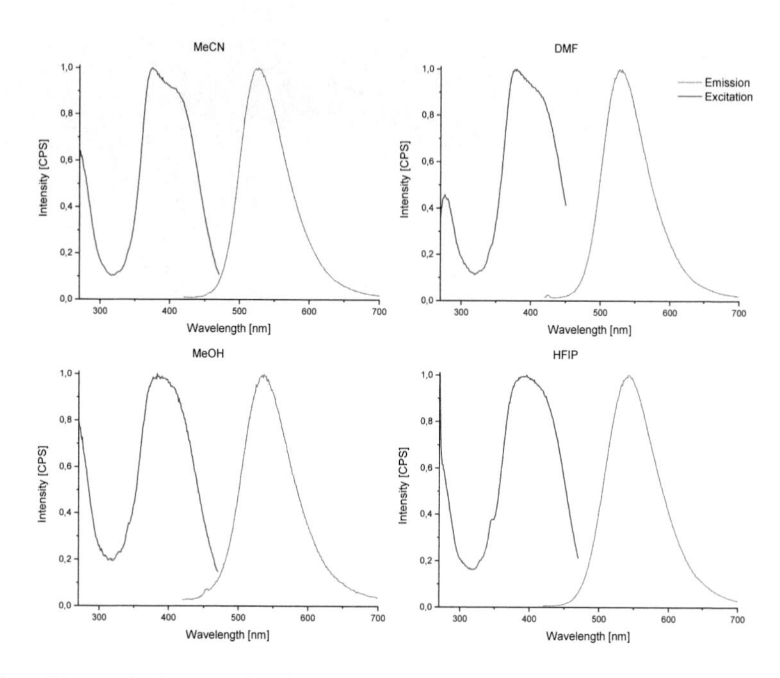

Figure 190: Normalized excitation (black) and emission spectrum (red, excited at 400 nm) of compound **94** in MeCN, DMF, MeOH, HFIP.

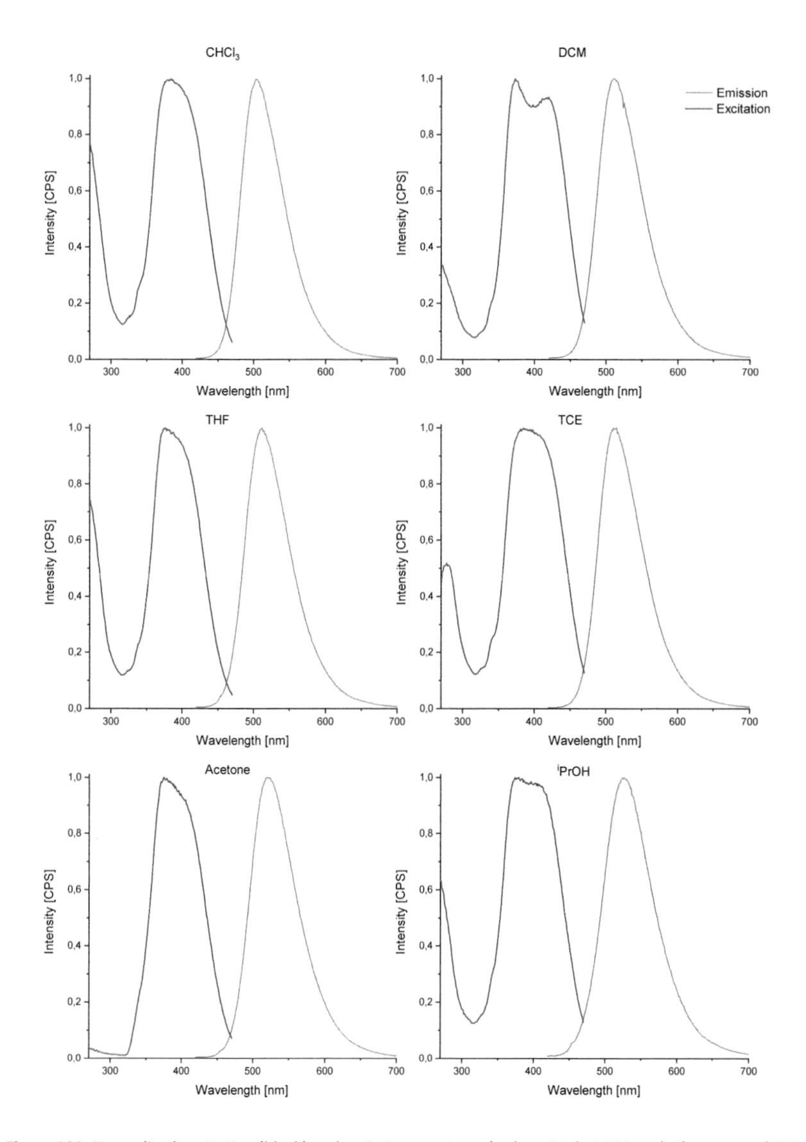

Figure 191: Normalized excitation (black) and emission spectrum (red, excited at 400 nm) of compound **94** in $CHCl_3$, DCM, THF, TCE, acetone, $^{i}PrOH$.

6 List of Abbreviations

3

3-NBA .... *m*-nitrobenzyl alcohol

A

ABs... azobenzene

ACQ....................... aggregation-caused quenching

AIE .... aggregation induced emission

AIEE. aggregation induced emission enhancement

AIE-gen .... AIE-luminogen

ATR .... attenuated total reflection

C

cH .... cyclohexane

CPDs... cyclophanedienes

CPMG.......................... Carr-Purcell-Meiboom-Gill

D

DAE .... diarylethene

DASAs .... Donor-Acceptor Stenhouse adducts

DBU................ 1,8-diazabicyclo[5.4.0]undec-7-ene

DCM.. dichloromethane

DHPs .... dihydropyrenes

DKP .... diketopiperazine

DMEM........... Dulbecco's Modified Eagle Medium

DMSO .... dimethyl sulfoxide

DNA .... deoxyribonucleic acid , ribonucleic acid

DPBS .... Dulbecco's Phosphate-Buffered Saline

E

EDG .... electron donating group

EI-MS........ electron ionization-mass spectrometry

ESPT......................... excited-state proton transfer

Et_3N .... triethylamine

EtOAc .... ethyl acetate

F

FA .... formic acid

FAB-MS................. fast atom bombardment-mass spectroscopy

Fmoc......................... 9-fluorenylmethoxycarbonyl

G

GSH .... glutathione

H

HFIP .... hexafluoroisopropanol

HPI.. hemipiperazine

I

ICT .... intramolecular charge transfer

IR ....infrared spectroscopy

L

LMWG .... low molecular weight gelator

LVE .... linear viscoelastic regime

M

MOF .... metal organic framework

N

NCS .... *N*-chlorosuccinimide

NMR...........................nuclear magnetic resonance

O

OLED .... organic light emitting diode

P

PCR .... polymerase chain reaction

pI .... isoelectric point

PSS .... photostationary states

PST .... photostatin

R

RIM........................... restriction of intramolecular motions/rotation

RNA .... ribonucleic acid

S

SURMOF .... surface-mounted-metal-organic framework

T

TCE 1,1,2,2-tetrachlorethane

THF .... tetrahydrofurane

TICT .... twisted intramolecular charge transfer

TLC.............. analytical thin layer chromatography

U

UV .... ultraviolet

7 Literature index

[1] W. A. Velema, W. Szymanski, B. L. Feringa, *J. Am. Chem. Soc.* **2014**, *136*, 2178-2191.

[2] C. Knie, M. Utecht, F. Zhao, H. Kulla, S. Kovalenko, A. M. Brouwer, P. Saalfrank, S. Hecht, D. Bleger, *Chem. Eur. J.* **2014**, *20*, 16492-16501.

[3] W. Szymanski, J. M. Beierle, H. A. Kistemaker, W. A. Velema, B. L. Feringa, *Chem. Rev.* **2013**, *113*, 6114-6178.

[4] A. L. Leistner, Z. L. Pianowski, *Eur. J. Org. Chem.* **2022**, *2022*, e202101271.

[5] I. R. Edwards, J. K. Aronson, *Lancet* **2000**, *356*, 1255-1259.

[6] G. Quandt, G. Hofner, J. Pabel, J. Dine, M. Eder, K. T. Wanner, *J. Med. Chem.* **2014**, *57*, 6809-6821.

[7] S. Kirchner, A. L. Leistner, P. Godtel, A. Seliwjorstow, S. Weber, J. Karcher, M. Nieger, Z. Pianowski, *Nat. Commun.* **2022**, *13*, 6066.

[8] A. Cembran, F. Bernardi, M. Garavelli, L. Gagliardi, G. Orlandi, *J Am Chem Soc* **2004**, *126*, 3234-3243.

[9] B. Heinz, S. Malkmus, S. Laimgruber, S. Dietrich, C. Schulz, K. Ruck-Braun, M. Braun, W. Zinth, P. Gilch, *J. Am. Chem. Soc.* **2007**, *129*, 8577-8584.

[10] Z. L. Pianowski, *Chem. Eur. J.* **2019**, *25*, 5128-5144.

[11] S. Aiken, R. J. L. Edgar, C. D. Gabbutt, B. M. Heron, P. A. Hobson, *Dyes Pigm.* **2018**, *149*, 92-121.

[12] R. S. Blackburn, T. Bechtold, P. John, *Color. Technol.* **2009**, *125*, 193-207.

[13] J. Pina, D. Sarmento, M. Accoto, P. L. Gentili, L. Vaccaro, A. Galvao, J. S. Seixas de Melo, *J. Phys. Chem. B* **2017**, *121*, 2308-2318.

[14] T. Arai, M. Ikegami, *Chem. Lett.* **1999**, *28*, 965-966.

[15] M. Hooper, W. N. Pitkethly, *J. Chem. Soc., Perkin Trans. 1* **1973**, 2804.

[16] C. Petermayer, S. Thumser, F. Kink, P. Mayer, H. Dube, *J. Am. Chem. Soc.* **2017**, *139*, 15060-15067.

[17] K. Ichimura, T. Seki, T. Tamaki, T. Yamaguchi, *Chem. Lett.* **1990**, *19*, 1645-1646.

[18] M. W. H. Hoorens, M. Medved, A. D. Laurent, M. Di Donato, S. Fanetti, L. Slappendel, M. Hilbers, B. L. Feringa, W. Jan Buma, W. Szymanski, *Nat. Commun.* **2019**, *10*, 2390.

[19] C. Petermayer, H. Dube, *Acc. Chem. Res.* **2018**, *51*, 1153-1163.

[20] A. R. Olson, *Trans. Faraday Soc.* **1931**, *27*, 69.

[21] D. H. Waldeck, *Chem. Rev.* **1991**, *91*, 415-436.

[22] a) M. Oelgemöller, R. Frank, P. Lemmen, D. Lenoir, J. Lex, Y. Inoue, *Tetrahedron* **2012**, *68*, 4048-4056; b) D. Villaron, S. J. Wezenberg, *Angew. Chem. Int. Ed.* **2020**, *59*, 13192-13202.
[23] M. Irie, *Chem. Rev.* **2000**, *100*, 1685-1716.
[24] M. Irie, M. Mohri, *J. Org. Chem.* **1988**, *53*, 803-808.
[25] R. M. Kellogg, M. B. Groen, H. Wynberg, *J. Org. Chem.* **1967**, *32*, 3093-&.
[26] M. Irie, T. Fukaminato, K. Matsuda, S. Kobatake, *Chem. Rev.* **2014**, *114*, 12174-12277.
[27] Z. Y. Li, C. J. He, Z. Q. Lu, P. S. Li, Y. P. Zhu, *Dyes Pigm.* **2020**, *182*, 108623.
[28] G. M. Tsivgoulis, J.-M. Lehn, *Adv. Mater.* **1997**, *9*, 627-630.
[29] N. M. Wu, M. Ng, W. H. Lam, H. L. Wong, V. W. Yam, *J. Am. Chem. Soc.* **2017**, *139*, 15142-15150.
[30] R. T. Jukes, V. Adamo, F. Hartl, P. Belser, L. De Cola, *Inorg. Chem.* **2004**, *43*, 2779-2792.
[31] C. J. Carling, J. C. Boyer, N. R. Branda, *J. Am. Chem. Soc.* **2009**, *131*, 10838-10839.
[32] K. Mori, Y. Ishibashi, H. Matsuda, S. Ito, Y. Nagasawa, H. Nakagawa, K. Uchida, S. Yokojima, S. Nakamura, M. Irie, H. Miyasaka, *J. Am. Chem. Soc.* **2011**, *133*, 2621-2625.
[33] J. H. Day, *Chem. Rev.* **2002**, *63*, 65-80.
[34] C. J. Martin, G. Rapenne, T. Nakashima, T. Kawai, *J. Photochem. Photobiol., C* **2018**, *34*, 41-51.
[35] M. Li, Q. Zhang, Y.-N. Zhou, S. Zhu, *Prog. Polym. Sci.* **2018**, *79*, 26-39.
[36] R. Klajn, *Chem. Soc. Rev.* **2014**, *43*, 148-184.
[37] L. Kortekaas, W. R. Browne, *Chem. Soc. Rev.* **2019**, *48*, 3406-3424.
[38] J. Buback, M. Kullmann, F. Langhojer, P. Nuernberger, R. Schmidt, F. Wurthner, T. Brixner, *J. Am. Chem. Soc.* **2010**, *132*, 16510-16519.
[39] O. Ivashenko, J. T. van Herpt, B. L. Feringa, P. Rudolf, W. R. Browne, *J. Phys. Chem. C* **2013**, *117*, 18567-18577.
[40] G. S. Hartley, R. J. W. Le Fèvre, *J. Chem. Soc.* **1939**, *0*, 531-535.
[41] a) R. J. Mart, R. K. Allemann, *Chem. Commun.* **2016**, *52*, 12262-12277; b) N. Preusske, W. Moormann, K. Bamberg, M. Lipfert, R. Herges, F. D. Sonnichsen, *Org. Biomol. Chem.* **2020**, *18*, 2650-2660; c) S. Samanta, C. Qin, A. J. Lough, G. A. Woolley, *Angew. Chem. Int. Ed.* **2012**, *51*, 6452-6455.
[42] a) T. Hugel, N. B. Holland, A. Cattani, L. Moroder, M. Seitz, H. E. Gaub, *Science* **2002**, *296*, 1103-1106; b) S. Muramatsu, K. Kinbara, H. Taguchi, N. Ishii, T. Aida, *J. Am. Chem. Soc.* **2006**, *128*, 3764-3769; c) A. S. Lubbe,

Q. Liu, S. J. Smith, J. W. de Vries, J. C. M. Kistemaker, A. H. de Vries, I. Faustino, Z. Meng, W. Szymanski, A. Herrmann, B. L. Feringa, *J. Am. Chem. Soc.* **2018**, *140*, 5069-5076.

[43] a) W. Wen, W. Q. Ouyang, S. Guan, A. H. Chen, *Polym. Chem.* **2021**, *12*, 458-465; b) X. Zheng, S. Guan, C. Zhang, T. Qu, W. Wen, Y. Zhao, A. Chen, *Small* **2019**, *15*, e1900110; c) A. Natansohn, P. Rochon, *Chem. Rev.* **2002**, *102*, 4139-4175; d) J. Park, D. Yuan, K. T. Pham, J. R. Li, A. Yakovenko, H. C. Zhou, *J. Am. Chem. Soc.* **2012**, *134*, 99-102.

[44] a) P. COURTOT, J. LE SAINT, R. PICHON, *Chemischer Informationsdienst* **1976**, *7*, no-no; b) P. Courtot, R. Pichon, J. Le Saint, *Tetrahedron Lett.* **1976**, *17*, 1181-1184; c) R. Pichon, J. Le Saint, P. Courtot, *Tetrahedron* **1981**, *37*, 1517-1524.

[45] L. A. Tatum, X. Su, I. Aprahamian, *Acc. Chem. Res.* **2014**, *47*, 2141-2149.

[46] a) S. M. Landge, E. Tkatchouk, D. Benitez, D. A. Lanfranchi, M. Elhabiri, W. A. Goddard, 3rd, I. Aprahamian, *J. Am. Chem. Soc.* **2011**, *133*, 9812-9823; b) H. Qian, S. Pramanik, I. Aprahamian, *J. Am. Chem. Soc.* **2017**, *139*, 9140-9143; c) Q. Qiu, S. Yang, M. A. Gerkman, H. Fu, I. Aprahamian, G. G. D. Han, *J. Am. Chem. Soc.* **2022**, *144*, 12627-12631; d) S. Yang, J. D. Harris, A. Lambai, L. L. Jeliazkov, G. Mohanty, H. Zeng, A. Priimagi, I. Aprahamian, *J. Am. Chem. Soc.* **2021**, *143*, 16348-16353.

[47] B. Shao, H. Qian, Q. Li, I. Aprahamian, *J. Am. Chem. Soc.* **2019**, *141*, 8364-8371.

[48] S. Helmy, F. A. Leibfarth, S. Oh, J. E. Poelma, C. J. Hawker, J. Read de Alaniz, *J. Am. Chem. Soc.* **2014**, *136*, 8169-8172.

[49] J. J. Bozell, G. R. Petersen, *Green Chem.* **2010**, *12*, 539.

[50] M. M. Lerch, S. J. Wezenberg, W. Szymanski, B. L. Feringa, *J. Am. Chem. Soc.* **2016**, *138*, 6344-6347.

[51] a) B. F. Lui, N. T. Tierce, F. Tong, M. M. Sroda, H. Lu, J. Read de Alaniz, C. J. Bardeen, *Photochem. Photobiol. Sci.* **2019**, *18*, 1587-1595; b) M. M. Lerch, M. Di Donato, A. D. Laurent, M. Medved, A. Iagatti, L. Bussotti, A. Lapini, W. J. Buma, P. Foggi, W. Szymanski, B. L. Feringa, *Angew. Chem. Int. Ed.* **2018**, *57*, 8063-8068.

[52] R. H. Mitchell, Y. S. Chen, *Tetrahedron Lett.* **1996**, *37*, 5239-5242.

[53] K. Klaue, Y. Garmshausen, S. Hecht, *Angew. Chem. Int. Ed.* **2018**, *57*, 1414-1417.

[54] S. Kirchner, A.-L. Leistner, Z. L. Pianowski, in *Molecular Photoswitches*, **2022**, pp. 987-1013.

[55] X. Liang, H. Asanuma, H. Kashida, A. Takasu, T. Sakamoto, G. Kawai, M. Komiyama, *J. Am. Chem. Soc.* **2003**, *125*, 16408-16415.

[56] a) Y. R. Shi, M. A. Gerkman, Q. F. Qiu, S. R. Zhang, G. G. D. Han, *J. Mater. Chem.* **2021**, *9*, 9798-9808; b) Z. Wang, A. Knebel, S. Grosjean, D. Wagner, S. Brase, C. Woll, J. Caro, L. Heinke, *Nat. Commun.* **2016**, *7*, 13872; c) D. Mutruc, A. Goulet-Hanssens, S. Fairman, S. Wahl, A. Zimathies, C. Knie, S. Hecht, *Angew. Chem. Int. Ed.* **2019**, *58*, 12862-12867; d) C. Santos Hurtado, G. Bastien, M. Masat, J. R. Stocek, M. Dracinsky, I. Roncevic, I. Cisarova, C. T. Rogers, J. Kaleta, *J. Am. Chem. Soc.* **2020**, *142*, 9337-9351.

[57] P. Griess, *Ann. Chem. Pharm.* **1858**, *106*, 123-125.

[58] I. Szele, H. Zollinger, in *Preparative Organic Chemistry*, Springer, **1983**, pp. 1-66.

[59] T. Cohen, R. J. Lewarchik, J. Z. Tarino, *J. Am. Chem. Soc.* **2002**, *96*, 7753-7760.

[60] S. Wawzonek, T. McIntyre, *J. Electrochem. Soc.* **1972**, *119*, 1350.

[61] a) C. Knie, M. Utecht, F. Zhao, H. Kulla, S. Kovalenko, A. M. Brouwer, P. Saalfrank, S. Hecht, D. Bleger, *Chem. Eur. J.* **2014**, *20*, 16492-16501; b) H. Firouzabadi, Z. Mostafavipoor, *Bull. Chem. Soc. Jpn.* **1983**, *56*, 914-917; c) M. Wang, J. Ma, M. Yu, Z. Zhang, F. Wang, *Catal. Sci. Technol.* **2016**, *6*, 1940-1945.

[62] Y. An, H. Tan, S. Zhao, *Chinese J. Org. Chem.* **2017**, *37*, 226.

[63] A. Grirrane, A. Corma, H. Garcia, *Science* **2008**, *322*, 1661-1664.

[64] C. Zhang, N. Jiao, *Angew. Chem. Int. Ed.* **2010**, *49*, 6174-6177.

[65] V. Koch, S. Bräse, in *Molecular Photoswitches*, **2022**, pp. 39-64.

[66] C. Mills, *J. Chem. Soc., Trans.* **1895**, *67*, 925-933.

[67] Y. Zhu, Y. Shi, *Org. Lett.* **2013**, *15*, 1942-1945.

[68] E. Merino, *Chem. Soc. Rev.* **2011**, *40*, 3835-3853.

[69] C. Bleasdale, M. K. Ellis, P. B. Farmer, B. T. Golding, K. F. Handley, P. Jones, W. McFarlane, *J. Labelled Compd. Radiopharm.* **1993**, *33*, 739-746.

[70] J. M. J. Tronchet, E. Jean, G. Galland-Barrera, *Cheminform* **2010**, *22*, no-no.

[71] B. G. Gowenlock, G. B. Richter-Addo, *Chem. Rev.* **2004**, *104*, 3315-3340.

[72] R. F. Nystrom, W. G. Brown, *J. Am. Chem. Soc.* **1948**, *70*, 3738-3740.

[73] R. O. Hutchins, D. W. Lamson, L. Rua, C. Milewski, B. Maryanoff, *J. Org. Chem.* **1971**, *36*, 803-806.

[74] K. Stolarczyk, R. Bilewicz, A. Skwierawska, J. F. Biernat, *J. Incl. Phenom. Macrocycl. Chem.* **2004**, *49*, 173-179.

[75] M. Malinowski, Ł. Kaczmarek, F. Rozpłoch, *J. Chem. Soc. Perk. Trans. 2* **1991**, 879-883.
[76] S.-k. Won, W.-j. Kim, H.-b. Kim, *Bull. Korean Chem. Soc.* **2006**, *27*, 195-196.
[77] M. G. Pizzolatti, R. A. Yunes, *Journal of the Chemical Society, Perkin Transactions 2: Physical Organic Chemistry*

1990, 759.
[78] M. J. Hansen, M. M. Lerch, W. Szymanski, B. L. Feringa, *Angew. Chem. Int. Ed.* **2016**, *55*, 13514-13518.
[79] B. Haag, Z. Peng, P. Knochel, *Org. Lett.* **2009**, *11*, 4270-4273.
[80] T. Wendler, C. Schutt, C. Nather, R. Herges, *J. Org. Chem.* **2012**, *77*, 3284-3287.
[81] A. Antoine John, Q. Lin, *J. Org. Chem.* **2017**, *82*, 9873-9876.
[82] W. C. Lin, M. C. Tsai, C. M. Davenport, C. M. Smith, J. Veit, N. M. Wilson, H. Adesnik, R. H. Kramer, *Neuron* **2015**, *88*, 879-891.
[83] H. Rau, *Angew. Chem.* **1973**, *85*, 248-258.
[84] A. Polosukhina, J. Litt, I. Tochitsky, J. Nemargut, Y. Sychev, I. De Kouchkovsky, T. Huang, K. Borges, D. Trauner, R. N. Van Gelder, R. H. Kramer, *Neuron* **2012**, *75*, 271-282.
[85] N. A. Wazzan, P. R. Richardson, A. C. Jones, *Photochem. Photobiol. Sci.* **2010**, *9*, 968-974.
[86] a) M. Dong, A. Babalhavaeji, S. Samanta, A. A. Beharry, G. A. Woolley, *Acc. Chem. Res.* **2015**, *48*, 2662-2670; b) S. Samanta, T. M. McCormick, S. K. Schmidt, D. S. Seferos, G. A. Woolley, *Chem. Commun.* **2013**, *49*, 10314-10316; c) K. Kuntze, J. Viljakka, E. Titov, Z. Ahmed, E. Kalenius, P. Saalfrank, A. Priimagi, *Photochem Photobiol Sci* **2022**, *21*, 159-173.
[87] O. Sadovski, A. A. Beharry, F. Zhang, G. A. Woolley, *Angew. Chem. Int. Ed.* **2009**, *48*, 1484-1486.
[88] A. A. Beharry, O. Sadovski, G. A. Woolley, *J. Am. Chem. Soc.* **2011**, *133*, 19684-19687.
[89] D. Bleger, J. Schwarz, A. M. Brouwer, S. Hecht, *J. Am. Chem. Soc.* **2012**, *134*, 20597-20600.
[90] S. Samanta, A. A. Beharry, O. Sadovski, T. M. McCormick, A. Babalhavaeji, V. Tropepe, G. A. Woolley, *J. Am. Chem. Soc.* **2013**, *135*, 9777-9784.
[91] A. L. Leistner, S. Kirchner, J. Karcher, T. Bantle, M. L. Schulte, P. Godtel, C. Fengler, Z. L. Pianowski, *Chem. Eur. J.* **2021**, *27*, 8094-8099.
[92] D. B. Konrad, G. Savasci, L. Allmendinger, D. Trauner, C. Ochsenfeld, A. M. Ali, *J. Am. Chem. Soc.* **2020**, *142*, 6538-6547.

[93] a) K. Aggarwal, T. P. Kuka, M. Banik, B. P. Medellin, C. Q. Ngo, D. Xie, Y. Fernandes, T. L. Dangerfield, E. Ye, B. Bouley, K. A. Johnson, Y. J. Zhang, J. K. Eberhart, E. L. Que, *J. Am. Chem. Soc.* **2020**, *142*, 14522-14531; b) C. Poloni, W. Szymanski, L. Hou, W. R. Browne, B. L. Feringa, *Chem. Eur. J.* **2014**, *20*, 946-951.

[94] K. Kumar, C. Knie, D. Bleger, M. A. Peletier, H. Friedrich, S. Hecht, D. J. Broer, M. G. Debije, A. P. Schenning, *Nat. Commun.* **2016**, *7*, 11975.

[95] S. Stoyanov, L. Antonov, T. Stoyanova, V. Petrova, *Dyes Pigm.* **1996**, *32*, 171-185.

[96] M. Dong, A. Babalhavaeji, M. J. Hansen, L. Kalman, G. A. Woolley, *Chem. Commun.* **2015**, *51*, 12981-12984.

[97] S. Samanta, A. Babalhavaeji, M. X. Dong, G. A. Woolley, *Angew. Chem. Int. Ed.* **2013**, *52*, 14127-14130.

[98] K. Kokkinos, R. Wizinger, *Helv. Chim. Acta* **1971**, *54*, 330-334.

[99] M. Dong, A. Babalhavaeji, C. V. Collins, K. Jarrah, O. Sadovski, Q. Dai, G. A. Woolley, *J. Am. Chem. Soc.* **2017**, *139*, 13483-13486.

[100] C. E. Weston, R. D. Richardson, P. R. Haycock, A. J. White, M. J. Fuchter, *J. Am. Chem. Soc.* **2014**, *136*, 11878-11881.

[101] N. A. Simeth, S. Crespi, M. Fagnoni, B. Konig, *J. Am. Chem. Soc.* **2018**, *140*, 2940-2946.

[102] J. Garcia-Amoros, M. Diaz-Lobo, S. Nonell, D. Velasco, *Angew. Chem. Int. Ed.* **2012**, *51*, 12820-12823.

[103] A. D. W. Kennedy, I. Sandler, J. Andreasson, J. Ho, J. E. Beves, *Chem. Eur. J.* **2020**, *26*, 1103-1110.

[104] R. Siewertsen, H. Neumann, B. Buchheim-Stehn, R. Herges, C. Nather, F. Renth, F. Temps, *J. Am. Chem. Soc.* **2009**, *131*, 15594-15595.

[105] M. Hammerich, C. Schutt, C. Stahler, P. Lentes, F. Rohricht, R. Hoppner, R. Herges, *J. Am. Chem. Soc.* **2016**, *138*, 13111-13114.

[106] P. Lentes, E. Stadler, F. Rohricht, A. Brahms, J. Grobner, F. D. Sonnichsen, G. Gescheidt, R. Herges, *J. Am. Chem. Soc.* **2019**, *141*, 13592-13600.

[107] P. Lentes, J. Rudtke, T. Griebenow, R. Herges, *Beilstein J. Org. Chem.* **2021**, *17*, 1503-1508.

[108] H.-A. Wagenknecht, W. Schmucker, *Synlett* **2012**, *23*, 2435-2448.

[109] H.-A. Wagenknecht, in *Molecular Photoswitches*, **2022**, pp. 973-986.

[110] H. Asanuma, T. Ishikawa, Y. Yamano, K. Murayama, X. Liang, *ChemPhotoChem* **2019**, *3*, 418-424.

[111] B. Cheng, H. Kashida, N. Shimada, A. Maruyama, H. Asanuma, *Chem. Commun.* **2019**, *55*, 1080-1083.

[112] A. Aemissegger, D. Hilvert, *Nat. Protoc.* **2007**, *2*, 161-167.
[113] A. Aemissegger, V. Krautler, W. F. van Gunsteren, D. Hilvert, *J. Am. Chem. Soc.* **2005**, *127*, 2929-2936.
[114] F. Nuti, C. Gellini, M. Larregola, L. Squillantini, R. Chelli, P. R. Salvi, O. Lequin, G. Pietraperzia, A. M. Papini, *Front. Chem.* **2019**, *7*, 180.
[115] C. Renner, U. Kusebauch, M. Loweneck, A. G. Milbradt, L. Moroder, *J. Pept. Res.* **2005**, *65*, 4-14.
[116] D. G. Flint, J. R. Kumita, O. S. Smart, G. A. Woolley, *Chem. Biol.* **2002**, *9*, 391-397.
[117] G. Despras, V. Poonthiyil, T. K. Lindhorst, in *Molecular Photoswitches*, **2022**, pp. 1015-1045.
[118] a) P. G. Alluri, M. M. Reddy, K. Bachhawat-Sikder, H. J. Olivos, T. Kodadek, *J. Am. Chem. Soc.* **2003**, *125*, 13995-14004; b) J. Hong, *Curr. Opin. Chem. Biol.* **2011**, *15*, 350-354.
[119] M. Borowiak, W. Nahaboo, M. Reynders, K. Nekolla, P. Jalinot, J. Hasserodt, M. Rehberg, M. Delattre, S. Zahler, A. Vollmar, D. Trauner, O. Thorn-Seshold, *Cell* **2015**, *162*, 403-411.
[120] A. H. Gelebart, D. J. Mulder, G. Vantomme, A. Schenning, D. J. Broer, *Angew. Chem. Int. Ed.* **2017**, *56*, 13436-13439.
[121] a) D. Wang, M. Wagner, H. J. Butt, S. Wu, *Soft Matter* **2015**, *11*, 7656-7662; b) Z. L. Pianowski, J. Karcher, K. Schneider, *Chem. Commun.* **2016**, *52*, 3143-3146.
[122] a) J. Kubitschke, C. Nather, R. Herges, *Eur. J. Org. Chem.* **2010**, *2010*, 5041-5055; b) R. Low, T. Rusch, T. Moje, F. Rohricht, O. M. Magnussen, R. Herges, *Beilstein J. Org. Chem.* **2019**, *15*, 1815-1821.
[123] Y. Zheng, H. Sato, P. Wu, H. J. Jeon, R. Matsuda, S. Kitagawa, *Nat. Commun.* **2017**, *8*, 100.
[124] Q. Wang, G. Ligorio, R. Schlesinger, V. Diez-Cabanes, D. Cornil, Y. Garmshausen, S. Hecht, J. Cornil, E. J. W. List-Kratochvil, N. Koch, *Adv. Mater. Interfaces* **2019**, *6*, 1900211.
[125] a) J. Orrego-Hernandez, A. Dreos, K. Moth-Poulsen, *Acc. Chem. Res.* **2020**, *53*, 1478-1487; b) Z. Wang, J. Udmark, K. Borjesson, R. Rodrigues, A. Roffey, M. Abrahamsson, M. B. Nielsen, K. Moth-Poulsen, *ChemSusChem* **2017**, *10*, 3049-3055.
[126] Z. Zhang, D. Chen, D. Mutruc, S. Hecht, L. Heinke, *Chem. Commun.* **2022**, *58*, 13963-13966.
[127] Z. Dang, L. Liu, Y. Li, Y. Xiang, G. Guo, *ACS Appl. Mater. Interfaces* **2016**, *8*, 31281-31288.

[128] D. Ahmad, I. van den Boogaert, J. Miller, R. Presswell, H. Jouhara, *Energy Sources, Part A* **2018**, *40*, 2686-2725.
[129] F. D. Jochum, P. Theato, *Chem. Soc. Rev.* **2013**, *42*, 7468-7483.
[130] T. Graham, *Philos. Trans. R. Soc. London, Ser. A* **1861**, *151*, 183-224.
[131] P. Hermans, *éd. HR Kruyt, New York* **1949**, 483-494.
[132] D. Wang, S. Zhao, R. Yin, L. Li, Z. Lou, G. Shen, *npj Flex. Electron.* **2021**, *5*, 13.
[133] S. Hong, Y. Yuan, C. Liu, W. Chen, L. Chen, H. Lian, H. Liimatainen, *J. Mater. Chem.* **2020**, *8*, 550-560.
[134] S. J. Carter, *Cooper and Gunn's Tutorial Pharmacy*, CBS Publishers & Distributors, **2021**.
[135] a) L. Wang, X. Shi, J. Zhang, Y. Zhu, J. Wang, *RSC Adv.* **2018**, *8*, 31581-31587; b) J. Karcher, S. Kirchner, A.-L. Leistner, C. Hald, P. Geng, T. Bantle, P. Gödtel, J. Pfeifer, Z. L. Pianowski, *RSC Adv.* **2021**, *11*, 8546-8551.
[136] a) Q. Wei, J. Duan, G. Ma, W. Zhang, Q. Wang, Z. Hu, *J. Mater. Chem.* **2019**, *7*, 2220-2225; b) Y. Liang, J. He, B. Guo, *ACS Nano* **2021**, *15*, 12687-12722.
[137] a) C. D. Spicer, *Polym. Chem.* **2020**, *11*, 184-219; b) S. Mantha, S. Pillai, P. Khayambashi, A. Upadhyay, Y. Zhang, O. Tao, H. M. Pham, S. D. Tran, *Materials* **2019**, *12*.
[138] S. Li, Y. Cong, J. Fu, *J. Mater. Chem.* **2021**, *9*, 4423-4443.
[139] M. W. Tibbitt, K. S. Anseth, *Biotechnol. Bioeng.* **2009**, *103*, 655-663.
[140] M. Patenaude, N. M. Smeets, T. Hoare, *Macromol. Rapid Commun.* **2014**, *35*, 598-617.
[141] D. Wang, S. Maharjan, X. Kuang, Z. Wang, L. S. Mille, M. Tao, P. Yu, X. Cao, L. Lian, L. Lv, J. J. He, G. Tang, H. Yuk, C. K. Ozaki, X. Zhao, Y. S. Zhang, *Sci. Adv.* **2022**, *8*, eabq6900.
[142] S. Van Vlierberghe, P. Dubruel, E. Schacht, *Biomacromolecules* **2011**, *12*, 1387-1408.
[143] C. Echeverria, S. N. Fernandes, M. H. Godinho, J. P. Borges, P. I. P. Soares, *Gels* **2018**, *4*.
[144] J. Mewis, *J. Non-Newtonian Fluid Mech.* **1979**, *6*, 1-20.
[145] J. Y. C. Lim, Q. Lin, K. Xue, X. J. Loh, *Mater. Today Adv.* **2019**, *3*, 100021.
[146] *Vol. 2023*, Nobel Prize Outreach AB 202, NobelPrize.org.
[147] F. Huang, E. V. Anslyn, *Chem. Rev.* **2015**, *115*, 6999-7000.
[148] in *Citation Report: supramolecular hydrogels 2000-2022, Vol. 20.02.2023*, https://www.webofscience.com/.

[149] a) S. Das, D. Das, *Front. Chem.* **2021**, *9*, 770102; b) O. Bellotto, S. Kralj, R. De Zorzi, S. Geremia, S. Marchesan, *Soft Matter* **2020**, *16*, 10151-10157; c) J. Shi, Y. Gao, Z. Yang, B. Xu, *Beilstein J. Org. Chem.* **2011**, *7*, 167-172.

[150] a) Q. Tang, T. N. Plank, T. Zhu, H. Yu, Z. Ge, Q. Li, L. Li, J. T. Davis, H. Pei, *ACS Appl. Mater. Interfaces* **2019**, *11*, 19743-19750; b) F. Tang, H. Feng, Y. Du, Y. Xiao, H. Dan, H. Zhao, Q. Chen, *Chem. - Asian J.* **2018**; c) K. J. Skilling, B. Kellam, M. Ashford, T. D. Bradshaw, M. Marlow, *Soft Matter* **2016**, *12*, 8950-8957.

[151] a) R. Van Lommel, L. A. J. Rutgeerts, W. M. De Borggraeve, F. De Proft, M. Alonso, *ChemPlusChem* **2020**, *85*, 267-276; b) V. Basavalingappa, T. Guterman, Y. Tang, S. Nir, J. Lei, P. Chakraborty, L. Schnaider, M. Reches, G. Wei, E. Gazit, *Adv. Sci. (Weinh.)* **2019**, *6*, 1900218; c) A. J. Kleinsmann, N. M. Weckenmann, B. J. Nachtsheim, *Chem. Eur. J.* **2014**, *20*, 9753-9761.

[152] J. Omar, D. Ponsford, C. A. Dreiss, T. C. Lee, X. J. Loh, *Chem. - Asian J.* **2022**, *17*, e202200081.

[153] L. Cao, W. Lu, A. Mata, K. Nishinari, Y. Fang, *Carbohydr. Polym.* **2020**, *242*, 116389.

[154] M. Hirsch, A. Charlet, E. Amstad, *Adv. Funct. Mater.* **2020**, *31*, 2005929.

[155] S. Hong, D. Sycks, H. F. Chan, S. Lin, G. P. Lopez, F. Guilak, K. W. Leong, X. Zhao, *Adv. Mater.* **2015**, *27*, 4035-4040.

[156] H. Goyal, S. Pachisia, R. Gupta, *Cryst. Growth Des.* **2020**, *20*, 6117-6128.

[157] a) C. Ou, J. Zhang, X. Zhang, Z. Yang, M. Chen, *Chem. Commun.* **2013**, *49*, 1853-1855; b) S. Marchesan, C. D. Easton, F. Kushkaki, L. Waddington, P. G. Hartley, *Chem. Commun.* **2012**, *48*, 2195-2197.

[158] A. J. Kleinsmann, B. J. Nachtsheim, *Chem. Commun.* **2013**, *49*, 7818-7820.

[159] a) T. Muraoka, C. Y. Koh, H. Cui, S. I. Stupp, *Angew. Chem. Int. Ed.* **2009**, *48*, 5946-5949; b) L. A. Haines, K. Rajagopal, B. Ozbas, D. A. Salick, D. J. Pochan, J. P. Schneider, *J. Am. Chem. Soc.* **2005**, *127*, 17025-17029.

[160] S. Matsumoto, S. Yamaguchi, S. Ueno, H. Komatsu, M. Ikeda, K. Ishizuka, Y. Iko, K. V. Tabata, H. Aoki, S. Ito, H. Noji, I. Hamachi, *Chem. Eur. J.* **2008**, *14*, 3977-3986.

[161] S. Matsumoto, S. Yamaguchi, A. Wada, T. Matsui, M. Ikeda, I. Hamachi, *Chem. Commun.* **2008**, 1545-1547.

[162] Z. Qiu, H. Yu, J. Li, Y. Wang, Y. Zhang, *Chem. Commun.* **2009**, 3342-3344.

[163] Y. Huang, Z. Qiu, Y. Xu, J. Shi, H. Lin, Y. Zhang, *Org. Biomol. Chem.* **2011**, *9*, 2149-2155.

[164] F. A. Larik, L. L. Fillbrook, S. S. Nurttila, A. D. Martin, R. P. Kuchel, K. Al Taief, M. Bhadbhade, J. E. Beves, P. Thordarson, *Angew. Chem.* **2021**, *133*, 6838-6844.

[165] J. Karcher, Z. L. Pianowski, *Chem. Eur. J.* **2018**, *24*, 11605-11610.

[166] Y. Hong, J. W. Lam, B. Z. Tang, *Chem. Soc. Rev.* **2011**, *40*, 5361-5388.

[167] J. Luo, Z. Xie, J. W. Lam, L. Cheng, H. Chen, C. Qiu, H. S. Kwok, X. Zhan, Y. Liu, D. Zhu, B. Z. Tang, *Chem. Commun.* **2001**, 1740-1741.

[168] Y. Hong, J. W. Lam, B. Z. Tang, *Chem. Commun.* **2009**, 4332-4353.

[169] J. Chen, C. C. W. Law, J. W. Y. Lam, Y. Dong, S. M. F. Lo, I. D. Williams, D. Zhu, B. Z. Tang, *Chem. Mater.* **2003**, *15*, 1535-1546.

[170] Z. Wu, Q. Yao, O. J. H. Chai, N. Ding, W. Xu, S. Zang, J. Xie, *Angew. Chem. Int. Ed.* **2020**, *59*, 9934-9939.

[171] N. B. Shustova, B. D. McCarthy, M. Dinca, *J. Am. Chem. Soc.* **2011**, *133*, 20126-20129.

[172] B. K. An, S. K. Kwon, S. D. Jung, S. Y. Park, *J. Am. Chem. Soc.* **2002**, *124*, 14410-14415.

[173] N. J. Hestand, F. C. Spano, *Chem. Rev.* **2018**, *118*, 7069-7163.

[174] J. Mei, N. L. Leung, R. T. Kwok, J. W. Lam, B. Z. Tang, *Chem. Rev.* **2015**, *115*, 11718-11940.

[175] J. Mei, Y. Hong, J. W. Lam, A. Qin, Y. Tang, B. Z. Tang, *Adv. Mater.* **2014**, *26*, 5429-5479.

[176] H.-H. Lin, Y.-C. Chan, J.-W. Chen, C.-C. Chang, *J. Mater. Chem.* **2011**, *21*, 3170.

[177] P. Gopikrishna, N. Meher, P. K. Iyer, *ACS Appl. Mater. Interfaces* **2018**, *10*, 12081-12111.

[178] M. Y. Yeh, C. T. Huang, T. S. Lai, F. Y. Chen, N. T. Chu, D. T. Tseng, S. C. Hung, H. C. Lin, *Langmuir* **2016**, *32*, 7630-7638.

[179] C. Balachandra, T. Govindaraju, *J. Org. Chem.* **2020**, *85*, 1525-1536.

[180] M. J. Frisch, G. W. Trucks, H. B. Schlegel, G. E. Scuseria, M. A. Robb, J. R. Cheeseman, G. Scalmani, V. Barone, G. A. Petersson, H. Nakatsuji, X. Li, M. Caricato, A. V. Marenich, J. Bloino, B. G. Janesko, R. Gomperts, B. Mennucci, H. P. Hratchian, J. V. Ortiz, A. F. Izmaylov, J. L. Sonnenberg, Williams, F. Ding, F. Lipparini, F. Egidi, J. Goings, B. Peng, A. Petrone, T. Henderson, D. Ranasinghe, V. G. Zakrzewski, J. Gao, N. Rega, G. Zheng, W. Liang, M. Hada, M. Ehara, K. Toyota, R. Fukuda, J. Hasegawa, M. Ishida, T. Nakajima, Y. Honda, O. Kitao, H. Nakai, T. Vreven, K. Throssell, J. A. Montgomery Jr., J. E. Peralta, F. Ogliaro, M. J. Bearpark, J. J. Heyd, E. N. Brothers, K. N. Kudin, V. N. Staroverov, T. A. Keith, R. Kobayashi, J.

Normand, K. Raghavachari, A. P. Rendell, J. C. Burant, S. S. Iyengar, J. Tomasi, M. Cossi, J. M. Millam, M. Klene, C. Adamo, R. Cammi, J. W. Ochterski, R. L. Martin, K. Morokuma, O. Farkas, J. B. Foresman, D. J. Fox, Wallingford, CT, **2016**.

[181] a) A. D. Becke, *J. Chem. Phys.* **1993**, *98*, 5648-5652; b) A. D. McLean, G. S. Chandler, *J. Chem. Phys.* **1980**, *72*, 5639-5648; c) R. Krishnan, J. S. Binkley, R. Seeger, J. A. Pople, *J. Chem. Phys.* **1980**, *72*, 650-654.

[182] A.-L. Leistner, Karlsruhe Institute of Technology, KIT library, **2019**.

[183] a) T. Kawakami, H. Suzuki, *Tetrahedron Lett.* **2000**, *41*, 7093-7096; b) Z. Zhang, Q. Tian, J. Qian, Q. Liu, T. Liu, L. Shi, G. Zhang, *J. Org. Chem.* **2014**, *79*, 8182-8188.

[184] C. Gallina, A. Liberatori, *Tetrahedron* **1974**, *30*, 667-673.

[185] H. E. Gottlieb, V. Kotlyar, A. Nudelman, *J. Org. Chem.* **1997**, *62*, 7512-7515.

[186] C. R. Martinez, B. L. Iverson, *Chem. Sci.* **2012**, *3*, 2191.

[187] A.-L. Leistner, D. G. Kistner, C. Fengler, Z. L. Pianowski, *RSC Adv.* **2022**, *12*, 4771-4776.

[188] Z. Zhang, D. C. Burns, J. R. Kumita, O. S. Smart, G. A. Woolley, *Bioconjugate Chem.* **2003**, *14*, 824-829.

[189] H. Ramli, N. F. A. Zainal, M. Hess, C. H. Chan, *CTI* **2022**, *4*, 307-326.

[190] a) K. Y. Lee, D. J. Mooney, *Prog. Polym. Sci.* **2012**, *37*, 106-126; b) S. K. Tam, S. Bilodeau, J. Dusseault, G. Langlois, J. P. Halle, L. H. Yahia, *Acta Biomater.* **2011**, *7*, 1683-1692; c) F. Abasalizadeh, S. V. Moghaddam, E. Alizadeh, E. Akbari, E. Kashani, S. M. B. Fazljou, M. Torbati, A. Akbarzadeh, *J. Biol. Eng.* **2020**, *14*, 8.

[191] G. Tan, J. Xu, Q. Yu, J. Zhang, X. Hu, C. Sun, H. Zhang, *Micromachines* **2022**, *13*.

[192] D. B. Konrad, J. A. Frank, D. Trauner, *Chem. Eur. J.* **2016**, *22*, 4364-4368.

[193] D. R. Fahey, *J. Chem. Soc. D* **1970**, 417a.

[194] X.-C. Wang, Y. Hu, S. Bonacorsi, Y. Hong, R. Burrell, J.-Q. Yu, *Cheminform* **2013**, *44*, no-no.

[195] Z. Yin, X. Jiang, P. Sun, *J. Org. Chem.* **2013**, *78*, 10002-10007.

[196] T. W. a. W. Greene, P.G.M., in *Protective Groups in Organic Synthesis*, **1999**, pp. 494-653.

[197] a) M. Goodman, A. Kossoy, *J. Am. Chem. Soc.* **2002**, *88*, 5010-5015; b) M. Sato, T. Kinoshita, A. Takizawa, Y. Tsujita, T. Osada, *Polym. J.* **1989**, *21*, 533-541; c) A. Fissi, O. Pieroni, G. Ruggeri, F. Ciardelli, *Macromolecules* **1995**, *28*, 302-309; d) N. Angelini, B. Corrias, A. Fissi, O. Pieroni, F. Lenci,

Biophys. J. **1998**, *74*, 2601-2610; e) A. Fissi, O. Pieroni, F. Ciardelli, D. Fabbri, C. Ruggeri, K. Umezawa, *Biopolymers* **1993**, *33*, 1505-1517; f) P. H. Vandewyer, G. Smets, *J. Polym. Sci., Part A-1: Polym. Chem.* **1970**, *8*, 2361-2374.
[198] D. Hermann, H. A. Schwartz, U. Ruschewitz, *ChemistrySelect* **2017**, *2*, 11846-11852.
[199] G. Indrayanto, G. S. Putra, F. Suhud, *Profiles Drug Subst. Excip. Relat. Methodol.* **2021**, *46*, 273-307.
[200] M. M. Most, Institute of Organic Chemistry KIT, KIT Library, **2022**.
[201] H. Dai, H. Xu, *Chin. J. Chem .* **2012**, *30*, 267-272.
[202] R. Beckert, E. Fanghänel, K. Schwetlick, *Organikum: organisch-chemisches Grundpraktikum*, 23. ed., Wiley VCH, Weinheim, **2009**.
[203] a) W. C. Still, M. Kahn, A. Mitra, *J. Org. Chem.* **1978**, *43*, 2923-2925; b) J. E. Silver, C. A. Bailey, R. L. Lewis, S. R. Paeschke, *Abstr. Pap. Am. Chem. Soc.* **2013**, *246*.
[204] H. E. Gottlieb, V. Kotlyar, A. Nudelman, *J. Org. Chem.* **1997**, *62*, 7512-7515.
[205] S. Manchineella, T. Govindaraju, *RSC Adv.* **2012**, *2*, 5539-5542.
[206] M. G. Anne, B. W. Nicholas, M. G. Lindsey, in *Rheology* (Ed.: V. Juan De), IntechOpen, Rijeka, **2012**, p. Ch. 3.
[207] a) A. Maus, C. Hertlein, K. Saalwächter, *Macromol. Chem. Phys.* **2006**, *207*, 1150-1158; b) S. Meiboom, D. Gill, *Rev. Sci. Instrum.* **1958**, *29*, 688-691.
[208] T. Gullion, D. B. Baker, M. S. Conradi, *J. Magn. Reson., Ser A* **1990**, *89*, 479-484.
[209] X. Y. Tian, J. W. Han, Q. Zhao, H. N. Wong, *Org. Biomol. Chem.* **2014**, *12*, 3686-3700.
[210] D. Mazzier, M. Maran, O. Polo Perucchin, M. Crisma, M. Zerbetto, V. Causin, C. Toniolo, A. Moretto, *Macromolecules* **2014**, *47*, 7272-7283.
[211] A. Szymańska, K. Wegner, L. Łankiewicz, *Helv. Chim. Acta* **2003**, *86*, 3326-3331.
[212] S. P. Ludwig, Institute of Organic Chemistry KIT, KIT library, **2022**.
[213] R. G. Vaswani, A. R. Chamberlin, *J. Org. Chem.* **2008**, *73*, 1661-1681.

8 Appendix

8.1 Curriculum Vitae

PERSONAL DATA

Name: Anna-Lena Leistner

Place of birth: 07.08.1996 / Neuenbürg

Adress: Adlerstraße 40, 76133 Karlsruhe, Germany

Annalena.Leistner@web.de

EDUCATIONAL BACKGROUND

01/2020 – 04/2023	PhD student Chemistry, Karlsruhe Institute of Technology, Topic: "Biocompatible Photochromic Diketopiperazine-based Supramolecular Hydrogelator Systems"
10/2017 – 09/2019	Master studies Chemical Biology, Karlsruhe Institute of Technology, Final grade: 1.0
10/2014 – 09/2017	Bachelor studies Chemical Biology, Karlsruhe Institute of Technology, Final grade: 1.6
09/2006 – 06/2014	Abitur, Enztalgymnasium, Bad Wildbad, Final grade: 1.4

ACADEMIC ACTIVITY

10/2019 – 12/2019	Research Assistant, Karlsruhe Institute of Technology, Pianowski group, cytotoxic assays, organic synthesis

FELLOWSHIPS AND AWARDS

01/2020 – 12/2022	LGF scholarship: Graduate Funding from the German State of Baden-Württemberg
2014	Graduation prize: Baetzner-Award (best of the year), Dr. Rommel-Award (science), award in biology, award in mathematics

LANGUAGES

English: fluently spoken and written, C1 level

French: B1 level

8.2 List of Publications

Peer-reviewed publications

Karcher, J.; Kirchner, S.; **Leistner, A.-L.**; Hald, C.; Geng, P.; Bantle, T.; Gödtel, P.; Pfeifer, J.; Pianowski, Z. L., Selective release of a potent anticancer agent from a supramolecular hydrogel using green light. RSC Adv. **2021,** 11 (15), 8546-8551.

Leistner, A. L. Kirchner, S.; Karcher, J. Bantle, T.; Schulte, M. L.; Godtel, P.; Fengler, C.; Pianowski, Z. L., Fluorinated Azobenzenes Switchable with Red Light. *Chem. Eur. J.* **2021**, *27*, 8094.

Leistner, A.-L.; Kistner, D. G.; Fengler, C.; Pianowski, Z. L., Reversible photodissipation of composite photochromic azobenzene-alginate supramolecular hydrogels. RSC Adv. **2022,** 12 (8), 4771-4776.

Leistner, A. L.; Pianowski, Z. L., Cover Feature: Smart Photochromic Materials Triggered with Visible Light (Eur. J. Org. Chem. 19/2022). *Eur. J. Org. Chem.* **2022**, e202101271.

Kirchner, S.; **Leistner, A. L.**; Godtel, P.; Seliwjorstow, A.; Weber, S.; Karcher, J.; Nieger, M.; Pianowski, Z., Hemipiperazines as peptide-derived molecular photoswitches with low-nanomolar cytotoxicity. Nat. Commun. **2022,** 13 (1), 6066.

Hoffmann, F.; **Leistner, A. L.**; Kirchner, S; Luy, B.; Muhle-Goll, C.*; Pianowski, Z.*, Cargo encapsulation in photochromic supramolecular hydrogels depends on specific guest-gelator supramolecular interactions. *Eur. J. Org. Chem.* 2023, e202300227.

Book chapter

Kirchner, S.; **Leistner, A.-L.**; Pianowski, Z. L., Photoswitchable Peptides and Proteins. In *Molecular Photoswitches*, **2022**; pp 987-1013.

Conference poster

Leistner, A. L.; Pianowski, Z. L., “Fluorinated Azobenzenes Switchable with Red Light” from November 29th to 30th, 2021, 3rd International Conference on Photopharmacology (PPIII)

Patent

„Diketopiperazine mit Licht-aktivierter Zytotoxizität” Pianowski, Z.; **Leistner, A.-L**.; Kirchner, S.; Weber, S.; Seliwjorstow, A.; Karcher, J.; DE 10 2019 005 005.3

8.3 Acknowledgements

At this point, I would like to thank all those who supported me during my doctorate and thus contributed to the success of this work.

First, I would like to cordially thank my doctoral supervisor, Priv.-Doz. Dr. Zbigniew L. Pianowski for the opportunity to do research in his group. His support, trust and especially freedom granted in the context of the preparation of this work allowed me to gain valuable experience for scientific growth and personal development.

My thanks also go to Prof. Dr. Stefan Bräse, who generously supports our group by providing infrastructural and financial help.

Prof. Dr. Hans-Achim Wagenknecht is kindly acknowledged for the acceptance of the co-reference of this thesis.

I would like to thank the Land of Baden-Württemberg for the financial support of this work in form of "Landesgraduiertenstipendium".

Special thanks to Dr. Christian Fengler and Maxi Hoffmann for measuring rheology of my samples, which gave us some troubles, but also ended in nice discussions and publications.

Dr. Heike Störmer and Mr. Volker Zibat I want to thank you for the assistance in electron microscopy data collection and the nice talks we had.

Dr. Martin Nieger, I thank you for the crystal structure analyses. I would also like to thank Dr. Norbert Foitzik, Angelika Mösle, Danny Wagner, Christoph Götz, Lara Hirsch. Andreas Rapp, Pia Lang, Tanja Ohmer-Scherrer, Richard von Budberg, Karolin Kohnle and Despina Savvidou-Kourmpidou for all their efforts.

Thanks to Christiane Lampert, Janine Bolz, and Dr. Christin Bednarek for their always friendly and quick support in bureaucratic matters.

I would like to thank my students Hendrik Kirchhoff, Ali Acan, David Kistner, Simon Ludwig, Dinh Tien Nguyen, and Mario Most, who chose to write their thesis supervised by me for their support in the lab, good ideas and motivation.

Johannes, you taught me all the skills I needed to pursue this doctorate from the beginning and awakened mine with your own enthusiasm. Through you I have chosen this path and I am infinitely grateful to you.

The past three years were not always easy, especially the Covid-19 pandemic cast its negative shadow on the time. Without the good atmosphere within the group, the funny conversations, wild parties, and other cozy get-togethers, I would not have survived this time. At this point, special thanks to you Simon, for the helpful conversations and the good time. I would also like to thank Lukas, Lisa, Henrik, Steffen, Lisa-Lou, Christoph, Philipp G. and Jens, who made this time special.

Many thanks to the whole Zibilab: Susanne, Peter, Valentin and Angelika for the good time we spent together, the helpful discussions and mutual support on a professional and personal level. Particular thanks to Angelika, Valentin and Peter for the correction of my thesis.

Finally, I want to thank my family. My mother and Johanna for the many ways of support and endless belief in me. Thank you, dad, for the enthusiasm and pride you had about my interest in natural sciences, you always knew that I would find my way.